Traditional Foods

Traditional Foods: Impact on Gut Health delves into the profound influence of traditional foods and dietary interventions on gut health, immune modulation, and disease prevention. This comprehensive book unites ancient culinary wisdom with contemporary scientific advancements, presenting a detailed exploration of traditional food practices, their phytochemical properties, and their profound impact on digestive and overall health.

Drawing from a multidisciplinary perspective, the chapters traverse diverse topics such as the historical and nutritional value of traditional foods, the gut health benefits of millets, phytochemicals and their chemistry, the dynamic interplay of fermented foods and microbiomes, and strategies for food allergy management. Special emphasis is placed on critical health concerns, including gut-associated cancers, liver disorders, leaky gut syndrome, and the emerging roles of biopolymers in cancer treatment.

By integrating state-of-the-art research with historical practices, this volume serves as an invaluable resource for researchers, health professionals, and anyone intrigued by the science of food and health. Real-world case studies, illustrative examples, and cutting-edge insights offer a bridge between cultural heritage and modern medicine, showcasing the transformative power of food as medicine.

Sapna Sharma serves as an assistant professor at Uttar Pradesh State Institute of Forensic Science, Lucknow, India. She completed her Ph.D. at CSIR–Indian Institute of Toxicology Research, focusing on probiotics in hepatotoxicity. With extensive academic and research experience, Dr. Sharma has worked in leading institutions worldwide, including Jawaharlal Nehru University and the University of Gothenburg. Her expertise encompasses gut-associated metabolic diseases, intestinal permeability, and vaccine development. Dr. Sharma has authored more than 25 publications and holds several patents.

Anand Prakash Singh is a research investigator at the Institute for Heart and Brain Health, University of Michigan, U.S.A. He earned his Ph.D. from Jawaharlal Nehru University, India, with significant contributions to understanding the disruption of tight junctions by bacterial effectors. With over seven years of postdoctoral experience, Dr. Singh has worked at renowned institutions, including Harvard Medical School and the Broad Institute. His research spans cardiac regeneration, cancer therapy-induced cardiovascular toxicity, and the gut-heart axis. He has authored over 24 research publications in esteemed journals.

Sustainable Industrial and Environmental Bioprocesses

This book series aims to provide a comprehensive collection of books focusing on bioprocesses in industrial and environmental biotechnology. The multidisciplinary content encompasses the chemical and biochemical engineering, industrial microbiology, and energy biosciences, all with a central theme of sustainable development and circular economy principles. The books align with the Sustainable Development Goals (SDGs) and offer state-of-the-art information and in-depth knowledge on the subject matter. The book in these series also emphasizes on the application of emerging tools, such as machine learning and artificial intelligence, for the advancement of bioprocesses. While primarily targeting academicians and researchers, the series is valuable for policy planners and industry professionals, with carefully tailored contents to cater to their specific needs and interests.

Series Editor: Dr Ashok Pandey

Professor Ashok Pandey is currently Distinguished Scientist at the Centre for Innovation and Translational Research, CSIR-Indian Institute of Toxicology Research, Lucknow, India. His major research and technological development interests are industrial & environmental biotechnology and energy biosciences, focusing on biomass to biofuels & chemicals, waste to wealth & energy, etc. He has 16 patents, 120 books, >1000 papers and book chapters, etc with h index of 131 and >74,000 citations. Professor Pandey is the recipient of many national and international awards and honours, which include Fellow, The World Academy of Sciences (TWAS), Highly Cited Researcher (Top 1% in the world), Clarivate Analytics (since 2018- till date), Rank #1 in India in Biology in India and Biochemistry and #417 in the world, Research.Com (2023), Rank #1 in India in Microbiology under Enabling and Strategic Technologies sector, Elsevier (2021) , Rank #1 in India in Biotechnology and #8 in the world, Stanford University Report (2020-2021-2022), etc.

Biodegradation of Toxic and Hazardous Chemicals: Detection and Mineralization
Kashyap K. Dubey, Kamal. K. Pant, Ashok Pandey and Maria Angeles Sanromán

Biodegradation of Toxic and Hazardous Chemicals: Remediation and Resource Recovery
Kashyap K. Dubey, Kamal K. Pant, Ashok Pandey and Maria Angeles Sanromán

Waste Management in Climate Change and Sustainability- Lignocellulosic Waste
Sunita Varjani, Izharul Haq, Ashok Pandey, Vijai Kumar Gupta, and Xuan-Thanh Bui

Ashwagandha for Quality of Life: Scientific Evidence
Sunil C Kaul & Renu Wadhwa

Microfluidics in Food Processing:Technologies and Applications
Ayon Tarafdar, Ranjna Sirohi, Barjinder Pal Kaur, Ashok Pandey, and Claude-Gilles Dussap

Sustainable technologies in Food waste management
Ranjna Sirohi, Ayon Tarafdar, Luciana Porto de Souza Vandenberghe, Mohammad J. Taherzadeh, and Ashok Pandey

Microplastic Pollution: Occurrence, Health Risk and Challenges
Navish Kataria, Vinod Kumar Garg, Changseok Han and Eldon Rene

Diatoms and Sustainable Agriculture
Archana Tiwari and Ashok Pandey

For more information about this series, please visit https://www.routledge.com/Sustainable-Industrial-and-Environmental-Bioprocesses/book-series/SIEB

Traditional Foods
Impact on Gut Health

Edited by
Sapna Sharma and Anand Prakash Singh

CRC Press
Taylor & Francis Group
Boca Raton London New York

CRC Press is an imprint of the
Taylor & Francis Group, an **informa** business

BRSI
The Biotech Research Society, India

Cover image: Getty Images

First edition published 2026
by CRC Press
2385 NW Executive Center Drive, Suite 320, Boca Raton FL 33431

and by CRC Press
4 Park Square, Milton Park, Abingdon, Oxon, OX14 4RN

CRC Press is an imprint of Taylor & Francis Group, LLC

ISBN: 9781032846378 (hbk)
ISBN: 9781032846408 (pbk)
ISBN: 9781003514244 (ebk)

DOI: 10.1201/9781003514244

Typeset in Times New Roman
by Deanta Global Publishing Services, Chennai, India

Contents

SECTION III Molecular Aspects in Gut-Associated Diseases and Their Management

About the Editors

Dr. Sapna Sharma works as an assistant professor at Uttar Pradesh State Institute of Forensic Science, Lucknow, UP, India. She completed her Ph.D. (2007–2012) at CSIR–Indian Institute of Toxicology Research, Lucknow, India. Dr. Sharma has vast academic experience as an assistant professor at the Institute of Biosciences and Biotechnology, SRM University, Lucknow, Uttar Pradesh. Dr. Sharma obtained a doctorate (Ph.D.) with a focus on the role of probiotics in hepatotoxicity. She completed her M.Phil in Microbiology and M.Sc in Applied Microbiology (gold medalist) at Ch. Charan Singh University, Meerut, India. She has more than eight years of post-Ph.D. research experience in international and national organizations. She worked as a research scientist (2015–2020) at the School of Biotechnology, Jawaharlal Nehru University, New Delhi, India. Her areas of focus during her tenure were metabolic diseases, the gut axis, and the role of intestinal permeability during early disease onset. Dr. Sharma worked as a senior post-doc fellow at the University of Gothenburg, Sweden, in vaccine development against myasthenia gravis, a rare neurological disorder. She also worked as a post-doc fellow at the University of Chicago, IL, U.S.A., studying the molecular mechanisms regulating anion exchangers, which play an important role in intestinal oxalate secretion in the presentation of hyperoxaluria. Dr. Sharma has more than 25 publications in international journals, several books, and has patents in the field of gut-associated metabolic disease, with a focus on membrane permeability. Dr. Sharma has received the award of best research paper at several international conferences/meetings. She has also received several national and international grants during her academic and research career.

Dr. Anand Prakash Singh is a research investigator at the Institute for Heart and Brain Health, University of Michigan, U.S.A.. He obtained his Ph.D. (2016) from the Special Center for Molecular Medicine, Jawaharlal Nehru University, India. During his Ph.D., he identified a previously unknown mechanism of tight junction disruption by the enteropathogenic E. coli (EPEC) effector Map. He has more than seven years of post-Ph.D. experience in national and international organizations. Dr. Singh worked as a postdoctoral fellow at Massachusetts General Hospital, Harvard Medical School, U.S.A., in the area of exercise and cardiac regeneration. He also worked as Visiting Scholar at the Broad Institute of MIT and Harvard Medical School, U.S.A., to learn and perform single-cell RNA sequencing of heart samples, as well as working as a postdoctoral fellow at the University of Alabama at Birmingham, and Vanderbilt University Medical Center, TN, U.S.A., in the area of cancer therapy-induced cardiovascular toxicity. His research interests encompass the alteration of cellular junctions in the gut and its role in cardiac diseases (heart-gut axis). He has more than 24 research publications in highly reputed journals.

List of Contributors

Ayman Younes Fathy Allam
Department of Food Science and Technology
Faculty of Agriculture, Minoufiya University
Shibin El-kom, Egypt

Amedeo Amedei
Department of Experimental and Clinical Medicine
University of Florence, 50134 Florence
Italy;

Joint Laboratory MIA-LAB (Microbiome-Immunity Axis research for a Circular Health)
University of Florence
Italy;

Network of Immunity in Infection, Malignancy and Autoimmunity (NIIMA), Universal Scientific Education and Research Network (USERN)
Florence
Italy

Ayushi
Faculty of Biotechnology, Institute of Biosciences and Technology
Shri RamSwaroop Memorial University
Lucknow-Deva Road, Barabanki, Uttar Pradesh, India

Nikita Basant
Faculty of Biotechnology, Institute of Biosciences and Technology
Shri RamSwaroop Memorial University
Lucknow-Deva Road, Barabanki, Uttar Pradesh, India

Sara Bertorello
Department of Experimental and Clinical Medicine
University of Florence, 50134 Florence, Italy

Shashi Lata Bharati
North Eastern Regional Institute of Science and Technology
Nirjuli, Arunachal Pradesh-791109, India

R. Chandrakanth
Department of Molecular Biology
Yuvaraja's College, University of Mysore
Mysuru-570006, India

Deepika S Channe
Assistant Chief Technical Officer, PME Cell
ICAR-Central Institute of Agricultural Engineering
Nabi Bagh Berasia Road, Bhopal, India

Pankaj Kumar Chaurasia
P.G. Department of Chemistry
L.S. College, B.R.A. Bihar University
Muzaffarpur Bihar, India

Francesca Cuffaro
Department of Experimental and Clinical Medicine
University of Florence
50134 Florence, Italy

Homa Fatma
Molecular Cancer Genetics and Translational Research Lab
Department of Zoology
Aligarh-202002, India

Shiwangi Gupta
Scientist-C
Department of Health Research MoHFW
Delhi, India

Kopal Jain
Faculty of Biotechnology
Institute of Biosciences and Technology
Shri RamSwaroop Memorial University, Lucknow-Deva Road, Barabanki, Uttar Pradesh, India

Mohd Jameel
Molecular Cancer Genetics and Translational Research Lab
Department of Zoology
Aligarh-202002, India

Narendra Kumar
Assistant Professor
Food Safety and Quality Assurance, College of Dairy and Food Technology
Rajasthan University of Veterinary and Animal Sciences, State Agricultural University
Bassi, Jaipur, India

Vipul Kumar
Department of Pharmaceutical Chemistry, Delhi Institute of Pharmaceutical Sciences and Research (DIPSAR)
Delhi Pharmaceutical Sciences and Research University
New Delhi, India

Samlesh Kumari
Scientist Dairy Microbiology, Centre of Excellence on Soybean Processing and Utilization
ICAR-Central Institute of Agricultural Engineering
Nabi Bagh Berasia Road, Bhopal, India

Vandana Kumari
Teaching Associate, Food Safety and Quality Assurance
College of Dairy and Food Technology
Rajasthan, India
University of Veterinary and Animal Sciences, State Agricultural University
Bassi, Jaipur, India

Yuxue Li
Department of Bioinformatics and Systems Biology, College of Life Science and Technology
Huazhong University of Science and Technology
Wuhan 430074, China

Smrutirekha Mallick
Scientist
ICAR-National Institute on Foot and Mouth Disease
Arugul, Bhubaneswar, India

Chandan Maurya
Department of Chemistry
Navyug Kanya Mahavidyalaya, University of Lucknow
Lucknow, Uttar Pradesh, India

Nishu Mittal
Faculty of Biosciences, Institute of Biosciences and Technology
Shri RamSwaroop Memorial University
Lucknow-Deva Road, Barabanki, Uttar Pradesh, India

Elena Niccolai
Department of Experimental and Clinical Medicine
University of Florence
50134 Florence, Italy;
Joint Laboratory MIA-LAB (Microbiome-Immunity Axis research for a Circular Health)
University of Florence
Italy

Kang Ning
Department of Bioinformatics and Systems Biology
College of Life Science and Technology, Huazhong University of Science and Technology
Wuhan 430074, China

Kritika Pandey
Faculty of Biotechnology
Institute of Biosciences and Technology, Shri RamSwaroop Memorial University
Lucknow-Deva Road, Barabanki, Uttar Pradesh, India

Shivam Pandey
Faculty of Biotechnology
Institute of Biosciences and Technology, Shri RamSwaroop Memorial University
Lucknow-Deva Road, Barabanki, Uttar Pradesh, India

Ramesh Pothuraju
Ramanujan Fellow, Scientist-C
Cancer Research Program, Rajiv Gandhi Centre for Biotechnology (RGCB)
Thycaud Post, Poojappura, Thiruvananthapuram, Kerala, India

Bettadahalli Sadashivaiah
Department of Molecular Biology
Yuvaraja's College, University of Mysore
Mysuru-570006, India

Avanish Chandra Sharma
Department of Chemistry
Navyug Kanya Mahavidyalaya, University of Lucknow
Lucknow, Uttar Pradesh, India

Jeetesh Sharma
Department of Clinical Science (k2)
University of Bergen
Bergen, Norway

Sapna Sharma
School of Forensic Science
Uttar Pradesh State Institute of Forensic Science
Lucknow, Uttar Pradesh, India

Mahak Shukla
Faculty of Biosciences
Institute of Biosciences and Technology, Shri RamSwaroop Memorial University
Lucknow-Deva Road, Barabanki, Uttar Pradesh, India

Hifzur R. Siddique
Molecular Cancer Genetics and Translational Research Lab
Department of Zoology
Aligarh-202002

Rachana Singh
Faculty of Biosciences
Institute of Biosciences and Technology, Shri RamSwaroop Memorial University
Lucknow-Deva Road, Barabanki, Uttar Pradesh, India

Sunita Singh
Department of Chemistry
Navyug Kanya Mahavidyalaya, University of Lucknow
Lucknow, Uttar Pradesh, India

Anand Prakash Singh
Institute for Heart and Brain Health
University of Michigan
Ann Arbor, MI, U.S.A.

Vimarsh Srivastava
Faculty of Biotechnology
Institute of Biosciences and Technology, Shri RamSwaroop Memorial University
Lucknow-Deva Road, Barabanki, Uttar Pradesh, India

L. Sunil
Department of Pharmacological & Pharmaceutical Sciences
College of Pharmacy, University of Houston
Houston, TX, U.S.A.

Neeraj Tiwari
Faculty of Biosciences
Institute of Biosciences and Technology, Shri RamSwaroop Memorial University
Lucknow-Deva Road, Barabanki, Uttar Pradesh, India

Section I

Introduction

1 Importance of Traditional Foods

Nikita Basant[1], Ayushi[1], Mahak Shukla[2], Rachana Singh[2], Sapna Sharma[3], and Neeraj Tiwari[2]

[1]Faculty of Biotechnology, Institute of Biosciences and Technology, Shri RamSwaroop Memorial University, Lucknow-Deva Road, Barabanki, Uttar Pradesh, India
[2]Faculty of Biosciences, Institute of Biosciences and Technology, Shri RamSwaroop Memorial University, Lucknow-Deva Road, Barabanki, Uttar Pradesh, India
[3]School of Forensic Science, Uttar Pradesh State Institute of Forensic Science, Lucknow, Uttar Pradesh, India

1.1 OVERVIEW OF TRADITIONAL FOODS, FUNCTIONAL FOODS, AND NUTRACEUTICALS

Tradition can be defined as a long-standing belief or custom that has been passed down through the years from one generation to the next. It includes following long-standing traditions or established conventions that have been passed down through the ages. It also includes customary ingredients and cooking techniques. These traditional eating customs reflect cultural diversity. Customary cuisines are manifestations of a people's identity, heritage, culture, and way of life (Fieldhouse 2013). Traditional foods represent cultural legacy and have a big impact on modern eating habits. They are essential elements of eating practices in many different countries, which makes them vital for accurately determining the amount of food consumed by the populace. These foods, which endure in our increasingly globalized world, are the embodiment of culture, history, and way of life (Rocillo-Aquino et al. 2021). Examining traditional foods offers important insights into how dietary norms have changed over time. Additionally, the long-standing use of traditional foods and dietary practices may validate their potential health benefits (Singh et al. 2020).

Food scientists, producers, and most importantly, consumers, have realized that foods have health benefits that go beyond supporting certain bodily functions to support general health and well-being. In the past few decades, a substantial amount of research has been done to gain a better understanding of the health-promoting properties of traditional food (Figure 1.1) (Witkamp 2022). The desire of consumers to take control of their health and the growing understanding of the relationship between diet and optimal health are the driving forces behind this increased research

DOI: 10.1201/9781003514244-2

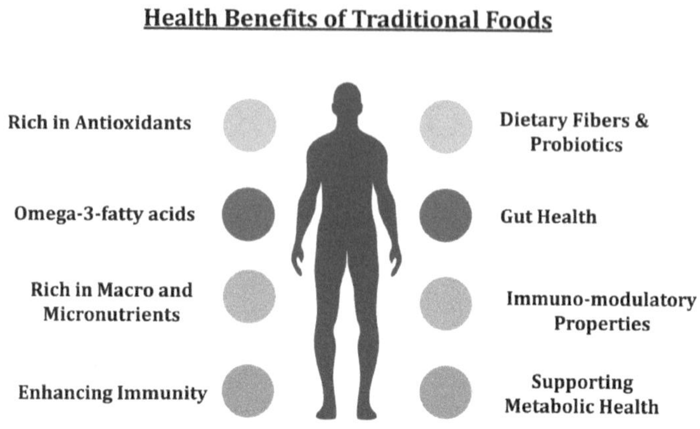

FIGURE 1.1 Health benefits of traditional foods.

effort. These beliefs are supported by mounting scientific evidence, which shows that foods and their ingredients can offer many general health benefits that are sometimes disregarded (Dillard and German 2000). Words like "functional foods," "traditional foods," and "nutraceuticals" are commonly used in this context. Since many people believe that anything considered traditional has inherent nutritional value and other health benefits, these terms are sometimes used interchangeably.

Traditional foods include a broad variety of culinary techniques or staple foods that are deeply embedded in customs from around the world. These are foods that have been ingested for many generations and are deeply ingrained in cultural and traditional practices (Ghosh, Meyer-Rochow, and Jung 2023). Traditional foods are generally regarded as having positive attributes and are valued for their genuineness, local origin, production, and lack of industrial processing. These foods are identified by particular characteristics that differentiate them from comparable products within the same category, such as the use of traditional primary products, compositions, production or processing techniques, or raw materials (Barska and Wojciechowska-Solis 2018).

Similarly, natural or processed foods containing either known or unknown biologically active compounds are known as functional foods. When ingested in specific, manageable, and non-toxic amounts, these foods provide clinically demonstrated and documented health benefits for the management or treatment of chronic illnesses (Arshad et al. 2021). Likewise, nutraceuticals are basically plant-based phytochemicals that give food functional properties. Their active ingredients are the source of these functional qualities. They interact with different targets at the physiological, cellular, and subcellular levels of the human body (Dixit et al. 2023). The effects of this interaction are associated with outcomes related to prevention as well as treatment. There's been a noticeable upsurge in interest lately in eating traditional fermented foods that are made with different ingredients and methods. The majority of these traditional fermented foods are produced in the Middle East, Asia, and Africa – often in homes, villages, or small-scale businesses (Bell et al. 2018). Several *in vitro* and *in vivo* studies have shown the health-promoting properties of fermented foods, highlighting a variety of advantages such as hypoglycemic, antihypertensive, antidiarrheal, and antithrombotic effects, among others (Ahn et al. 2006; Kwon et al. 2009). For instance, in India, traditional fermented dairy products like dahi, lassi, and chach are widely consumed due to their capacity to decrease serum cholesterol and lessen gastrointestinal infections (Rasane, Kailey, and Singh 2017). Moreover, it has been shown that fermented milk products have anticarcinogenic properties and modulate human brain activity (Tillisch et al. 2013; García-Burgos et al. 2020).

Traditional Japanese fermented foods like natto, miso, and tempeh are made with a combination of beneficial microorganism cultures, using age-old techniques. Due to their link to longer lifespans, these products have become well-known worldwide and are frequently considered functional foods (do Prado et al. 2022). Hence, fermented foods fall under the category of functional foods since they offer consumers health benefits in addition to essential nutrients. Prebiotics, probiotics, proteins, and peptides that have particular biological activities are commonly found in fermented foods. Though other bacteria, fungi, and yeasts also contribute to fermentation, lactic acid bacteria (LAB) from various genera – most notably *Lactobacillus*, *Streptococcus*, and *Leuconostoc* – usually dominate fermentation in foods that are fermented. Therefore, probiotics, when consumed in sufficient amounts, enhance the host's microbiota and thereby promote health.

1.2 CATEGORIES OF TRADITIONAL FOODS

Foods classified as traditional are those that have been consumed for many generations and have become essential parts of cultures and traditions. There are various categories into which these foods can be divided. Traditional foods can be divided into six categories: (1) cereal-based foods; (2) pulse- and legume-based foods; (3) fruit- and vegetable-based foods; (4) dairy-based foods; (5) meat- and fish-based foods; and (6) traditional foods enhanced with spices. Fermentation, curing/brining (salting), syruping (sugaring), drying, and smoking are common techniques for preserving traditional foods. These methods' main goals are to control the activity of water in order to prevent the growth of microorganisms, offer antibacterial qualities, and reduce chemical and biological reactions.

1.2.1 Traditional Foods Based on Cereals

For centuries, cereal-based traditional foods have been an essential part of diets around the world. They have cultural and historical significance that reflects the dietary customs and culinary traditions of various regions. Rice (including parboiled, flaked, hand-pounded, puffed, and rice wine), whole wheat flour (often used in bread, biscuits, naan, dosa, idli, vermicelli, noodles, and sweet dishes), maize, finger millet, and sorghum are some of the ingredients commonly used in these dishes. These dishes influence the regional culinary traditions by reflecting agricultural methods, weather patterns, and ingredient availability (Das, Raychaudhuri, and Chakraborty 2012).

Throughout Europe, wheat flour bread is a staple meal that varies in taste and texture from nation to nation. Pasta made from durum wheat is popular in Italy, where it is used in a variety of shapes and forms for recipes like lasagna, ravioli, and spaghetti (Elias 1995). France is home to the world-famous baguette, a popular side dish for a variety of meals, with each variety having a distinct cultural significance and being enjoyed most often during holidays or family get-togethers (Dryansky and Dryansky 2012). In traditional cuisine, rice holds a prominent place throughout Asia. Japan is a showcase of sushi artistry, where rice, vinegar, seafood, and vegetables are skillfully blended. In India, rice is the main ingredient in meals like dosa, pulao, and biryani. Rice can be prepared in a variety of ways, highlighting its versatility in terms of flavor profiles and cooking techniques.

Maize is a staple crop in Africa and is used to make traditional dishes like ugali in Kenya, sadza in Zimbabwe, and pap in South Africa. These porridges made from maize are usually served with stews, vegetables, or meats, giving people all throughout the continent a filling and healthy meal. Even though they are straightforward, sharing these foods with others creates a sense of community and has great cultural significance (Ekpa et al. 2019).

A wide variety of organic acids, alcohols, aldehydes, and ketones, as well as carbonyl compounds, are produced during cereal fermentation. Furthermore, during cereal fermentations, a few volatile compounds are produced that add to the unique flavor of fermented foods (Sharma et al. 2020). Fermented cereal-based products get their aroma from the presence of diacetyl, butyric acid, and acetic acid. However, conventional processing techniques like cooking, fermentation, and soaking help remove trypsin inhibitors from foods made with cereal.

A balanced diet requires a variety of essential nutrients, many of which are found in traditional cereal-based foods. Whole grains, such as brown rice, whole wheat, and oats, are high in fiber, vitamins, minerals, and antioxidants that promote overall health and reduce the risk of chronic illnesses (Wei et al. 2022). On the other hand, because the bran and germ layers are removed during processing, refined grains like white rice and processed wheat flour might be deficient in certain nutrients that are found in whole grains (Joye 2020). Cereals are essential to a balanced diet because they are a vital source of carbohydrates, the body's main energy source. By adding a range of cereals to customary meals, people can take advantage of a wide range of nutrients that promote general health and well-being. Additionally, the cultural significance of these foods gives eating a social and emotional component that helps people feel like they belong in their communities and have an identity. Finally, traditional foods made from cereals serve as a source of nutrition as well as a reflection of cultural history, culinary prowess, and community ties.

1.2.2 Traditional Foods Based on Pulses and Legumes

For centuries, pulses and legumes have played a crucial role in traditional diets across the globe, providing essential nutrients along with a variety of flavors and cultural significance. Brazil, Australia, Bangladesh, India, Russia, China, Canada, and the United States are some of the world's top countries producing pulses. Pulses are superfoods that are high in protein, fiber, vitamins, and minerals. Pulses include lentils, chickpeas, beans, and peas as well as legumes like peanuts and soybeans (Kumar and Pandey 2020). They are known to be able to support populations in an economical and sustainable way, which makes them essential sources of plant-based protein for vegans and vegetarians.

Dal, or lentil soup, is a staple dish in Indian cuisine that varies by region in terms of preparation and flavour. Dal is a complete protein source that is tasty and nourishing. It is typically seasoned with a mixture of spices called "masala," and it can include different types of lentils such as masoor (red lentils), moong (mung beans), or chana (chickpeas). Dal is typically served with rice or flatbread. Pulses have historically been processed using techniques like germination and malting, which improve iron absorption by raising vitamin C levels or lowering phytic acid and tannin levels (Zhou et al. 2023). Additionally, by lowering the tannin content, legume sprouting increases iron bioavailability, and fermentation helps to lower oligosaccharide fractions, which relieves digestive problems. For example, in beans, tempeh fermentation deactivates trypsin inhibitors and reduces oligosaccharides (Verni et al. 2022).

In terms of nutrition, legumes and pulses are great providers of fiber, complex carbohydrates, plant-based protein, and micronutrients like folate, zinc, and iron (Semwal et al. 2023). Their low fat and cholesterol content helps to reduce the occurrence of chronic diseases like diabetes, obesity, and heart disease, as well as promote heart health (Lukus, Doma, and Duncan 2020). Because it increases satiety, their high fiber content helps with blood sugar regulation, weight management, and digestive health. Furthermore, the adaptability of pulses and legumes enables a broad range of culinary creations, from side dishes to soups, stews, salads, and dips, adding flavor, texture, and nutrition to meals. This adaptability helps to create a sustainable food system that promotes food security and environmental stewardship.

1.2.3 Traditional Foods Based on Fruits and Vegetables

Fruits and vegetables are staples in many traditional diets around the world and are integral parts of many cultures' culinary traditions. These plant-based foods offer a multitude of flavors, textures, and colors that enhance and diversify traditional cuisines, in addition to serving as a source of essential nutrients. They also serve as the foundation for many famous dishes that showcase local climates, agricultural methods, and cultural preferences. There are many different ways to use fruits and vegetables to create delicious and nutritious meals, ranging from robust stews to colorful salads and aromatic side dishes. This is best demonstrated by the Mediterranean diet, which places a strong focus on lean proteins, whole grains, olive oil, and fresh produce. Classic recipes like tabbouleh, ratatouille, and Greek salad highlight how fruits and vegetables can be combined to make energizing and colorful meals.

Similar to this, fruits and vegetables are a common feature of Asian cuisines, which highlight the wide variety of flavors and ingredients available in dishes like stir-fries, curries, soups, and condiments. In Latin American traditional dishes like guacamole, salsa, and ceviche, fruits like mangoes, pineapples, and papayas, and vegetables like avocados, tomatoes, corn, and peppers take center stage, contributing to their flavor and vibrancy. In addition to being delicious foods, fruits and vegetables are an abundant source of flavonoids, carotenoids, vitamins, minerals, and other phytochemicals that are good for human health. Mulberry fruit juice, for example, contains a wealth of organic acids, minerals, tannins, and anthocyanins that may help prevent heart disease (Kristbergsson and Otles 2016). Similarly, persimmon and its leaves – which are widely used in traditional Chinese, Japanese, and Korean fruit and tea – contain flavonoids, tannins, vitamins, and minerals that may have therapeutic benefits (Hossain, Moon, and Kim 2017). Asia's staple fruit, jackfruit, is rich in carotenoids, phenolic compounds, antioxidants, vitamins, and minerals (Ranasinghe, Maduwanthi, and Marapana 2019). Globally used in cuisine, cucumbers are rich in cucurbitacins, lignans, glucosides, and apigenin (Mukherjee et al. 2013). In Korea, kimchi, a fermented cabbage dish, is highly valued for its numerous health benefits, which include probiotic and anticancer effects (Kwak et al. 2014).

Traditional foods made primarily of fruits and vegetables are, in essence, more than just culinary creations; they are symbols of cultural identity, inventiveness in the kitchen, and a deep connection to the land. By including a wide variety of fruits and vegetables in our diets, we can embrace the timeless knowledge and flavors of traditional cuisines from around the world while also enjoying the complex flavors and textures of plant-based foods.

1.2.4 Traditional Foods Based on Milk

Dairy products are essential to many traditional diets and can be found in a wide variety of savory and sweet dishes. Examples of these products are milk, cheese, yogurt, and butter (Górska-Warsewicz et al. 2019). The pastoral traditions of many cultures, where raising livestock and farming dairy have been important components of agriculture and culinary practices, are reflected in the use of dairy in cooking. Foods made with milk are prized for their protein content, creamy texture, and capacity to intensify flavor in cooking. Dairy products are staples in many traditional European dishes, such as French béchamel sauce, Italian risotto, and British scones.

Dairy is a staple of Middle Eastern and South Asian cuisines, as seen in dishes like Lebanese labneh, Afghan firnee pudding, and Indian paneer (Roufs and Roufs 2014). Paneer is a versatile ingredient that works well in curries, kebabs, and desserts. Its mild flavor and creamy texture go well with vegetables and spices (Dubey 2010). Enjoyed on special occasions like weddings and holidays, firnee is a delicately sweet dessert that is a traditional Afghan milk pudding flavored with cardamom and rosewater (Kopka 2011). The majority of traditional milk-based foods are fermented milk products, such as cheese, yogurt/dahi, curds, buttermilk, ghee (clarified butter), and sweet dishes. These products have excellent nutritional and medicinal qualities. The lactic acid bacteria (LAB) strain and species have a significant impact on the texture and flavor of fermented milk (Sarkar 2008). Probiotics like fermented milk products help prevent colon cancer, lower cholesterol and triacylglycerols, treat gastroenteritis, and fight infections (Hendler and Zhang 2018; Liu, Tran, and Rhoads 2018). Yogurt, for instance, lowers cholesterol, strengthens the immune system, increases appetite and vitality, and improves lactose intolerance. Yogurt contains a wide range of volatile substances, including hydrocarbons, alcohols, esters, acids, sulfur-containing compounds, heterocyclic compounds, aromatic compounds, and carbonyl compounds. These substances, which include lactic acid, diacetyl and acetoin, and acetaldehyde, are what give yogurt its final scent (Cheng 2010). Likewise, butter includes minerals, vitamins (primarily A, D, E, and K), conjugated linoleic acid, and butyric acid (Mehta 2015). In traditional Indian medicine, i.e., Ayurveda, ghee has long been utilized for numerous medical applications, including the treatment of allergy, skin, and respiratory diseases (Sharma, Zhang, and Dwivedi 2010).

1.2.5 Traditional Foods Based on Meat and Fish

The incorporation of meat and fish in traditional cuisines mirrors the historical, geographical, and cultural factors at play, giving rise to unique culinary customs across various areas. The cooking techniques applied to meat and fish in traditional dishes exhibit a broad spectrum, encompassing grilling, roasting, stewing, frying, and curing. These cooking methods not only enrich the taste and consistency of the ingredients but also serve as reflections of the culinary proficiency and heritage of the communities from which these culinary delights emerge.

For traditional meat products, methods like drying, salting, fermenting, and fish preservation are frequently used. For example, Shutiki, Lonailish, and Shidal are traditional fermented fish products from Northeast India that are prized for their high nutritional content (Majumdar et al. 2016). Because dried fish has high protein content and a low-calorie profile, its omega-3 fatty acid content helps lower the risk of strokes and heart attacks.

1.2.6 Traditional Foods Based on Spices

Spices are essential to creating vibrant and flavorful Indian dishes like masala chai, biryani, and curry. A traditional Indian curry is aromatic and filling, made with a combination of ingredients like meat, vegetables, or legumes and a medley of spices like cumin, coriander, turmeric, and garam masala (Shukla and Yadav 2018). Spices like paprika, saffron, cumin, and cinnamon are vital ingredients in Moroccan dishes like tagine and ras el hanout. Moroccan cuisine is known for its distinct depth of flavor, which is achieved through the use of a blend of spices that range from savory to sweet in tagine, a stew that is cooked slowly and named after the earthenware pot in which it is cooked (Koehler 2012).

Similarly, Mexican cooking uses a lot of spices, including paprika, cumin, chili, and oregano, in recipes like tamales, salsa, and mole. The complex flavors and rich cultural legacy of Mexican cuisine are exemplified by the rich sauce known as "mole," which is made with a blend of chocolate, chiles, and spices. A zesty addition to dishes like tacos and enchiladas is salsa, a fresh condiment made with tomatoes, onions, cilantro, and chili peppers. Along with adding flavor and aroma to traditional foods, spices have symbolic and cultural value in many cultures. They are frequently linked to customs, festivals, and rituals, signifying wealth, friendliness, and a link to the land. Spices have long been used in trade, exploration, and cross-cultural exchanges that have shaped cuisines around the world and culinary customs on many continents.

Furthermore, since ancient times, people have used spices extensively for their flavor-enhancing, colorant, preservative, and possibly therapeutic qualities. They contain a variety of phytochemicals with antimicrobial, antioxidant, anticancer, anti-inflammatory, and cardioprotective properties, including curcumin, piperine, capsaicin, and cinnamondehyde (Srinivasan 2014). For instance, studies have shown that turmeric and its curcuminoids have antimicrobial qualities against a range of pathogens and may have anti-cancer effects against a number of cancer types (Adamczak, Ożarowski, and Karpiński 2020; Tomeh, Hadianamrei, and Zhao 2019). It has been demonstrated that cumin can lessen oxidative stress and elevated blood sugar (Miah et al. 2021). These medicinal qualities draw attention to the many advantages of adding spices to traditional foods, enhancing flavor and health.

1.3 REVITALIZING TRADITIONAL FOODS THROUGH INNOVATION

The recent innovation of traditional foods has been greatly aided by the development of food biotechnology and fermentation techniques. These techniques could enhance traditional foods and drinks' nutritional value, shelf life, and therapeutic qualities (Xing et al. 2023). For example, multiple research studies have demonstrated that traditional fermented foods naturally increase in antioxidant and vitamin content through fermentation, improving their quality and medicinal value. Furthermore, by generating antifungal metabolites, fermentation biotechnology employing lactic acid bacteria can increase shelf life (Walther and Schmid 2017).

It has been noted that fermentation technology is being used for traditional food products in Africa, China, India, the Middle East, and the Mediterranean (Siddiqui et al. 2023). This includes the development of microbial production technology, the design of bioreactors, and the metabolic engineering of bacteria. Apart from food biotechnology and fermentation, advanced packaging techniques have also been employed as means to extend the food's shelf life (Drago et al. 2020). The current global trend in the food industry towards safer and longer-lasting food production is in line with the efforts made by food scientists and industries to modernize traditional foods. Modern techniques have been used to produce traditional foods on a larger scale, and creative positioning strategies have been employed in their marketing. For instance, creative packaging ideas have been effectively applied for traditional Indonesian foods like "canned gudeg," "hygienic tempeh," "sambal in a jar," and "powdered coconut milk" (Fibri and Frøst 2019).

The risk of modernizing and increasing production of traditional foods must be recognized, though, as it could result in the loss of their rich cultural legacy. When traditional products are overly altered, they may lose their authenticity and become less popular with customers. In the long run, this may result in diminished consumption of traditional, healthful products and a sense of cultural loss (Fibri and Frøst 2020). Standardization and registration are two other ways to keep traditional foods alive. But given the close relationship that exists between people and food, especially when it comes to traditional foods, it is imperative that modernized versions of these foods have qualities that go beyond their sensory appeal. Showcasing superior qualities can improve a product's standing in the market and boost repurchase probability. This strategy can be extremely helpful in maintaining traditional foods for future generations and protecting local biodiversity. Therefore, in order for the food industry to engage consumers beyond sensory experiences and into a deeper level of engagement, it is imperative that it capture the emotional attachment associated with traditional foods. This all-encompassing strategy can support customers' adherence to customs while safeguarding their legacy for a sustainable future.

1.4 TRADITIONAL FOODS/COMPONENTS EXHIBITING IMMUNO-MODULATORY EFFECTS

A healthy lifestyle can be maintained by utilizing the diverse immunological properties offered by traditional foods and their ingredients. Traditional foods are naturally high in nutrients because they have undergone little processing and have retained their natural nutrients. Moreover, some traditional foods, such as fermented foods like sauerkraut, kimchi, and yogurt, can help maintain a healthy gut microbiome. A diverse and well-balanced gut microbiome has been linked to better immune system performance, mental health, and digestion. Thus, eating a variety of traditional foods can provide essential vitamins, minerals, antioxidants, and other beneficial compounds that promote general health and vitality.

1.4.1 Omega-3-Fatty Acids

Omega-3 polyunsaturated fatty acids (n-3 PUFAs) are a class of compounds important for human diet and physiology. The most well-known ones are α-linolenic acid (ALA), eicosapentaenoic acid (EPA), and docosahexaenoic acid (DHA) (Saini and Keum 2018). Significantly, higher blood levels of n-3 PUFAs have been connected to a lower risk of death from a number of different diseases. n-3 PUFAs have been shown in numerous studies to have immunomodulatory effects, which can be attributed to their direct action on immune and non-immune cells or their indirect influence on the gut microbiome (Oppedisano et al. 2020; Jayapala and Lim 2023).

1.4.2 Dietary Fiber

Dietary fiber is a non-essential component of mammal diets. It is composed of carbohydrate polymers with ten or more monomeric units that can withstand digestion by pancreatic enzymes or the intestinal brush border. Rather, they enter the colon undigested, where bacteria break them down. Some forms of dietary fiber cause changes in the luminal microbiota, which benefits the host's health (Holscher 2017). Within the colon, gut microbiota ferments dietary fiber, producing large amounts of short-chain fatty acids (SCFAs), including propionate, butyrate, and acetate (Nogal, Valdes, and Menni 2021). By improving epithelial integrity, these SCFAs and dietary fiber help to lessen systemic endotoxemia. Furthermore, SCFAs influence gene transcription by inhibiting histone deacetylases (HDACs). Additionally, SCFAs support immune homeostasis by binding to G-protein coupled receptors (GPCRs), also referred to as "metabolite-sensing" receptors, such as GPR43 (Qian et al. 2022). Significant amounts of SCFAs produced in the colon are sufficient to reduce pH, which affects the activity of proton sensors like GPR65 that are expressed on immune cells and the gut epithelium. Inflammatory cytokines are suppressed by GPR65, an anti-inflammatory Gαs-coupled receptor (Xie et al. 2023).

1.4.3 Vitamins

Vitamins can modify immunological responses and are essential for immune system maintenance. Many vitamins, including the B complex vitamins, as well as A, C, E, and D, have been demonstrated to be important for immune regulation, which supports overall health (Mora, Iwata, and Von Andrian 2008). Vitamin A plays a critical role in preserving the integrity of mucosal barriers in the digestive and respiratory systems. It controls immune cell differentiation and function and promotes the synthesis of mucins and antimicrobial peptides, which help prevent infections (Filipe Rosa et al. 2023; Fan et al. 2015). Similarly, research has shown that vitamin C, especially at concentrated doses, has antimicrobial qualities that reduce the risk of infections (Mousavi, Bereswill, and Heimesaat 2019). As a potent antioxidant, vitamin C boosts the generation of white blood cells, fortifies immune cell barriers against infections, and protects immune cells from the damaging effects of free radicals (Carr and Maggini 2017).

Vitamin D has the power to influence the innate immune system, boost immune cell phagocytic capacity, and strengthen the physical barrier function of epithelial cells. Additionally, it affects the function of immune cells like T-cells and macrophages, strengthening the body's resistance to infections. A low level of vitamin D has been associated with heightened susceptibility to infections and autoimmune diseases (Sassi, Tamone, and D'Amelio 2018). Additionally, vitamin E stimulates T-cell differentiation and proliferation, lowers the production of inflammatory cytokines, and modifies inflammatory responses in different tissues, including the lungs (Moriguchi and Kaneyasu 2004). The synthesis of cytokines and antibodies, which are necessary for directing immune responses, depends on vitamin B6. By adjusting the size of the thymus and the maturation, proliferation, and function of B and T lymphocytes, it controls the adaptive immune response (Islas-Weinstein and Maldonado-García 2022). Vitamin B6 deficiency can weaken the immune system and make a person more prone to infections (Stach, Stach, and Augoff 2021). Last but not least, vitamin B12 is important for cellular immunity, especially for $CD8^{+}$ cells and the NK cell system (Tamura et al. 1999). It is essential for red blood cell and DNA production, critical for immune cell proliferation and function, and contributes to regulating inflammation and supporting immune cell activity (Ge, Zadeh, and Mohamadzadeh 2022).

1.4.4 Minerals

Minerals are necessary ingredients in every kind of food and perform a variety of bodily functions. They are important for the development of bones, affect the function of muscles and nerves, control the body's water balance, and aid in the synthesis of hormones, enzymes, and other molecules that are biologically active. Furthermore, minerals strengthen the body's immune system by acting as immunomodulators (Weyh et al. 2022). As a cofactor for many different enzymes, magnesium is essential for the activation of membrane function, enzymes, and intracellular signaling. Magnesium is necessary for the synthesis of DNA and RNA, the secretion of hormones and enzymes, oxidative phosphorylation, and muscle contraction. Additionally, it influences the acute phase response and macrophage function in response to cytokines, stabilizes membrane structures, and modifies ion transporters (Volpe 2013). As an antioxidant, zinc affects how multiprotein complexes like the T-cell receptor are arranged and how stable biological membranes are. In addition, it influences the production of cytokines, balances the function of the immune system, controls the development of hormones and receptor activity, and affects the production of antibodies and T-cell differentiation (Chasapis et al. 2012; Bonaventura et al. 2015). Likewise, iron is a dietary mineral that is necessary for many essential human processes, such as erythropoiesis, the metabolism of cellular energy, and the growth and operation of the immune system. It plays a multifaceted and intricately complex role in immune system modulation, which emphasizes the significance of maintaining healthy iron stores. It has been demonstrated that low iron inhibits T-cell function, B-cell proliferation, and adaptive antibody responses; these effects can be effectively reversed with iron supplementation (Jiang et al. 2019; Mullick et al. 2006).

1.4.5 Probiotics

Probiotic bacteria make up a substantial portion of the large gut microbiota and are an important component of gut-associated lymphoid tissue (GALT). GALT is a key component of the immune system because it includes 75% of the lymphoid cells of the overall immune system and around 50% of the lymphocytes. Probiotic bacteria have the ability to influence specific immune functions by interacting with and stimulating commensal microflora and intestinal immune cells (Mazziotta et al. 2023). Their ability to promote the proper function and differentiation of T and B lymphocytes, dendritic cells, and macrophages is thought to be the reason behind their anti-inflammatory qualities. Probiotic bacteria support this increased activity, which increases the presentation of antigens to B lymphocytes and the production of IgA (Latif et al. 2023). Furthermore, research suggests that probiotic bacteria can control the production of cytokines. Some Lactobacillus strains cause Th1 lymphocytes to secrete more IFN-γ and IL-12, which is linked to NK cell cytotoxic activity. Lactobacillus and Bifidobacterium affect the expression of regulatory and anti-inflammatory cytokines like IL-10 and TGF-β, which inhibit Treg lymphocytes (Hardy et al. 2013). Supplements and fermented foods, such as sour dairy products or fermented vegetables, are good sources of probiotics. Studies have indicated that foods that have undergone fermentation provide a number of health advantages, such as immune system-related antioxidant, antimicrobial, antifungal, anti-inflammatory, antidiabetic, and anti-atherosclerotic properties.

1.4.6 Nigella sativa

Nigella sativa is highly valued for its practical qualities and is well-known for its many uses in both the culinary and medical domains. The benefits of N. sativa seeds include increased immunity, reduced pain and inflammation, and lowered blood sugar. They have an impressive nutritional profile that includes 26% proteins, 28% fats, 25% carbohydrates, 8.4% crude fiber, and 4.8% ash. They also contain high levels of minerals like iron, copper, zinc, and phosphorus, as well as carotene (Ciesielska-Figlon et al. 2023).

This plant is native to Southwest Asia, Southern Europe, and North Africa, and is known by many names in different languages. In English, it is called black seed, black cumin, and black caraway; in Arabic, it is called Habbah Al-Sauda, or "seed of blessing"; in Russian, it is called chernushka; in Turkish, it is called çörek otu; and in Persian, it is called Cyah-daneh. Its seeds are an essential part of traditional medicine and cuisine due to their wide range of biological qualities, which incorporate anticancer, antibacterial, antihypertensive, hepatoprotective, anti-inflammatory, antipyretic, analgesic, and antioxidative effects (Amin and Hosseinzadeh 2015). Thymoquinone (TQ), in particular, is one of the components of N. sativa that exhibits immunomodulatory effects by influencing T cell and natural killer (NK) cell activity (Salem and Hossain 2000). Majdalawieh et al. investigated how N. sativa affected NK cell activity, macrophage function, and splenocyte proliferation in cell cultures. According to their findings, aqueous extracts of N. sativa seeds suppressed the secretion of proinflammatory mediators by macrophages, such as tumor necrosis factor-alpha (TNF-α), interleukin-6 (IL-6), and nitric oxide (NO), while concurrently increasing splenocyte proliferation and the secretion of Th2 cytokines responsible for humoral immune responses in a dose-dependent manner (Majdalawieh, Hmaidan, and Carr 2010).

Additionally, a three-month study on asthmatic patients assessed the preventive benefits of water extracts from N. sativa seeds. When compared to baseline measurements, the group that received boiled N. sativa seed extract demonstrated significant improvements in wheezing, frequency of symptoms, asthma severity, and pulmonary function test (PFT) values (Boskabady et al. 2007). Finally, the unique qualities of N. sativa seeds emphasize their value in both conventional and alternative medicine, underscoring their potential as a natural treatment for a range of illnesses.

1.4.7 Garlic

The versatile Allium sativum L., or garlic, has been an essential ingredient in traditional cuisines around the world for centuries due to its strong aroma and distinct flavor. Beyond just food, its cultural significance extends to historical, social, and even mythological realms. Garlic has been shown in numerous studies to have immunomodulatory qualities. This means that it may improve immune system performance by activating different cell types, including NK cells, dendritic cells, macrophages, lymphocytes, and eosinophils through the activation of macrophages, phagocytosis, modulating immunoglobulin production, and stimulating cytokine secretion (Arreola et al. 2015).

Garlic compounds have different compositions based on how they are prepared and extracted. S-allyl-L-cysteine sulfoxide (alliin) and γ-glutamyl cysteine derivatives are the main ingredients in fresh bulbs; steam-distilled oils are abundant in sulfide family compounds; diallyl disulfide (DADS) and alliin are found in crushed and dried garlic powder; macerates (ground garlic) are enhanced with sulfide family compounds, dithiines, and (E–Z)-ajoene compounds; and S-allyl mercaptocysteine (SAC) and S-allyl mercaptocysteine (SAMC) are found in aged garlic extract (AGE) (Trio et al. 2014).

A study by Feng et al. concluded that the administration of allicin to Balb/c mice infected with Plasmodium yoelii significantly resulted in decreased parasitemia and extended survival by increasing proinflammatory mediators like interferon-gamma (IFN-γ). Furthermore, treatment with allicin promoted the growth of macrophages and CD4+ T cells (Feng et al. 2012). In a similar vein, Hanieh et al. discovered that feeding white Leghorn chickens low doses of garlic (*Allium sativum)* and onion (*Allium cepa*) after they had been immunized against Brucella abortus (BA), Newcastle disease virus (NDV), and sheep red blood cells (SRBC) increased the production of anti-NDV, anti-SRBC, and anti-BA antibodies. The increased T cell proliferation associated with garlic consumption, according to the authors, may either directly or indirectly promote B-cell differentiation and proliferation (Hanieh et al. 2010).

1.4.8 Cinnamon

Two species in this genus are Cinnamomum zeylanicum and C. cassia, the only parts of which are used in food or medicine. Major ingredients in cinnamon bark include coumarin, eugenol, cinnamaldehyde, cinnamic acid, and cinnamyl alcohol. Thanks to its antioxidant content and bioactive components, cinnamon has a wide range of health benefits, including immunomodulatory properties. Potential therapeutic benefits for ailments like diabetes, neurological disorders, microbial infections, and cardiovascular diseases have been associated with these properties.

One important ingredient, cinnamaldehyde, has been shown to have anti-neuro-inflammatory effects by lowering the expression and synthesis of nitric oxide (NO), interleukin (IL)-1b, IL-6, and tumor necrosis factor (TNF)-a in lipopolysaccharide (LPS)-activated BV2 microglia (Ho, Chang, and Chang 2013). Furthermore, it may help protect neurons by preventing tau protein from aggregating, which is a characteristic of Alzheimer's disease (AD) (Błaszczyk, Rosiak, and Kałużna-Czaplińska 2021).

1.4.9 Licorice Root

Glycyrrhiza uralensis Fisch., Glycyrrhiza inflata Bat., and Glycyrrhiza glabra L. are the plants that yield dried roots and rhizomes that are referred to as licorice, or "gancao." It has a variety of benefits, such as the ability to eliminate toxins and heat, relieve pain and cough, expel phlegm, and work well with different medications (Wang et al. 2013). Due to its many therapeutic benefits, which include antiviral, anti-inflammatory, antibacterial, and immunomodulatory qualities, as well as possible hypoglycemic, anti-obesity, and detoxifying effects, licorice has a long history of use in medicine in different cultures worldwide (Zhang et al. 2021).

In a study by Ayeka et al., immunomodulatory activities of licorice polysaccharides were observed in BALB/c mice with CT 26 tumors. These polysaccharides raised the immune organ index and significantly inhibited the growth of tumors. The activation of $CD4^+$ and $CD8^+$ immune cell populations further demonstrated the immunomodulatory effect in CT 26 tumor-bearing mice (Ayeka et al. 2017). Similarly, Adianti et al. (2014) found that by suppressing the expression of HCV full-length viral particles and their core gene, licorice root containing glycyrrhizin (GL) may be able to significantly help in the treatment of patients with chronic hepatitis C (Adianti et al. 2014).

1.4.10 Black Pepper

Black pepper (Piper nigrum L.) contains many bioactive substances, such as those with anti-inflammatory, antiviral, antipyretic, immune-boosting, and bioavailability-enhancing effects (Takooree et al. 2019). Alkaloids, terpenes, flavonoids, phenolics, steroids, special amino acids, and polyoses are just a few of the phytochemicals found in pepper that contribute to its health-promoting properties. Black pepper has been found to have immunomodulatory effects on human health. Piper nigrum has been demonstrated to significantly lower body weight, triglycerides, total cholesterol, LDL, VLDL, and fat mass while maintaining constant HDL levels. Due to their hypolipidemic and antiatherogenic properties, black pepper and piperine can lower the risk of atherosclerosis through diet (Shah et al. 2011). Furthermore, piperine prevents human umbilical vein endothelial cells from proliferating and from going through the G1/S transition (Doucette et al. 2013).

Dietary Piper nigrum improves digestion and shortens the duration of gastrointestinal transit by inducing pancreatic enzyme secretion. The liver's stimulation of bile acid secretion upon oral administration of piperine or Piper nigrum is important for fat digestion and absorption (Saleem et al. 2022). Moreover, black pepper can increase white blood cell production and effectiveness, strengthening the body's defenses against invasive microorganisms. According to studies, piperine can improve immune competence and neutralize the negative effects of medications like 5-FU that might weaken the immune system (Bezerra et al. 2008). Furthermore, piperine has been observed to suppress B cell immunoglobulin α-IgM and lipopolysaccharide (LPS)-induced proliferative responses, which impact humoral immune responses to T-un/dependent antigens (Bernardo et al. 2015). Additionally, piperine has been shown to up-regulate NF-κB and IL-10. These two factors, along with gamma-aminobutyric acid (GABA), activate the p38, JNK, and MAPK pathways, which in turn increase the expression of EPO and EPO-R (Lee et al. 2018).

1.4.11 Honey

Honey has been utilized for food and medicinal purposes since ancient times. Its primary ingredient is sugar, which primarily consists of fructose and glucose. Honey contains proline as an amino acid, in addition to organic acids like gluconic acid, contributing to its diverse properties (da Silva et al. 2016). The body naturally stimulates the immune system with honey. Numerous lines of research have demonstrated honey's effectiveness as an immunomodulatory substance. According to a study by Tonks et al., published almost 20 years ago, there was a significant reduction in reactive oxygen species (ROS) production when the MM6 monocytic cell line was exposed to pasture and manuka honeys (Tonks et al. 2001). The administration of manuka honey to rats that bore gastric ulcers led to a notable rise in the levels of glutathione (GSH), nitric oxide (NO), and superoxide dismutase (SOD) in the gastric mucosa. Additionally, the plasma levels of pro-inflammatory cytokines TNF-α, IL-6, and IL-1β were found to decrease (Almasaudi et al. 2016). In a carcinogen-induced breast tumor model in Sprague Dawley rats, it was demonstrated that oral administration of relatively high concentrations (1.0 g/kg body weight per day) of Malaysian Tualang honey (TH) or MH would slow the growth of cancer after tumors became palpable. This led to smaller tumor weights and volumes as well as a better histological grade in comparison to the untreated group. In parallel, the expression of anti-apoptotic proteins TNF-α, COX-2, and Bcl-xL declined, while that of pro-apoptotic proteins Apaf-1, Caspase-9, and IFN-γ increased (Ahmed, Sulaiman, and Othman 2017).

1.4.12 Mushroom

For centuries, people have valued mushrooms as a staple food and a source of health supplements. Enriched with essential amino acids, dietary fiber, minerals, and a multitude of bioactive compounds – particularly those connected to immune system processes – they are valuable sources of natural immunomodulators (Bhambri, Srivastava, and Karn 2022; Valverde, Hernández-Pérez, and Paredes-López 2015). Terpenes, lectins, terpenoids, fungal immunomodulatory proteins (FIPs), and polysaccharides are just a few of the many immunoregulatory substances found in mushrooms. These compounds are distributed differently in different species of mushrooms, and their strong immune-modulating effects depend on changes in fraction composition and core structures (Zhao et al. 2020). Products made from mushrooms have many health advantages, including immune-stimulating and anticancer qualities, as well as antioxidant, antihypertensive, antihyperglycemic, neuroprotective, hepatoprotective, antidiabetic, antifungal, antibacterial, and antiviral effects (Venturella et al. 2021).

Furthermore, a variety of mushrooms have immune-modulating properties that have aided in the creation of immune-modulating medications. As an example, FIO-a-β and FA-2-b-Md immune modulators are found in Agaricus and Himematsutake mushrooms, while KS-2, Lentinan, and LEM immune stimulators are found in Shiitake and Golden Oak mushrooms (Mizuno, Kawakami, and Fujitake 2003).

Additionally, mushroom proteins and protein–conjugate complexes are also well-known immunomodulatory compounds. These protein-based immunomodulatory substances found in medicinal mushrooms fall into two main categories, much like the polysaccharide-based compounds: lectins and fungal immunomodulatory proteins (FIPs). According to a number of research studies, these FIPs exhibit antitumor activities, such as the suppression of cell growth and proliferation, the promotion of autophagy and apoptosis, and the decrease in tumor cell invasion and migration (Zhao et al. 2020).

Many bioactive substances found in mushrooms are linked to these effects, including proteins, polysaccharides, dietary fiber, minerals, lipopolysaccharides, glycoproteins, and secondary metabolites. It has been demonstrated that among them, complex organic compounds have immunomodulatory effects. For instance, the polysaccharides in mushrooms can activate innate immune cytokines and interleukins, as well as neutrophils, macrophages, and natural killer cells (Davis et al. 2020). Furthermore, by maintaining essential metabolic processes, secondary metabolites like phenols, terpenes, and sterols contribute to improving host survival (Chugh et al. 2022).

1.5 SUMMARY

In conclusion, traditional food is a blend of regional ingredients and savory legacy that has been passed down through the ages. It has a richness of nutritional content and culinary wisdom that has sustained communities for centuries. Traditional foods frequently demonstrate a thorough awareness of local ecosystems and sustainable farming practices that promote biodiversity and environmental harmony. Furthermore, traditional foods have been linked to numerous health benefits, thus providing a foundation for balanced and wholesome nutrition. The consumption of traditional foods offers a wide array of health advantages due to their reliance on natural, unprocessed ingredients that provide a well-rounded and wholesome dietary base. The appreciation and adoption of traditional foods can also contribute to agricultural biodiversity, given that numerous traditional crops and ingredients are typically sourced locally.

Modern-day changes in social and cultural standards push food scientists and industry innovators to reimagine classic recipes, leading to longer shelf lives and better distribution. However, there's a significant risk that the modernization process, with its increased production and changes to traditional recipes, will destroy the rich cultural legacy ingrained in food. In addition to essential vitamins (A, C, D, E, B6, B12) and minerals (zinc, copper, iron, selenium), many traditional foods are rich in phytochemical compounds (phenolics, flavonoids, alkaloids, polysaccharides, amino acids) that have been shown to have immune-boosting properties through a variety of pathways. This chapter demonstrates traditional foods, their categories, and how some traditional foods can strengthen immunity, thus helping in achieving a balanced and healthy life.

REFERENCES

Adamczak, Artur, Marcin Ożarowski, and Tomasz M. Karpiński. 2020. Curcumin, a natural antimicrobial agent with strain-specific activity. *Pharmaceuticals* 13 (7):153.

Adianti, Myrna, Chie Aoki, Mari Komoto, Lin Deng, Ikuo Shoji, Tutik Sri Wahyuni, Maria Inge Lusida, Soetjipto, Hiroyuki Fuchino, and Nobuo Kawahara. 2014. Anti-hepatitis C virus compounds obtained from Glycyrrhiza uralensis and other Glycyrrhiza species. *Microbiology and immunology* 58 (3):180–187.

Ahmed, Sarfraz, Siti Amrah Sulaiman, and Nor Hayati Othman. 2017. Oral administration of Tualang and Manuka honeys modulates breast cancer progression in Sprague-Dawley rats model. *Evidence-Based Complementary and Alternative Medicine* 1:5904361.

Ahn, In-Sook, Myoung-Sool Do, Su-Ok Kim, Hun-Soon Jung, Young-In Kim, Hye-Jung Kim, and Kun-Young Park. 2006. Antiobesity effect of Kochujang (Korean fermented red pepper paste) extract in 3T3-L1 adipocytes. *Journal of Medicinal Food* 9 (1):15–21.

Almasaudi, Saad B., Nagla A. El-Shitany, Aymn T. Abbas, Umama A. Abdel-Dayem, Soad S. Ali, Soad K. Al Jaouni, and Steve Harakeh. 2016. Antioxidant, anti-inflammatory, and antiulcer potential of manuka honey against gastric ulcer in rats. *Oxidative Medicine and Cellular Longevity* 1:3643824.

Amin, Bahareh, and Hossein Hosseinzadeh. 2015. Black cumin (Nigella sativa) and its active constituent, thymoquinone: An overview on the analgesic and anti-inflammatory effects. *Planta Medica*: 82(01/01):8–16.

Arreola, Rodrigo, Saray Quintero-Fabián, Rocío Ivette López-Roa, Enrique Octavio Flores-Gutiérrez, Juan Pablo Reyes-Grajeda, Lucrecia Carrera-Quintanar, and Daniel Ortuño-Sahagún. 2015. Immunomodulation and anti-inflammatory effects of garlic compounds. *Journal of Immunology Research* 1:401630.

Arshad, Muhammad Sajid, Waseem Khalid, Rabia Shabir Ahmad, Muhammad Kamran Khan, Muhammad Haseeb Ahmad, Saira Safdar, Safura Kousar, Haroon Munir, Umair Shabbir, and Muhammad Zafarullah. 2021. Functional foods and human health: An overview. *Functional Foods Phytochem Health Promoting Potential* 3.

Ayeka, Peter Amwoga, YuHong Bian, Peter Mwitari Githaiga, and Ying Zhao. 2017. The immunomodulatory activities of licorice polysaccharides (Glycyrrhiza uralensis Fisch.) in CT 26 tumor-bearing mice. *BMC Complementary and Alternative Medicine* 17:1–9.

Barska, Anetta, and Julia Wojciechowska-Solis. 2018. Traditional and regional food as seen by consumers–research results: The case of Poland. *British Food Journal* 120 (9):1994–2004.

Bell, Victoria, Jorge Ferrão, Lígia Pimentel, Manuela Pintado, and Tito Fernandes. 2018. One health, fermented foods, and gut microbiota. *Foods* 7 (12):195.

Bernardo, Aline Rodrigues, Juliana Dutra Barbosa da Rocha, Marco Edilson Freire de Lima, Débora Decote Ricardo, Lúcia Helena Pinto da Silva, Lígia Maria Torres Peçanha, and Maria das Graças Miranda Danelli. 2015. Modulating effect of the piperine, the main alkaloid from Piper nigrum Linn., on murine B lymphocyte function. *Brazilian Journal of Veterinary Medicine* 37 (3):209–216.

Bezerra, Daniel P., Fernanda O. de Castro, Ana Paula N. N. Alves, Cláudia Pessoa, Manoel Odorico de Moraes, Edilberto R. Silveira, Mary Anne S. Lima, Francisca Juliana M. Elmiro, Nylane M. N. de Alencar, and Rodney O. Mesquita. 2008. In vitro and in vivo antitumor effect of 5-FU combined with piplartine and piperine. *Journal of Applied Toxicology: An International Journal* 28 (2):156–163.

Bhambri, Anne, Malay Srivastava, and Santosh Kumar Karn. 2022. Mushrooms as potential sources of active metabolites and medicines. *Frontiers in Microbiology* 13:837266.

Błaszczyk, Natalia, Angelina Rosiak, and Joanna Kałużna-Czaplińska. 2021. The potential role of cinnamon in human health. *Forests* 12 (5):648.

Bonaventura, Paola, Giulia Benedetti, Francis Albarède, and Pierre Miossec. 2015. Zinc and its role in immunity and inflammation. *Autoimmunity Reviews* 14 (4):277–285.

Boskabady, Mohammad Hossein, H. Javan, M. Sajady, and H. Rakhshandeh. 2007. The possible prophylactic effect of Nigella sativa seed extract in asthmatic patients. *Fundamental & Clinical Pharmacology* 21 (5):559–566.

Carr, Anitra C., and Silvia Maggini. 2017. Vitamin C and immune function. *Nutrients* 9 (11):1211.

Chasapis, Christos T., Ariadni C. Loutsidou, Chara A. Spiliopoulou, and Maria E. Stefanidou. 2012. Zinc and human health: An update. *Archives of Toxicology* 86:521–534.

Cheng, Hefa. 2010. Volatile flavor compounds in yogurt: A review. *Critical Reviews in Food Science and Nutrition* 50 (10):938–950.

Chugh, Rishi Man, Pooja Mittal, Namratha Mp, Tanu Arora, Tanima Bhattacharya, Hitesh Chopra, Simona Cavalu, and Rupesh K. Gautam. 2022. Fungal mushrooms: A natural compound with therapeutic applications. *Frontiers in Pharmacology* 13:925387.

Ciesielska-Figlon, Klaudia, Karolina Wojciechowicz, Anna Wardowska, and Katarzyna Aleksandra Lisowska. 2023. The immunomodulatory effect of Nigella sativa. *Antioxidants* 12 (7):1340.

da Silva, Priscila Missio, Cony Gauche, Luciano Valdemiro Gonzaga, Ana Carolina Oliveira Costa, and Roseane Fett. 2016. Honey: Chemical composition, stability and authenticity. *Food Chemistry* 196:309–323.

Das, Arpita, Utpal Raychaudhuri, and Runu Chakraborty. 2012. Cereal based functional food of Indian subcontinent: A review. *Journal of Food Science and Technology* 49:665–672.

Davis, Renee, Alex Taylor, Regan Nally, Kathleen F. Benson, Paul Stamets, and Gitte S. Jensen. 2020. Differential immune activating, anti-inflammatory, and regenerative properties of the aqueous, ethanol, and solid fractions of a medicinal mushroom blend. *Journal of Inflammation Research* 13:117–131.

Dillard, Cora J., and J. Bruce German. 2000. Phytochemicals: nutraceuticals and human health. *Journal of the Science of Food and Agriculture* 80 (12):1744–1756.

Dixit, Versha, S. William Joseph Kamal, Pranjali Bajrang Chole, Deen Dayal, Kundan Kumar Chaubey, Anish Kumar Pal, Jobi Xavier, B. T. Manjunath, and Rakesh Kumar Bachheti. 2023. Functional foods: Exploring the health benefits of bioactive compounds from plant and animal sources. *Journal of Food Quality* 1:5546753.

do Prado, Fernanda Guilherme, Maria Giovana Binder Pagnoncelli, Gilberto Vinícius de Melo Pereira, Susan Grace Karp, and Carlos Ricardo Soccol. 2022. Fermented soy products and their potential health benefits: A review. *Microorganisms* 10 (8):1606.

Doucette, Carolyn D., Ashley L. Hilchie, Robert Liwski, and David W. Hoskin. 2013. Piperine, a dietary phytochemical, inhibits angiogenesis. *The Journal of Nutritional Biochemistry* 24 (1):231–239.

Drago, Emanuela, Roberta Campardelli, Margherita Pettinato, and Patrizia Perego. 2020. Innovations in smart packaging concepts for food: An extensive review. *Foods* 9 (11):1628.

Dryansky, Gerry Y., and Joanne Dryansky. 2012. *Coquilles, Calva, & Crème: Exploring France's Culinary Heritage: A Love Affair with French Food*. Open Road Media.

Dubey, Krishna Gopal. 2010. *The Indian cuisine*. PHI Learning Pvt. Ltd.

Ekpa, Onu, Natalia Palacios-Rojas, Gideon Kruseman, Vincenzo Fogliano, and Anita R. Linnemann. 2019. Sub-Saharan African maize-based foods-processing practices, challenges and opportunities. *Food Reviews International* 35 (7):609–639.

Elias, E. M. 1995. Durum wheat products. *Durum Wheat Improvement in the Mediterranean Region: New Challenges, Serie A: Séminaires Méditerranéennes* 40:23–31.

Fan, Xiaoxiao, Shaoqiong Liu, Guanhua Liu, Jingpeng Zhao, Hongchao Jiao, Xiaojuan Wang, Zhigang Song, and Hai Lin. 2015. Vitamin A deficiency impairs mucin expression and suppresses the mucosal immune function of the respiratory tract in chicks. *PLoS One* 10 (9):e0139131.

Feng, Yonghui, Xiaotong Zhu, Qinghui Wang, Yongjun Jiang, Hong Shang, Liwang Cui, and Yaming Cao. 2012. Allicin enhances host pro-inflammatory immune responses and protects against acute murine malaria infection. *Malaria Journal* 11:1–9.

Fibri, Dwi Larasatie Nur, and Michael Bom Frøst. 2019. Consumer perception of original and modernised traditional foods of Indonesia. *Appetite* 133:61–69.

Fibri, Dwi Larasatie Nur, and Michael Bom Frøst. 2020. Indonesian millennial consumers' perception of tempe–And how it is affected by product information and consumer psychographic traits. *Food Quality and Preference* 80:103798.

Fieldhouse, Paul. 2013. *Food and Nutrition: Customs and Culture*. Springer.

Filipe Rosa, Louisa, Patricia P. Petersen, Lisa F. Görtz, Iris Stolzer, Valentina Kaden-Volynets, Claudia Günther, and Stephan C. Bischoff. 2023. Vitamin A-and D-deficient diets disrupt intestinal antimicrobial peptide defense involving Wnt and STAT5 signaling pathways in mice. *Nutrients* 15 (2):376.

García-Burgos, María, Jorge Moreno-Fernández, María J. M. Alférez, Javier Díaz-Castro, and Inmaculada López-Aliaga. 2020. New perspectives in fermented dairy products and their health relevance. *Journal of Functional Foods* 72:104059.

Ge, Yong, Mojgan Zadeh, and Mansour Mohamadzadeh. 2022. Vitamin B12 regulates the transcriptional, metabolic, and epigenetic programing in human ileal epithelial cells. *Nutrients* 14 (14):2825.

Ghosh, Sampat, Victor Benno Meyer-Rochow, and Chuleui Jung. 2023. *Embracing Tradition: The Vital Role of Traditional Foods in Achieving Nutrition Security*. MDPI.

Górska-Warsewicz, Hanna, Krystyna Rejman, Wacław Laskowski, and Maksymilian Czeczotko. 2019. Milk and dairy products and their nutritional contribution to the average polish diet. *Nutrients* 11 (8):1771.

Hanieh, Hamza, Kiyoaki Narabara, Mingzi Piao, Chaogetu Gerile, Asaki Abe, and Yasuhiro Kondo. 2010. Modulatory effects of two levels of dietary Alliums on immune response and certain immunological variables, following immunization, in White Leghorn chickens. *Animal Science Journal* 81 (6):673–680.

Hardy, H., J. Harris, E. Lyon, J. Beal, and A. D. Foey. 2013. Probiotics, prebiotics and immunomodulation of gut mucosal defences: homeostasis and immunopathology. *Nutrients* 5: 1869–1912.

Hendler, Robert, and Yue Zhang. 2018. Probiotics in the treatment of colorectal cancer. *Medicines* 5 (3):101.

Ho, Su-Chen, Ku-Shang Chang, and Pei-Wen Chang. 2013. Inhibition of neuroinflammation by cinnamon and its main components. *Food Chemistry* 138 (4):2275–2282.

Holscher, Hannah D. 2017. Dietary fiber and prebiotics and the gastrointestinal microbiota. *Gut Microbes* 8 (2):172–184.

Hossain, Abul, Hey Kyung Moon, and Jong-Kuk Kim. 2017. Effect of pre-treatment and extraction conditions on the antioxidant properties of persimmon (Diospyros kaki) leaves. *Bioscience, Biotechnology, and Biochemistry* 81 (11):2079–2085.

Islas-Weinstein, Leon, and José Luis Maldonado-García. 2022. Immunomodulatory supplements. In *Encyclopedia of Infection and Immunity*, edited by N. Rezaei. Oxford: Elsevier.

Jayapala, Hettiarachchige P. S., and Sun Y. Lim. 2023. N-3 polyunsaturated fatty acids and gut microbiota. *Combinatorial Chemistry & High Throughput Screening* 26 (5):892–905.

Jiang, Yuhang, Cuifeng Li, Qian Wu, Peng An, Laiquan Huang, Jia Wang, Chen Chen, Xi Chen, Fan Zhang, and Li Ma. 2019. Iron-dependent histone 3 lysine 9 demethylation controls B cell proliferation and humoral immune responses. *Nature Communications* 10 (1):2935.

Joye, Iris J. 2020. Dietary fibre from whole grains and their benefits on metabolic health. *Nutrients* 12 (10):3045.

Koehler, Jeff. 2012. *Morocco: A Culinary Journey with Recipes from the Spice-Scented Markets of Marrakech to the Date-Filled Oasis of Zagora*. Chronicle Books.

Kopka, Deborah. 2011. *Passport Series: Middle East*. Milliken Publishing Company.

Kristbergsson, Kristberg, and Semih Otles. 2016. *Functional Properties of Traditional Foods*. Vol. 12. Springer.

Kumar, Shishir, and Geetanjali Pandey. 2020. Biofortification of pulses and legumes to enhance nutrition. *Heliyon* 6 (3).

Kwak, Shin-Hye, Young-Mi Cho, Geon-Min Noh, and Ae-Son Om. 2014. Cancer preventive potential of kimchi lactic acid bacteria (Weissella cibaria, Lactobacillus plantarum). *Journal of Cancer Prevention* 19 (4):253.

Kwon, Dae Young, Sang Mee Hong, Il Sung Ahn, Young Suk Kim, Dong Wha Shin, and Sunmin Park. 2009. Kochujang, a Korean fermented red pepper plus soybean paste, improves glucose homeostasis in 90% pancreatectomized diabetic rats. *Nutrition* 25 (7–8):790–799.

Latif, Anam, Aamir Shehzad, Sobia Niazi, Asna Zahid, Waqas Ashraf, Muhammad Waheed Iqbal, Abdur Rehman, Tahreem Riaz, Rana Muhammad Aadil, and Imran Mahmood Khan. 2023. Probiotics: Mechanism of action, health benefits and their application in food industries. *Frontiers in Microbiology* 14:1216674.

Lee, Yoon-Mi, Jun-Ha Choi, Wan-Kwon Min, Jong-Kwon Han, and Jae-Wook Oh. 2018. Induction of functional erythropoietin and erythropoietin receptor gene expression by gamma-aminobutyric acid and piperine in kidney epithelial cells. *Life Sciences* 215:207–215.

Liu, Yuying, Dat Q. Tran, and J. Marc Rhoads. 2018. Probiotics in disease prevention and treatment. *The Journal of Clinical Pharmacology* 58:S164–S179.

Lukus, Patricia K., Katarina M. Doma, and Alison M. Duncan. 2020. The role of pulses in cardiovascular disease risk for adults with diabetes. *American Journal of Lifestyle Medicine* 14 (6):571–584.

Majdalawieh, Amin F., Reem Hmaidan, and Ronald I. Carr. 2010. Nigella sativa modulates splenocyte proliferation, Th1/Th2 cytokine profile, macrophage function and NK anti-tumor activity. *Journal of Ethnopharmacology* 131 (2):268–275.

Majumdar, Ranendra K., Deepayan Roy, Sandeep Bejjanki, and N. Bhaskar. 2016. Chemical and microbial properties of shidal, a traditional fermented fish of Northeast India. *Journal of Food Science and Technology* 53:401–410.

Mazziotta, Chiara, Mauro Tognon, Fernanda Martini, Elena Torreggiani, and John Charles Rotondo. 2023. Probiotics mechanism of action on immune cells and beneficial effects on human health. *Cells* 12 (1):184.

Mehta, Bhavbhuti M. 2015. Chemical composition of milk and milk products. *Handbook of Food Chemistry*:511–553.

Miah, Pintu, Surovi Binte Sharmin Mohona, Md Mizanur Rahman, Nusrat Subhan, Ferdous Khan, Hemayet Hossain, Shazid Md Sharker, and Md Ashraful Alam. 2021. Supplementation of cumin seed powder prevents oxidative stress, hyperlipidemia and non-alcoholic fatty liver in high fat diet fed rats. *Biomedicine & Pharmacotherapy* 141:111908.

Mizuno, Masashi, Sachiko Kawakami, and Nobuhide Fujitake. 2003. Macrophages stimulated by polysaccharide purified from Agaricus brasiliensis S. Wasser et al.(Agaricomycetideae) enhance mRNA expression of Th1 cytokine including IL-12 and 18. *International Journal of Medicinal Mushrooms* 5 (4).

Mora, J. Rodrigo, Makoto Iwata, and Ulrich H. Von Andrian. 2008. Vitamin effects on the immune system: Vitamins A and D take centre stage. *Nature Reviews Immunology* 8 (9):685–698.

Moriguchi, Satoru, and Mayumi Kaneyasu. 2004. Role of vitamin E in immune system. *Journal of Clinical Biochemistry and Nutrition* 34 (3):97–109.

Mousavi, Soraya, Stefan Bereswill, and Markus M. Heimesaat. 2019. Immunomodulatory and antimicrobial effects of vitamin C. *European Journal of Microbiology and Immunology* 9 (3):73–79.

Mukherjee, Pulok K., Neelesh K. Nema, Niladri Maity, and Birendra K. Sarkar. 2013. Phytochemical and therapeutic potential of cucumber. *Fitoterapia* 84:227–236.
Mullick, Shalini, Usha Rusia, Meera Sikka, and M. A. Faridi. 2006. Impact of iron deficiency anaemia on T lymphocytes & their subsets in children. *Indian Journal of Medical Research* 124 (6):647–654.
Nogal, Ana, Ana M. Valdes, and Cristina Menni. 2021. The role of short-chain fatty acids in the interplay between gut microbiota and diet in cardio-metabolic health. *Gut Microbes* 13 (1):1897212.
Oppedisano, Francesca, Roberta Macrì, Micaela Gliozzi, Vincenzo Musolino, Cristina Carresi, Jessica Maiuolo, Francesca Bosco, Saverio Nucera, Maria Caterina Zito, and Lorenza Guarnieri. 2020. The anti-inflammatory and antioxidant properties of n-3 PUFAs: Their role in cardiovascular protection. *Biomedicines* 8 (9):306.
Qian, Xiao-hang, Ru-yan Xie, Xiao-li Liu, and Hui-dong Tang. 2022. Mechanisms of short-chain fatty acids derived from gut microbiota in Alzheimer's disease. *Aging and Disease* 13 (4):1252.
Ranasinghe, R. A. S. N., S. D. T. Maduwanthi, and R. A. U. J. Marapana. 2019. Nutritional and health benefits of jackfruit (Artocarpus heterophyllus Lam.): A review. *International Journal of Food Science* 1:4327183.
Rasane, Prasad, Rekha Kailey, and Shailesh Kumar Singh. 2017. Fermented indigenous Indian dairy products: standards, nutrition, technological significance and opportunities for its processing. *Journal of Pure & Applied Microbiology* 11 (2):1199–1213.
Rocillo-Aquino, Zeltzin, Fernando Cervantes-Escoto, Juan Antonio Leos-Rodríguez, Daniela Cruz-Delgado, and Angélica Espinoza-Ortega. 2021. What is a traditional food? Conceptual evolution from four dimensions. *Journal of Ethnic Foods* 8:1–10.
Roufs, Kathleen Smyth, and Timothy G. Roufs. 2014. *Sweet treats around the world*. California: ABC-CLIO.
Saini, Ramesh Kumar, and Young-Soo Keum. 2018. Omega-3 and omega-6 polyunsaturated fatty acids: Dietary sources, metabolism, and significance—A review. *Life Sciences* 203:255–267.
Saleem, Aisha, Irum Naureen, Muhammad Naeem, Gulnaz Tasleem, Hassam Ahmed, and Umar Farooq. 2022. Therapeutic role of Piper Nigrum L (Black Pepper) and pharmacological activities. *Scholars International Journal of Biochemistry* 5 (1):15–21.
Salem, Mohamed Labib, and Mohammad Sohrab Hossain. 2000. Protective effect of black seed oil from Nigella sativa against murine cytomegalovirus infection. *International Journal of Immunopharmacology* 22 (9):729–740.
Sarkar, S. 2008. Innovations in Indian fermented milk products—a review. *Food Biotechnology* 22 (1):78–97.
Sassi, Francesca, Cristina Tamone, and Patrizia D'Amelio. 2018. Vitamin D: nutrient, hormone, and immunomodulator. *Nutrients* 10 (11):1656.
Semwal, Prabhakar, Sakshi Painuli, Shabaaz Begum J. P., Abhishek Jamloki, Abdur Rauf, Ahmed Olatunde, Md Mominur Rahman, Nobendu Mukerjee, Anees Ahmed Khalil, and Abdullah S. M. Aljohani. 2023. Exploring the nutritional and health benefits of pulses from the Indian Himalayan region: A glimpse into the region's rich agricultural heritage. *Food Chemistry* 422:136259.
Shah, Shreya S., Gaurang B. Shah, Satbeer D. Singh, Priyanshi V. Gohil, Kajal Chauhan, Khyati A. Shah, and Mehul Chorawala. 2011. Effect of piperine in the regulation of obesity-induced dyslipidemia in high-fat diet rats. *Indian Journal of Pharmacology* 43 (3):296–299.
Sharma, Hari, Xiaoying Zhang, and Chandradhar Dwivedi. 2010. The effect of ghee (clarified butter) on serum lipid levels and microsomal lipid peroxidation. *AYU (An international Quarterly Journal of Research in Ayurveda)* 31 (2):134–140.

Sharma, Ranjana, Prakrati Garg, Pradeep Kumar, Shashi Kant Bhatia, and Saurabh Kulshrestha. 2020. Microbial fermentation and its role in quality improvement of fermented foods. *Fermentation* 6 (4):106.

Shukla, Ankit, and Nagendra Yadav. 2018. Role of Indian spices in Indian history. *International Journal of Management Research and Reviews* 8 (11):1.

Siddiqui, Shahida Anusha, Zeki Erol, Jerina Rugji, Fulya Taşçı, Hatice Ahu Kahraman, Valeria Toppi, Laura Musa, Giacomo Di Giacinto, Nur Alim Bahmid, and Mohammad Mehdizadeh. 2023. An overview of fermentation in the food industry-looking back from a new perspective. *Bioresources and Bioprocessing* 10 (1):85.

Singh, Aditi, Pallabi Banerjee, Mohammad Anas, Nagendra Singh, and Imteyaz Qamar. 2020. Traditional nutritional and health practices targeting lifestyle behavioral changes in humans. *Journal of Lifestyle Medicine* 10 (2):67.

Srinivasan, K. 2014. Antioxidant potential of spices and their active constituents. *Critical Reviews in Food Science and Nutrition* 54 (3):352–372.

Stach, Kamilla, Wojciech Stach, and Katarzyna Augoff. 2021. Vitamin B6 in health and disease. *Nutrients* 13 (9):3229.

Takooree, Heerasing, Muhammad Z. Aumeeruddy, Kannan R. R. Rengasamy, Katharigatta N. Venugopala, Rajesh Jeewon, Gokhan Zengin, and Mohamad F. Mahomoodally. 2019. A systematic review on black pepper (Piper nigrum L.): From folk uses to pharmacological applications. *Critical Reviews in Food Science and Nutrition* 59 (supl):S210–S243.

Tamura, J., K. Kubota, H. Murakami, M. Sawamura, T. Matsushima, T. Tamura, T. Saitoh, H. Kurabayshi, and T. Naruse. 1999. Immunomodulation by vitamin B12: Augmentation of CD8+ T lymphocytes and natural killer (NK) cell activity in vitamin B12-deficient patients by methyl-B12 treatment. *Clinical & Experimental Immunology* 116 (1):28–32.

Tillisch, Kirsten, Jennifer Labus, Lisa Kilpatrick, Zhiguo Jiang, Jean Stains, Bahar Ebrat, Denis Guyonnet, Sophie Legrain–Raspaud, Beatrice Trotin, and Bruce Naliboff. 2013. Consumption of fermented milk product with probiotic modulates brain activity. *Gastroenterology* 144 (7):1394–1401. e4.

Tomeh, Mhd Anas, Roja Hadianamrei, and Xiubo Zhao. 2019. A review of curcumin and its derivatives as anticancer agents. *International Journal of Molecular Sciences* 20 (5):1033.

Tonks, A., Rose A. Cooper, A. J. Price, Peter C. Molan, and K. P. Jones. 2001. Stimulation of TNF-α release in monocytes by honey. *Cytokine* 14 (4):240–242.

Trio, Phoebe Zapanta, Sixiang You, Xi He, Jianhua He, Kozue Sakao, and De-Xing Hou. 2014. Chemopreventive functions and molecular mechanisms of garlic organosulfur compounds. *Food & Function* 5 (5):833–844.

Valverde, María Elena, Talía Hernández-Pérez, and Octavio Paredes-López. 2015. Edible mushrooms: improving human health and promoting quality life. *International Journal of Microbiology* 1:376–387.

Venturella, Giuseppe, Valeria Ferraro, Fortunato Cirlincione, and Maria Letizia Gargano. 2021. Medicinal mushrooms: bioactive compounds, use, and clinical trials. *International Journal of Molecular Sciences* 22 (2):634.

Verni, Michela, Erica Pontonio, Marco Montemurro, and Carlo Giuseppe Rizzello. 2022. Fermentation as strategy for improving nutritional, functional, technological, and sensory properties of legumes. In J.C. Jimenez-Lopez and A. Clemente. *Legumes research-volume 2*. IntechOpen.

Volpe, Stella Lucia. 2013. Magnesium in disease prevention and overall health. *Advances in Nutrition* 4 (3):378S–383S.

Walther, B., and A. Schmid. 2017. Effect of fermentation on vitamin content in food. In J. Frias, C. Martinez-Villaluenga and E. Peñas. *Fermented foods in health and disease prevention*: Elsevier.

Wang, Xiaoying, Han Zhang, Lili Chen, Lihua Shan, Guanwei Fan, and Xiumei Gao. 2013. Liquorice, a unique "guide drug" of traditional Chinese medicine: A review of its role in drug interactions. *Journal of Ethnopharmacology* 150 (3):781–790.

Wei, Xun, Wei Yang, Jianhui Wang, Yong Zhang, Yaxuan Wang, Yan Long, Bin Tan, and Xiangyuan Wan. 2022. Health effects of whole grains: A bibliometric analysis. *Foods* 11 (24):4094.

Weyh, Christopher, Karsten Krüger, Peter Peeling, and Lindy Castell. 2022. The role of minerals in the optimal functioning of the immune system. *Nutrients* 14 (3):644.

Witkamp, Renger F. 2022. Bioactive components in traditional foods aimed at health promotion: A route to novel mechanistic insights and lead molecules? *Annual Review of Food Science and Technology* 13:315–336.

Xie, Liang, Md Jahangir Alam, Francine Z. Marques, and Charles R. Mackay. 2023. A major mechanism for immunomodulation: Dietary fibres and acid metabolites. Paper read at Seminars in Immunology.

Xing, Yanxia, Mengzhen Huang, Chinasa V. Olovo, Chiamaka L. Mgbechidinma, Yu Yang, Jing Liu, Bo Li, Mengliu Zhu, Kexue Yu, and He Zhu. 2023. Traditional fermented foods: Challenges, sources, and health benefits of fatty acids. *Fermentation* 9 (2):110.

Zhang, Qian-hui, Hao-zhou Huang, Zhen-feng Wu, Jun-zhi Lin, and Ding-kun Zhang. 2021. Traditional uses, pharmacological effects, and molecular mechanisms of licorice in potential therapy of COVID–19. *Frontiers in Pharmacology* 12:719758.

Zhao, Shuang, Qi Gao, Chengbo Rong, Shouxian Wang, Zhekun Zhao, Yu Liu, and Jianping Xu. 2020. Immunomodulatory effects of edible and medicinal mushrooms and their bioactive immunoregulatory products. *Journal of Fungi* 6 (4):269.

Zhou, Jiajing, Minhao Li, Qian Bai, Thaiza S. P. de Souza, Colin Barrow, Frank Dunshea, and Hafiz A. R. Suleria. 2023. Effects of different processing methods on pulses phytochemicals: An overview. *Food Reviews International* 40 (4):1138–1195.

2 Phytochemical Chemistry and its Benefits

Sunita Singh[1], Avanish Chandra Sharma[1], Chandan Maurya[1], Pankaj Kumar Chaurasia[2], Vipul Kumar[3], Shashi Lata Bharati[4], and Ayman Younes Fathy Allam[5]

[1]Department of Chemistry, Navyug Kanya Mahavidyalaya, University of Lucknow, Lucknow, Uttar Pradesh, India
[2]P.G. Department of Chemistry, L.S. College, B.R.A. Bihar University, Muzaffarpur Bihar, India
[3]Department of Pharmaceutical Chemistry, Delhi Institute of Pharmaceutical Sciences and Research (DIPSAR), Delhi Pharmaceutical Sciences and Research University, New Delhi, India
[4]North Eastern Regional Institute of Science and Technology, Nirjuli, Arunachal Pradesh-791109, India
[5]Department of Food Science and Technology, Faculty of Agriculture, Minoufiya University, Shibin El-kom, Egypt

2.1 INTRODUCTION

Desiderius Erasmus, who is responsible for the well-known phrase "Prevention is better than cure," emphasized the role of a healthy lifestyle for achieving good and sound health. In various civilizations, from Egyptian to Chinese, the importance of plants and their different formulations in the management of various ailments have been mentioned. Recently, plant-derived compounds, or phytochemicals, have attained prominence as many people have developed a positive attitude towards nature-based healthy lifestyles. Scientific advancements in dietary options have highlighted the effectiveness of these phytochemicals in the treatment of many diseases. Phytochemical compounds, which have been traditionally used in medicines for many years, are now being used as food supplements. These phytochemical compounds are now being combined with modern science to produce significant health benefits for humans. Phytochemicals are non-nutritive parts of plants with preventive properties against various diseases and external factors. They are biologically active components produced as a result of primary and secondary metabolism

DOI: 10.1201/9781003514244-3

in plants. Phytochemicals are widely present in fruits, vegetables, legumes, whole grains, dark chocolate, seeds and nuts, and tea. It is believed that the consumption of these dietary phytochemicals reduces the risk of many chronic diseases (Vasmehjani et al., 2021). Recently, functional food, or nutraceuticals, which is food or processed food containing bioactive ingredients that impart health benefits in various ways, has become widespread.

Phytochemicals are compounds naturally produced by plants, each with various effects on the human body. Some of these are antioxidants that protect cells from oxidative stress damage and lower the risk of cancer (Li et al., 2023). Some have hormonal effects and influence metabolic pathways related to cancer and cardiovascular diseases (Liu et al., 2019). Some stimulate various enzymes that help the body function better (Yang, 2022). Some have antimicrobial effects and restrict pathogens from sticking to the cell wall (Yang, 2022). Anthocyanins, flavonoids of the phenolic class, tannins, ellagitannins, dietary fibers, and polysaccharides have all shown the ability to reduce LDL cholesterol and prevent blood clots (Islam et al., 2021).

2.1.1 Role of Phytochemicals in Plant Physiology

Plants use phytochemicals to protect themselves from various threats. Many phytochemicals can kill or stop the growth of bacteria, fungi, and other harmful microbes (Wani et al., 2023). These compounds also help plants resist the attack of pathogens. Phytochemicals also help plants defend themselves from herbivores. Some phytochemicals make plants taste bad or poisonous to herbivores, so they avoid eating them. For example, glucosinolates and terpenoids are two groups of phytochemicals that deter herbivores with their smell, taste, and toxicity (Divekar, 2022).

Moreover, some phytochemicals affect the interactions between plants and other living things in their surroundings. These phytochemicals are called allelochemicals, and they can have positive or negative effects on other plants. For example, some plants release phenolic compounds that prevent other plants from growing near them, giving them an advantage. However, some plants also release compounds that help other plants grow better (Dai et al., 2010).

Phytochemicals also help plants cope with environmental stresses, such as UV radiation, drought, and extreme temperatures. Some phytochemicals, especially flavonoids, act as sunscreens, shielding plants from UV damage (Faccio et al., 2021). Some phytochemicals also act as antioxidants, removing harmful molecules and preventing oxidative damage in plant cells. These antioxidant properties help plants survive and thrive under stress.

Interestingly, plants can produce more phytochemicals when they sense stress or damage from pathogens or herbivores. This is called the plant's secondary metabolite pathway, and it is controlled by various signals and regulators. When plants detect a threat, they turn on specific genes that make phytochemicals, and they store these compounds for defense. This flexible response allows plants to adjust their defense strategies to different situations and use their resources wisely.

2.2 ROLE OF PHYTOCHEMICALS IN LIVING ORGANISMS

Phytochemicals have many functions in living organisms, but their exact mechanisms are not fully known. Some of the functions of phytochemicals are:

1. Protecting important biomolecules like nucleic acids, proteins, and fats from oxidative stress conditions.
2. Acting as antimicrobial agents against bacteria, fungi, viruses, and trypanosomes.
3. Boosting the immune system.
4. Regulating the enzymes that detoxify harmful substances.
5. Reducing inflammation.
6. Preventing the clumping of blood platelets.
7. Affecting physiological processes such as blocking the attachment of pathogens to cell receptors.

About 80% of the world's most useful drugs come from plants. Phytochemicals also have other functions, such as fighting malaria, diarrhea, worms, liver damage, atherosclerosis, allergies, diabetes, mutations, and high blood pressure. They can also help with sore throat, cough, toothache, ulcers, menstrual bleeding, sperm count, dysentery, stomach problems, vertigo, and appetite (Mahomoodally, 2013).

2.3 CLASSIFICATION AND BENEFITS OF PHYTOCHEMICALS

Phytochemicals, due to their diverse forms and structures, have not been definitively classified. Traditionally, they are categorized as primary or secondary metabolites based on their function in plant metabolism. Primary metabolites encompass common sugars, amino acids, proteins, purines, pyrimidines of nucleic acids, and chlorophylls. Secondary metabolites include other plant chemicals such as alkaloids, terpenes, flavonoids, lignans, plant steroids, saponins, phenolics, and glucosides. Phenolics are the most prevalent and structurally diverse plant chemicals. Figure 2.1 gives the generalized classification of phytochemicals (Santhiravel et al., 2022)

2.4 CHEMISTRY AND HEALTH BENEFITS OF PHYTOCHEMICALS

In recent years, phytochemicals have attracted much attention, as they have been shown to have various beneficial effects on human health. Phytochemicals can modulate various cellular and molecular pathways that are involved in the development and progression of chronic diseases. The chemistry and health benefits of carotenoids, polyphenols, terpenoids, phytosterols, saponins, dietary fibers, and polysaccharides depend on their structure, bioavailability, metabolism, and interactions with other components in the food matrix (Roy and Dutta, 2020; Kumar et al., 2023). This section will explore some of the major classes of phytochemicals and their health benefits.

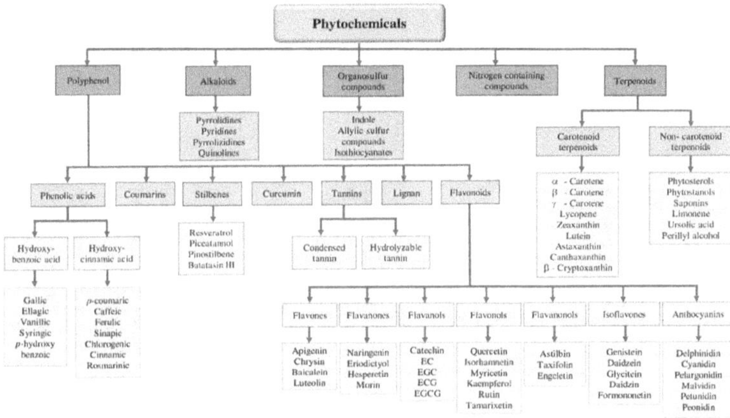

FIGURE 2.1 Generalized classification of phytochemicals.

2.4.1 Phenolic Compounds

Phenolic phytochemicals contain one or more aromatic rings with at least one hydroxyl group. They provide a protective function in plants by reducing the impact of predators, parasites, and ultraviolet radiation. Phenolics are found in fruits, cereals, legumes, and vegetables. Plant phenolics are further classified as flavonoids, phenolic acids, stilbenes, lignans, and tannins.

2.4.1.1 Flavonoids

Flavonoids are a class of polyphenolic compounds that have a 15-carbon skeleton, consisting of two benzene rings (A and B) and a heterocyclic pyran ring. These flavonoids are further classified as Anthocyanidins, Flavan-3-ols, Flavonols, Flavones, Flavanone, and Isoflavones (Roy and Dutta, 2020). Spinach, broccoli, flax, apple, plum, and chicory are good sources of these flavonoids (D' Archivio et al., 2007). In beverages such as tea, olive oil, and red wine, a good amount of these components are present.

There are many health benefits of flavonoid classes containing at least one hydroxyl group, such as free radical scavenging activity, protective effects against cardiovascular diseases, cancers, and other age-related diseases, and prevention of inflammation and allergies (Ballard et al., 2019). Table 2.1 provides an account of various health benefits of compounds belonging to the classes of Anthocyanidins, Flavan-3-ols, Flavonols, Flavones, Flavanone, and Isoflavones. They have been found to be beneficial in the management of diabetes, gastrointestinal diseases, lymphocytic leukemia, rhinitis, traumatic cerebral infarction, and as inhibitors of NF-κB (Roy and Datta, 2020). The mode of action involves their pro-oxidant potential, ability to prevent oxidative stress at the cellular level, inhibition of different inflammatory factors, and regulation of enzymes such as cyclooxygenase, lipo-oxygenase, and xanthine-oxidase. Figure 2.2 shows some important flavonoids of different classes possessing health benefits.

TABLE 2.1
Types of Flavonoid Compounds with Their Health Benefits

Type	Subtype	Example	Sources	Health Benefits
Anthocyanidins	Malvidin, pelargondin, peoidin, cyanidin, apigeninidin, delphinidin, diosmetinidin, guibourtinidin, aurentinidin, columnidin, hirsutinidin, and rosinidin	Cyanidin	Berries, plums, grapes, pomegranates	Antioxidant, antimicrobial, anticancer, anti-inflammatory, obesity control
Flavan-3-ols	Monomers (catechins), dimers, and polymers	Catechin	Bananas, apples, tea, cocoa, grapes, fava beans, blueberries, peaches, pears	Anti-inflammatory, antioxidant, anticancer
Flavonols	Quercetin, kaempferol, myricetin, rutin, morin, fisetin, galangin, rhamnetin, robinin, and spirenoside	Quercetin	Onions, leeks, lettuce, Brussels sprouts, kale, broccoli, tea, berries, beans, apples	Anti-inflammatory, antioxidant, antihistamine, anticancer
Flavones	Apigenin, tangeretin, baicalein, chrysin, diosmetin, diosmin, luteolin, etc.	Apigenin	Celery, parsley, broccoli, thyme, dandelion, chamomile tea, oregano, rosemary	Antioxidant, anticancer, anti-inflammatory, apoptosis induction
Flavanones	Eriodictyol, hesperidin, hesperetin, likvirtin, naringin, naringenin, and pinocembrin	Naringenin	Oranges, lemons, grapes, grapefruits	Antioxidant, anti-inflammatory, cardiovascular health, nervous system health
Isoflavones	Genistein, glycitein, irilone, luteone, prunetin, pratensein, and daidzein	Genistein	Soybeans, peas, peanuts, chickpeas, navy beans, alfalfa	Phytoestrogen, anticancer, menstrual cycle regulation

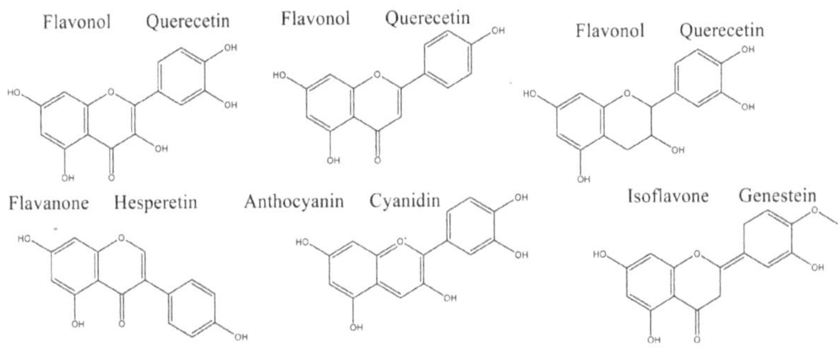

FIGURE 2.2 Different flavonoid components possessing health benefits.

2.4.1.2 Phenolic Acids

Phenolic acids contain a carboxylic group as well as an aromatic phenolic group. Cereals, wheat flours, onions, coffee, kiwis, berries, apples, and citrus fruits are good sources of phenolic acids. They have anti-inflammatory, antioxidant, antimutagenic, and anticarcinogenic properties (Roy and Datta, 2020). They occur in plants as esters or glycosides, which are linked to flavonoids, alcohols, sterols, etc. Phenolic acids have the same basic structure, but they differ in the number and position of hydroxyl groups. There are two main groups of phenolic acids: hydroxybenzoic acids and hydroxycinnamic acids. The hydroxybenzoic group includes various hydroxybenzoic and salicylic acids, while the hydroxycinnamic group includes caffeic, ferulic, and sinapic acids and their esters. Phenolic acids can be easily absorbed by the human body and act as antioxidants due to their phenolic group. They can also donate hydrogen atoms and scavenge free radicals (Kumar et al., 2017). Moreover, phenolic acids such as ferulic, isovanillic, p-hydroxycinnamic, p-hydroxybenzoic, and caffeic acids exhibit antibacterial properties. Caffeic acid and its esters have shown potential to fight against colon cancer (Olthof et al., 2001). Some phenolic acids, such as gallic acid, can also reduce allergic reactions and inflammation. Caffeic acid and its derivatives can inhibit the enzyme that integrates HIV-1 DNA into the host genome (Falsaperlaa et al., 2005). Furthermore, phenolic acids can prevent cell proliferation and cancer development.

2.4.1.3 Stilbenes

Stilbenes are non-flavonoid polyphenolic compounds found in plants of the *Cyperaceae*, *Dipterocarpaceae*, *Gnetaceae*, *Leguminaceae*, and *Vitaceae* families. Grapes are a good source of stilbenes. The basic structure of stilbenes contains a bridge of ethylenic moiety between two aromatic rings. It exists in two diastereoisomeric forms, namely, E and Z-1,2-diphenyethylene. They are generally biosynthesized under stress conditions in plants such as high temperature, oxidation, insect attacks, and fungal and bacterial infections. Compounds like resveratrol, pterostilbene, and combrestatin A-4 are well known for their significant role in drug development.

Many studies have reported the wide application of the stilbene scaffold in the synthesis of compounds utilizing the biological properties of being antioxidant, anti-cancer, antimicrobial, neuroprotective, antidiabetic, and anti-inflammatory (Pecyna et al., 2020). Raloxifene and tamoxifen are examples of FDA-approved drugs that are derived from stilbene. Ongoing clinical trials have supported the use of resveratrol, combretastatin, and tapinarof as chemopreventive agents (Sale, 2004). Stilbene based Ramizol® has been found to be effective against 100 isolates of Clostridium difficile species (Wolfe et al., 2018). Structure activity relationship studies showed that the presence of a hydroxyl group is responsible for anti-inflammatory properties, including inhibitory effects against TNF-induced NF-kappa B factor. Resveratrol with curcumin and ellagic acid showed antiproliferative activity against human leukemia cells and human colon cancer cells, respectively.

2.4.1.4 Lignans

Dietary lignans are generally a group of phytoestrogens containing aglycones or glycoside units (Li et al., 2018). They are found in roots, rhizomes, stems, leaves, flowers, fruits, seeds, xylem, and resins of plants. Some plants of the genera *Machilus*, *Ocotea*, and *Nectandra* have good percentages of lignans. *Annonaceae*, *Orchidaceae*, *Berberidaceae*, and *Schisandraceae* families also have many lignans and neolignans in their chemical composition (Cui et al., 2020). In more than 70 plant families, approximately 200 classical lignans and 100 neolignans have been identified (Pan et al., 2009). They are usually in pairs, but some are in groups of three or four. Most of the lignans in plants are free, but some can join with sugars and make glycosides and other compounds. Lignans possess many pharmacological effects, such as antitumor, antioxidant, antibacterial, immunosuppressive, and antiasthmatic activities (Pan et al., 2009; Marcotullio et al., 2018). Structurally, lignans and neolignans are dimers of phenylpropane and are generally classified into lignans, neolignans, and oxyneolignans based on the C–C bond and oxygen bridge joining the two typical phenylpropane units. These phenylpropanoid units are derived from cinnamic acid, cinnamyl alcohol, propenyl benzene, and allyl benzene. These units are linked in different ways to form different types of lignans. If the linkage is between the β-β ′ positions (or 8-8′), they are called "classical lignans." If the linkage is not between the β-β ′ positions, they are called "neolignans." Structurally, neolignans are more diverse in nature (Cui et al., 2020).

Flaxseed is one of the best dietary sources of lignans, as it contains a high percentage of secoisolariciresinol, lariciresinol, pinoresinol, and matairesinol in high amounts. Sesame seeds, whole grains, fruits like apricots, pears, peaches, and strawberries, and vegetables such as garlic, asparagus, and carrots, as well as lentils, are also good sources of lignans.

With such structural diversity found in lignans, it is not surprising that many attractive pharmacological activities of the lignan family have been reported, such as being antioxidant (Lu et al., 1992), antibacterial (Kawazoe et al., 2001), antiviral (Cui et al., 2020), antitumor (Capilla et al., 2001), and anti-inflammatory (Nyamai et al., 2016; Cui et al., 2020). Lignans are known to reduce cell proliferation activity in colon cancer cells and exhibit anticancer effects in *in vivo* models. They inhibit the formation of metastatic secondary tumors and decrease levels of colon cancer markers in rat models. Lignans also suppress the receptor binding of platelet-activating factor and inhibit the replication of the human immunodeficiency virus at the integration stage. Furthermore, they inhibit tumor necrosis factor-alpha from lipopolysaccharide-triggered murine macrophages.

2.4.1.5 Tannins

Tannins are a diverse group of polyphenolic compounds of high molecular weight having various biological properties (Roy and Datta, 2020). They are generally present in coffee, tea, wine, grapes, pomegranate, and berries. They are found in leaves, wood, bark, fruit, and roots. In fact, tannin makes up 5–10% of the dry weight of plant leaves. Tannins are broadly divided into hydrolyzable tannin (HT) and condensed tannin (CT) (Santhiravel et al.,2022). HTs are polyol units of d-glucose with phenolic groups attached and include gallic acid and ellagic acid. Gallotannins and ellagitannins are examples of HT. Condensed tannins are polyphenolic bioflavonoids that are comprised of polyhydroxy flavan-3-ol units and a few examples of condensed tannins include catechin, epicatechin, gallocatechin, epigallocatechin, and luteolin.

Tannins have various biological roles in plants, such as defense against herbivores, pathogens, and UV radiation, as well as regulation of growth, development, and reproduction (Lara et al., 2020). Lopez-Corona et al. (2022) found that raspberry extracts had excellent ROS-lowering, anti-inflammatory, and cytotoxic effects due to the presence of ellagic acid, anthocyanins, and ellagitannins. Epigallocatechin-3-gallate (EGCG), a component present in green tea, reduced the ROS level and activated the Nrf2 transcription factor and antioxidant proteins (Xie et al., 2020). Moreover, Luo et al. (2022) confirmed that corilagin, a type of ellagitannin, could lower the expression of inflammatory cytokines and decrease pyroptosis. All these mechanisms lowered ROS production by mitochondria (Luo et al., 2022). Virgen-Carrillo et al. (2022), using a computational method to evaluate the energy interaction between molecules, suggested that phytochemicals present in West Mexican berries containing ellagitannins could block enzymes related to obesity and diabetes.

2.4.2 Alkaloids and Their Health Benefits

Alkaloids contain heterocyclic nitrogen atoms and are fundamentally basic in nature. Their name originates from their "alkaline" properties and is used to describe any nitrogen-containing base. Most alkaloids are known for their bitter taste, with quinine being one of the most bitter substances known. Due to their diverse molecular structures, classifying alkaloids is challenging. However, they can be grouped into families based on the type of heterocyclic ring system present in the molecule (Table 2.2) (Saxena et al., 2013).

Since the beginning of civilization, the use of alkaloid-containing plants can be seen in their use as dyes, spices, drugs, or poisons. Alkaloids play a very significant role in plant defense mechanisms as they provide protection against microorganism infections and insect attacks. Alkaloids derived from different plant species have

TABLE 2.2
Classification of Alkaloids

Family	Ring System	Examples	Sources	Uses
Pyrrolidine alkaloids	Pyrrolidine (tetrahydropyrrole)	Hygrine	*Erythroxylum coca* leaves	Stimulant, local anesthetic
Pyridine alkaloids	Piperidine (hexahydropyridine)	Coniine, piperine, isopelletierine	Hemlock, black pepper, Punica granatum	Neurotoxin, flavoring, anthelmintic
Pyrrolidine-pyridine alkaloids	Pyrrolidine-pyridine	Myosmine, nicotine	Tobacco (Nicotiana tabacum) plant	Stimulant, addictive, insecticide
Pyridine-piperidine alkaloids	Pyridine-piperidine	Anabasine	Anabasis aphylla	Nicotinic agonist, insecticide
Quinoline alkaloids	Quinoline	Quinine, primaquine	Cinchona tree	Antimalarial, antipyretic, fluorescent
Isoquinoline alkaloids	Isoquinoline	Narcotine, papaverine, morphine, codeine, heroin	Opium poppy (Papaver somniferum)	Analgesic, antispasmodic, narcotic

many biological properties such as being antioxidative, anticancer, antimalarial, antiparasitic, antibacterial, anti-hypertensive, anti-arrhythmic, anti-plasmodial, and insecticidal (Gutiérrez-Grijalva et al., 2020).

Various studies reported that these naturally occurring compounds have various effects on cancer, the nervous system, and inflammation (Shi et al.,2012; Farias et al.,2012; Huang et al., 2012; Qiu et al., 2014). Vinblastine, berberine, and coronaridine showed antitumor activity, which can interfere with cell division, signaling pathways, migration, and survival of cancer cells. The alkaloid fraction from *Crassocephalum bauchiense* and *huperzine* A showed effects on the central nervous system. Strictosidinic acid, an alkaloid extracted from *Psychotria myriantha* Müll.-Arg. (Rubiaceae) leaves, can modulate behavior, cognition, oxidative stress, and neurotransmitter levels. Oxymatrine, lycorine, and acetylcorynoline are effective as anti-inflammatory agents, as they can inhibit inflammatory mediators, oxidative damage, and immune cell activation. Some studies suggest that alkaloids can protect the brain, stimulate the nervous system, and fight oxidative damage in Alzheimer's disease (O'Brien et al., 2006). These substances can also improve memory and cognition in Alzheimer's patients. This was thought to be the case due to the effects of reducing inflammation and oxidative stress, boosting cholinergic signaling, increasing estrogen and other neurotrophic factors, and preventing the formation and toxicity of β-amyloid (O'Brien et al., 2006). These substances block the enzyme acetylcholinesterase, which enhances the activity of acetylcholine, a key neurotransmitter in Alzheimer's treatment. An amide alkaloid called tetramethylpyrazine can lower blood pressure by preventing platelet clumping and blood vessel narrowing (Guruvayoorappan et al., 2014). This alkaloid can also affect the contraction and relaxation of the heart muscle. Tetramethylpyrazine is used to treat diseases that block blood flow to the brain because of its ability to dilate blood vessels. Alkaloids can also kill microbes, cancer cells, and parasites. They do this by inserting themselves into DNA and disrupting its copying and reading processes, leading to genetic errors. Alkaloids may also interfere with protein synthesis and interact with nerve receptors to exert their antimicrobial and antiparasitic effects (Nyamai et al., 2016).

2.4.3 Terpenoids

Terpenoids are natural products mainly comprising five-carbon isoprene units. They are one of the largest groups of naturally occurring secondary metabolites. Terpenoids can be divided into several groups based on the number of isoprene units present in the compound (Saxena et al., 2013; Masyita et al., 2022) (Table 2.3).

Terpenoids play a vital role in the defense physiology of plants, as they provide protection against various external factors including microorganisms, insects, herbivores, and other plants. They have many biological effects, such as antioxidant, anti-inflammatory, and anticancer activities (Masyita et al., 2022). They can also

TABLE 2.3
Different Subtypes of Terpene Classes

Category	Isoprene Units	Examples	Sources
Hemiterpenoids	1	Isoprene, isovaleric acid, prenol	Various plants
Monoterpenoids	2	Geranyl pyrophosphate, eucalyptol, limonene, citral, camphor, pinene	Essential oils, citrus fruits, herbs, spices
Sesquiterpenes	3	Artemisinin, bisabolol, farnesol, eudesmol	Artemisia plants, chamomile, flowers, eucalyptus
Diterpenes	4	Cembrene, kahweol, taxadiene, cafestol, retinol, retinal, phytol	Corals, coffee beans, yew trees, animalia, chlorophyll
Triterpenes	6	Lanosterol, squalene	Wheat germ, olives, cholesterol
Tetraterpenoids	8	Lycopene, gamma-carotene, alpha-carotene, beta-carotene	Tomatoes, carrots, pumpkins, algae

Source: Saxena et al., 2013; Masyita et al., 2022.

affect the NF-kB system, which is involved in immune response and cell survival. Some terpenoids can block the activation or function of NF-kB in different ways, such as aucubin, limonene, α-pinene, and helenalin A. Other terpenoids, such as artemisinin, have additional benefits, such as antimalarial and antifungal properties. Terpenes and terpenoids can inhibit the growth, proliferation, invasion, and metastasis of cancer cells by modulating various signaling pathways, such as apoptosis, cell cycle, angiogenesis, and inflammation. Some examples of anticancer terpenes and terpenoids are β-caryophyllene, limonene, perillyl alcohol, α-pinene, β-pinene, myrcene, linalool, geraniol, nerolidol, citronellol, and thymoquinone. They can also enhance the efficacy of chemotherapy and radiotherapy by sensitizing cancer cells to these treatments. They can modulate the immune response by regulating the balance between Th1 and Th2 cells, which are involved in allergic reactions. Some examples of antiallergic terpenes and terpenoids are α-bisabolol, β-caryophyllene, eugenol, farnesol, geraniol, and thymol. Terpenes and terpenoids can reduce inflammation by inhibiting the activation of NF-κB factors, a key transcription factor that regulates the expression of inflammatory genes. They can also scavenge free radicals and reactive oxygen species, which are involved in oxidative stress and tissue damage. They can modulate the gut microbiota, which plays a role in maintaining the intestinal barrier and immune homeostasis. Some examples of anti-inflammatory terpenes and terpenoids are α-bisabolol, β-caryophyllene, curcumin, eugenol, farnesol, geraniol, limonene, linalool, myrcene, nerolidol, and thymol. Terpenes and terpenoids are important for human health, according to many studies. Figure 2.3 shows some important examples of terpenes and terpenoids having health benefits. These are the largest group of organic compounds that plants make in their essential oils (EOs). They have several isoprene units in their structure. They can help treat different diseases by acting as anticancer, antimicrobial, anti-inflammatory, antioxidant, antiallergic, neuroprotective, anticoagulant, sedative, and analgesic agents. They belong to different classes, such as monoterpenes, sesquiterpenes, diterpenes,

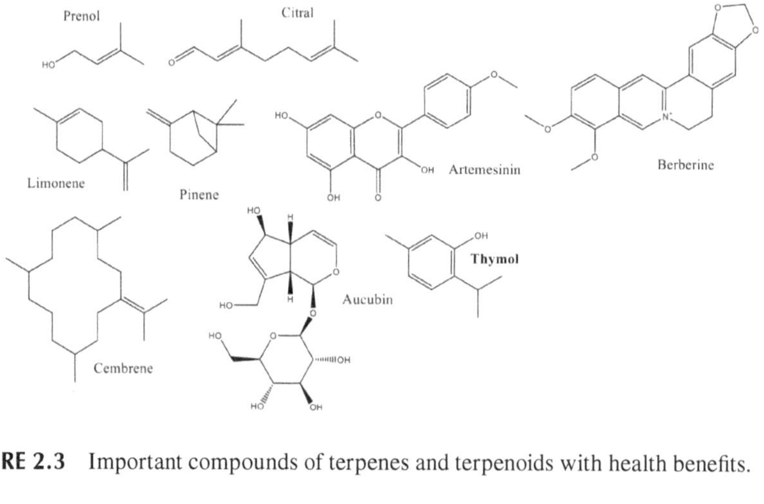

FIGURE 2.3 Important compounds of terpenes and terpenoids with health benefits.

triterpenes, tetraterpenes, and glycosides (Zhao et al., 2016). These compounds are also found in many human food and health products, as they provide vitamins A, E, K, and coenzyme Q10. Carotenoids and tocopherols from this group are also vital sources of vitamins for animals, including humans (Rodriguez-Concepcion et al., 2018). Terpenes are widely used in various industries, such as pharmaceutical, nutraceutical, food and beverage, cosmetic, perfume, synthetic chemical, aroma and flavor, rubber, and biofuel industries (Tetali, 2019).

2.4.4 Saponins

Saponins are surface-active glycosides having foam-forming properties. The term "saponin" was taken from the Saponaria plant (Soapwort), in which "sapo" means soap. The combination of fat-soluble and water-soluble parts enhances the foaming properties of saponin. On hydrolysis, saponins yield sugar and non-sugar components, namely glycone and aglycone (Desai et al., 2016).

Soybeans, peas, and some herbs such as soapwort, soapberry, soapbark, and soap root are good sources of saponin. Commercially available saponins are extracted from *Quillaja saponaria* and *Yucca schidigera*. These saponins often occur in plants as complex mixtures and have a wide range of properties such as hypoglycemic, antimicrobial, hyperlipidaemic, and anti-inflammatory properties. The reason behind the biological efficacies of saponins is their ability to penetrate cell membranes (Desai et al., 2016). These saponins showed antimicrobial activities due to the presence of aglycone moieties. Fenugreek seeds are a rich source of saponins and play significant role in hypoglycemic activity (Singh et al., 2023). These saponins stimulate beta cells to slow down the transfer of glucose from the stomach to small intestine. The ability of saponins to form micelles with bile salts, in turn, helps in reducing the cholesterol levels. Oleanolic acid showed antiviral activity against HIV-1 protease activity. The anti-inflammatory property of saponins is reported against inflammation mediators such as serotonin and prostaglandins (Ayertey et al., 2021). There are reports that indicated that saponins have the ability to reduce the risk of chronic diseases (Castro et al., 2023). Saponins stimulate cytochrome C-caspase and induce apoptosis by stimulation. The saponins also affect tumor-specific activity in cytotoxic action. Saponins are effective against hormone- or non-hormone dependent cancers.

2.4.5 Cardiac Glycosides (CG)

CGs are a large group of secondary compounds widely distributed in nature from different sources with various applications (Botelho, 2019). Chemically, they have a steroid ring attached to an unsaturated lactone ring at position 17, and most of them are attached to a sugar moiety at position 3 (Botelho et al., 2019). Diversity of these molecules comes from variation in structure, especially the number of carbons present in the lactone ring and associated sugar moieties. Commonly found cardenolides have a butyrolactone ring with five carbons, while bufodienolides have a pyrone ring with six carbons (Prassas and Diamandis, 2008).

Commonly known CGs are digoxin, ouabain, and oleandrin, which belong to cardenolides, and bufalin, which belongs to bufodienolides. The common sugar moieties present in CGs include glucose, galactose, mannose, rhamnose, and digitalose. Ouabain is found in *Acokanthera* and *Strophanthus* (Cassels, 1985) species, digoxin in the *Digitalis* genus, namely *D. lanata* and *D. purpurea* (Braga et al., 1997), and oleandrin in the plant *Nerium oleander* (Pedroza et al., 2015).

CGs have shown cytotoxicity towards tumor cells, and research into their antineoplastic potential has risen over the last decade (Calderón-Montaño et al., 2014; Osman et al., 2017). Digitalis, by initiating apoptosis and increasing concentration of intracellular Ca^{2+}, directly inhibits the proliferation of prostate cancer cell lines (Yeh et al., 2003; Newman et al., 2007). Studies have shown that ouabain, oleandrin, and bufalin have an apoptotic effect on normal leukocytes (Newman et al., 2007).

2.4.6 Polysaccharides and Dietary Fibres

Polysaccharides are a collection of sugar monomers linked by glycosidic bonds. They can serve as energy storage, like starch and glycogen, or as non-digestible components such as cellulose, pectin, beta-glucan, hemicellulose, resistant starch, and lignin, collectively known as dietary fiber. These components are not broken down by human digestive enzymes but are digested by gut microbiota in the large intestine, where they selectively promote the growth of beneficial microorganisms.

All plant-based foods are good sources of dietary fiber, with particularly rich sources including carob beans, chicory, tamarind, Jerusalem artichoke, barley, corn, oats, wheat, and green beans. Regular consumption of dietary fiber can help prevent a range of health conditions, including cancer, inflammation, hypertension, hyperlipidemia, hypercholesterolemia, obesity, and cardiovascular diseases. It can also improve insulin sensitivity and promote a healthy gut microbiota. Additionally, dietary fibers can enhance the texture and health properties of food, reduce cooking loss and production costs, and serve as a fat substitute in health food.

2.4.7 Organosulphur Compounds (OSCs)

OSCs are compounds that contain sulfur atoms bonded to carbon or cyanate groups. They are mainly found in plants of the *Allium* and *Brassica* families, such as garlic, onion, broccoli, and cabbage. Allicin, s-allyl cysteine, sulforaphane, cysteine, methionine, and lipoic acid are some examples of OSCs with therapeutic properties. OSCs have many health benefits, such as antioxidant, antimicrobial, anticancer, antiviral, anti-inflammatory, antidiabetic, and immunomodulatory effects (Ruhee et al., 2020; Zhao et al., 2021). They can affect different metabolic pathways, including detoxification, inflammation, apoptosis, cell cycle, and signal transduction, which are directly related to oxidative stress and chronic diseases (Mithen, 2008). OSCs can also activate Nrf2, which is a key gene regulator that controls the expression of antioxidant and cytoprotective genes.

2.4.8 Carotenoids

Carotenoids are plant-made fat-soluble pigments that have yellow, orange, and red colors. They are produced by bacteria, fungi, and algae but not by animals (Sugawara et al., 2016). Out of the approximately 700 natural carotenoids, around 50 are important for human nutrition (Leitzmann, 2003). Carotenoids belong to two main classes: the first class has only hydrocarbons without oxygen (e.g., β-carotene, α-carotene, and lycopene) (Aizawa and Inakuma, 2015), and the second class has hydrocarbons with oxygen (e.g., xanthophylls) (PerezRodriguez, 2015) (Figure 2.4). Carotenoids are of great interest because of their various roles in human health. Most nutrition research studies have concentrated on the six carotenoids that are found in the highest levels in blood: β-carotene, lycopene, α-carotene, lutein, zeaxanthin and β-cryptoxanthin (Basu et al., 2001). α-Carotene and β-carotene act as important antioxidants in the lipid phase by neutralizing singlet oxygen and reactive oxygen species (Cvetkovic et al., 2013). However, the main biological function of these carotenoids is having provitamin A activity. These carotenes are mainly found in orange-colored fruits and vegetables, such as carrots, pumpkin, sweet potato, and apricot, and dark green leafy vegetables, such as spinach and kale (Sale et al., 2004). β-cryptoxanthin is a common carotenoid present in fruit as well as human blood and tissues. Foods that are rich in β-cryptoxanthin include peaches, papaya, tangerines, persimmons, and oranges (Sugiura et al., 2012).

2.5 BIOSYNTHETIC PATHWAYS OF PHYTOCHEMICALS

Phytochemicals are generally classified into three major groups based on their biosynthetic origins: terpenoids, alkaloids, and phenolics. Each group has different biosynthetic pathways that involve specific enzymes and precursors. The mevalonate pathway or the methylerythritol phosphate pathway and the shikimate pathway are among the common biosynthetic pathways by which phytochemical compounds are synthesized.

In the shikimate pathway, phosphoenolpyruvate and erythrose-4-phosphate are converted into chorismate, a precursor for phenylalanine and tyrosine. Phenylalanine and tyrosine are then converted to cinnamic acid and p-coumaric acid by the enzymes phenylalanine ammonia-lyase and tyrosine ammonia-lyase, respectively. These acids are the initial substrates for the phenylpropanoid pathway, which produces

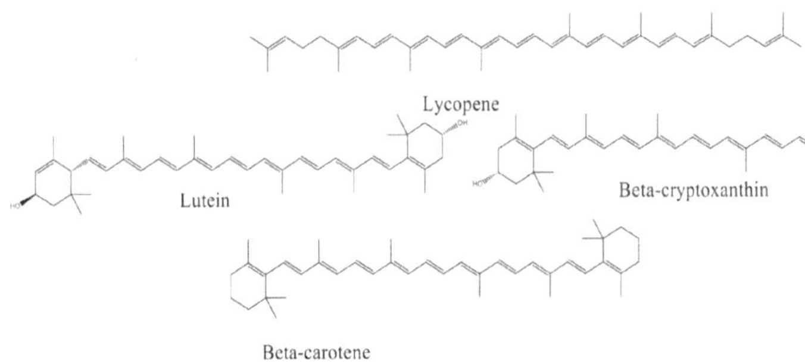

FIGURE 2.4 Commonly known carotenoids with health benefits.

various phenolic compounds, such as lignin, flavonoids, anthocyanins, and tannins. Terpenoids are derived from the condensation of five-carbon isoprene units, which are synthesized from acetyl-CoA via the mevalonate pathway or the methylerythritol phosphate pathway in plants. Table 2.4 accounts for the different phytochemicals and pathways involved in their synthesis along with the precursor and enzyme details, which convert different precursors into phytochemical compounds (Tijjani et al., 2018; Eljounaidi et al., 2020)

TABLE 2.4
Biosynthetic Pathways for Different Phytochemicals

Phytochemical	Biosynthetic Pathway	Precursor	Enzyme
Alkaloids	Amino acid-derived	Amino acids (e.g., tyrosine, tryptophan, lysine)	Various (e.g., decarboxylase, hydroxylase, methyltransferase, oxidase)
Anthocyanidins	Shikimate pathway	Phenylalanine	Phenylalanine ammonia-lyase, chalcone synthase, flavanone 3-hydroxylase (F3H), dihydroflavonol 4-reductase, anthocyanidin synthase
Anthraquinones	Shikimate pathway	Phenylalanine	Flavonoid 3'-hydroxylase, flavonoid 3',5'-hydroxylase (F3'5'H), anthraquinone synthase (AQS)
Flavonoids	Shikimate pathway	Phenylalanine	Flavone synthase, flavanone 4-reductase, flavan-3,4-diol synthase, isoflavone synthase, flavonol synthase
Glycosides	Various	Aglycone + sugar	Glycosyltransferase (GT)
Lignans	Shikimate pathway	Phenylalanine	Cinnamate 4-hydroxylase, 4-coumarate:CoA ligase, cinnamoyl-CoA reductase, cinnamyl alcohol dehydrogenase, dirigent protein
Phenolics	Shikimate pathway	Phenylalanine	Various hydroxylases and methyltransferases
Saponins	Mevalonate pathway	Isopentenyl pyrophosphate	Cytochrome P450 monooxygenase, glycosyltransferase
Steroids	Mevalonate pathway	Isopentenyl pyrophosphate	Cytochrome P450 monooxygenase, hydroxysteroid dehydrogenase
Terpenoids	Mevalonate pathway or methylerythritol phosphate pathway	Isopentenyl pyrophosphate and its isomer dimethylallyl pyrophosphate	Phenyltransferase, terpene synthase, cytochrome P450 monooxygenase, reductase, glycosyltransferase

2.6 ANALYTICAL TECHNIQUES FOR PHYTOCHEMICALS

Phytochemicals are extracted from plants using a process that involves cell lysis, collection of extract, isolation, purification using chromatographic techniques, and identification of phytochemicals using spectrophotometry. The efficiency of this extraction can be enhanced by optimizing conditions such as the choice of solvent, temperature, and time.

Categorically, solvents used for extracting phytochemicals are divided into green solvents and other organic solvents. Green solvents are generally non-toxic, biodegradable, recyclable, and renewable in nature. Water, ethanol, ionic liquids, supercritical CO_2, acetic acid, fatty oils, deep eutectic solvents, and natural deep eutectic solvents are considered green solvents. Acetone, chloroform, butanol, methanol, ethyl acetate, methyl acetate, benzene, hexane, and cyclohexane are examples of other organic solvents. Each solvent has its own unique properties and uses. For instance, water is the most used solvent due to its non-selective nature and ability to separate all hydrophilic substances. Ethanol is selective and used for the extraction of polyphenols and triterpenes. Glycerol is a selective solvent for polyphenolic extraction, while fatty oils are good for extracting hydrophobic substances. Acetic acid buffer can be used for the extraction of phenolics and anthocyanins, and isopropanol is a green alternative to n-hexane used for the extraction of oils, alkaloids, gums, and natural resins. Supercritical CO_2 is used for the extraction of lignans, anthocyanins, and essential oils. Ionic liquids and deep eutectic solvents are used for the extraction of a wide range of organic and inorganic compounds.

Natural deep eutectic solvents (NADES) are composed of natural metabolites such as organic acids, amino acids, sugars, polyols, and choline derivatives. These solvents are characterized by their low volatility and melting point, and a wide polarity range, making them suitable for extracting compounds that are poorly soluble in water.

Among the organic solvents, acetone is suitable for extracting alkaloids and oils due to its low toxicity. Ethyl acetate is used for extracting flavonoids and total phenolics, and it is non-toxic in nature. Despite the toxicity factor, methanol and chloroform are used for extracting flavonoids, saponins, and tannins. Butanol is primarily used for purifying fractions of individual compounds. Due to low acute toxicity, n-hexane can be used for extracting flavonoids, carbohydrates, anthraglycosides, and saponins. Methyl acetate, a by-product of esterification and carboxylation, is mainly used for extracting phytosterols and tocopherols. Cyclohexanes are generally used for extracting fats, waxes, and oils. Table 2.5 describes the common and recent methods employed for the extraction of phytochemicals, with their pros and cons (Kumar et al., 2023).

2.6.1 Qualitative and Quantitative Analysis of Phytochemicals

Phytochemical analysis is the investigation of the chemical constituents and activities of plant-derived substances. Phytochemical analysis can assist in the identification, isolation, quantification, and characterization of the phytochemicals found in

TABLE 2.5
Conventional and Non-Conventional Methods of Phytochemical Extraction

Extraction Method	Solvent Used	Pros	Cons	Phytochemical
Maceration	Water, ethanol, methanol, etc.	Simple, low cost, highly effecient	Long extraction time, solvent evaporation, thermal degradation	Flavonoids, alkaloids, phenolic acids, etc.
Percolation	Water, ethanol, methanol, etc.	Continuous extraction, high yield, less solvent consumption	Requires special equipment, solvent evaporation, thermal degradation	Flavonoids, alkaloids, phenolic acids, etc.
Decoction	Water	Simple, low cost, suitable for heat-stable compounds	Long extraction time, solvent evaporation, thermal degradation, low selectivity	Saponins, tannins, glycosides, etc.
Reflux extraction	Water, ethanol, methanol, etc.	High yield, less solvent consumption, suitable for heat-stable compounds	Requires special equipment, solvent evaporation, thermal degradation, low selectivity	Flavonoids, alkaloids, phenolic acids
Soxhlet extraction	Water, ethanol, methanol, etc.	High yield, less solvent consumption, suitable for heat-stable compounds	Requires special equipment, solvent evaporation, thermal degradation, low selectivity	Flavonoids, alkaloids, phenolic acids
Pressurized liquid extraction (PLE)	Water, ethanol, methanol, etc.	Fast extraction, high yield, high selectivity, less solvent consumption	Requires special equipment, high pressure, high temperature	Anthocyanins, flavonoids, phenolic acids
High hydrostatic pressure extraction (HHP)	Water, ethanol, methanol, etc.	Fast extraction, high yield, high selectivity, less solvent consumption, preserves bioactivity	Requires special equipment, high pressure, high cost	Anthocyanins, flavonoids, phenolic acids
Microwave-assisted extraction (MAE)	Water, ethanol, methanol, etc.	Fast extraction, high yield, high selectivity, less solvent consumption, less energy consumption	Requires special equipment, high temperature, thermal degradation	Flavonoids, alkaloids, phenolic acids

(Continued)

TABLE 2.5 (CONTINUED)
Conventional and Non-Conventional Methods of Phytochemical Extraction

Extraction Method	Solvent Used	Pros	Cons	Phytochemical
Ultrasound-assisted extraction (UAE)	Water, ethanol, methanol, etc.	Fast and efficient extraction, high yield, high selectivity, less solvent consumption, less energy consumption	Requires special equipment, can cause noise pollution, thermal degradation	Flavonoids, alkaloids, phenolic acids
Pulsed electric field extraction (PEF)	Water, ethanol, methanol, etc.	Fast and efficient extraction, high yield, high selectivity, less solvent consumption, preserves bioactivity	Requires special equipment, high voltage, high cost	Flavonoids, alkaloids, phenolic acids
Supercritical fluid extraction (SFE)	Carbon dioxide, propane, etc.	Fast and efficient extraction, high yield, high selectivity, less solvent consumption, no solvent residue	Requires special equipment, high pressure, high temperature, high cost	Carotenoids, terpenoids, essential oils
Natural deep eutectic solvent extraction (NADES)	Choline chloride, urea, glycerol, etc.	Green extraction, high yield, high selectivity, biodegradable, low cost	Requires special equipment, high viscosity, low volatility	Flavonoids, alkaloids, phenolic acids

different plant parts, such as leaves, flowers, fruits, seeds, roots, barks, and stems. Phytochemical analysis can also reveal the biosynthesis, metabolism, and distribution of phytochemicals in plants, as well as their interactions with other molecules and biological systems (Khandelwal, 2006). There are two main types of phytochemical analysis: qualitative and quantitative. Qualitative phytochemical analysis is the initial screening of plant extracts or fractions to determine the presence or absence of certain classes of phytochemicals, such as alkaloids, flavonoids, phenolics, tannins, saponins, steroids, glycosides, and terpenoids (Kumar and Sharma, 2020). Qualitative phytochemical analysis can be done by using simple and fast methods, such as colorimetric tests, precipitation tests, chromatographic tests, and spectroscopic tests. These methods rely on the specific reactions or interactions of phytochemicals with certain reagents, solvents, or detectors, which result in characteristic changes in color, precipitate formation, spot appearance, or spectral patterns. Qualitative phytochemical analysis can provide useful information about the chemical diversity and complexity of plant extracts or fractions, as well as their potential pharmacological activities. Quantitative phytochemical analysis can be done by using more advanced and precise methods, such as high-performance liquid chromatography (HPLC), gas chromatography (GC), mass spectrometry (MS), nuclear magnetic resonance spectroscopy (NMR), and Fourier transform infrared spectroscopy (FTIR). Table 2.6 accounts for various techniques employed for the analysis of phytochemicals (Singh and Singh, 2015; Harborne, 1998). These methods depend on the separation, identification, and quantification of phytochemicals by using different parameters, such as retention time, peak area, molecular weight, structural features, or functional groups. Quantitative phytochemical analysis can provide detailed information about the chemical composition and structure of plant extracts or fractions, as well as their quality, purity, and stability. Some of the common qualitative tests for phytochemicals include Mayer's test, Wagner's test, Dragendorff's test, and Hager's test for alkaloids; Shinoda test, $AlCl_3$ test, NaOH test, and lead acetate test for flavonoids; ferric chloride test, Folin-Ciocalteu reagent test, and vanillin-HCl test for phenolics; gelatin test, ferric chloride test, and bromine water test for tannins; Liebermann-Burchard test, Salkowski test, and Noller's test for steroids; Legal test, Baljet test, and Keller-Killiani test for glycosides; Salkowski test, Noller's test, and Liebermann-Burchard test for terpenoids; Molisch test, Benedict's test, and Fehling's test for carbohydrates; and foam test, hemolytic test, and vanillin-sulphuric acid test for saponins.

Phytochemical analysis is an important and challenging area of research, as it requires the application of various techniques and instruments, as well as the interpretation and integration of complex data. Phytochemical analysis can also encounter some limitations and difficulties, such as the variability of plant materials, the extraction and isolation methods, the standardization and validation of analytical methods, the availability and accessibility of reference standards, the identification and characterization of unknown or novel phytochemicals, the interference and interaction of phytochemicals with other components, the reproducibility and comparability of results, and the ethical and legal issues related to the use and exploitation of plant resources.

TABLE 2.6
Various Techniques Employed for the Analysis of Phytochemicals

Technique	Advantages	Disadvantages	Phytochemicals
Thin layer chromatography (TLC)	Simple, low cost, rapid, qualitative, and semi-quantitative, suitable for various phytochemicals	Low resolution, low sensitivity, solvent consumption, manual operation	Flavonoids, alkaloids, phenolic acids, terpenoids
High performance liquid chromatography (HPLC)	High resolution, high sensitivity, quantitative, automated, suitable for various phytochemicals	High cost, complex operation, solvent consumption, sample preparation	Flavonoids, alkaloids, phenolic acids, terpenoids
Gas chromatography (GC)	High resolution, high sensitivity, quantitative, automated, suitable for volatile phytochemicals	High cost, complex operation, sample preparation, unsuitable for non-volatile phytochemicals	Essential oils, terpenoids, fatty acids
Fourier transform infrared spectroscopy (FTIR)	Simple, rapid, non-destructive, qualitative, and semi-quantitative, suitable for various phytochemicals	Low sensitivity, interference from solvent and water, limited information	Flavonoids, alkaloids, phenolic acids, terpenoids, etc
Nuclear magnetic resonance spectroscopy (NMR)	Non-destructive, qualitative, and quantitative, comprehensive information, suitable for various phytochemicals	High cost, complex operation, low sensitivity, sample preparation	Flavonoids, alkaloids, phenolic acids, terpenoids
Mass spectrometry (MS)	High sensitivity, high specificity, qualitative and quantitative, comprehensive information, suitable for various phytochemicals	High cost, complex operation, sample preparation, ionization methods	Flavonoids, alkaloids, phenolic acids, terpenoids

2.7 CHALLENGES AND FUTURE PROSPECTS IN PHYTOCHEMICALS RESEARCH

Plants produce compounds called phytochemicals that can benefit human health in various ways, such as fighting and preventing chronic diseases, boosting immunity, and improving well-being. They have different chemical structures, such as phenolics, terpenoids, alkaloids, flavonoids, and glucosinolates, and different biological effects, such as antioxidant, anti-inflammatory, antimicrobial, anticancer, and antidiabetic properties. They can be applied in medicine, food, cosmetics, and agriculture, as they can affect cellular functions, protect against oxidative damage, kill harmful microorganisms, trigger cell death in cancer cells, and control glucose levels. Phytochemicals are also an economic opportunity for many developing countries, as they have a high market demand and value, and can be obtained from native plants where there is traditional knowledge of their health effects. Consumers are more interested in functional foods that contain phytochemicals, as they are willing to pay more for foods that have health claims and functional ingredients. However, the field of phytochemicals also faces many challenges, such as the lack of standardized methods and criteria for the isolation, extraction, purification, identification, and quantification of phytochemicals. The composition and activity of phytochemicals vary depending on source, processing methods, and environmental factors, which reduce their applicability. Their low bioavailability, poor stability, and solubility in the human body are also a concern in drug development processes. Producing enough phytochemicals for industrial and clinical applications, developing effective delivery systems, and identifying the main active phytochemicals and their mechanisms of action are also some of the goals for the scientific community in exploring the vast world of phytochemicals. These challenges require interdisciplinary collaboration and innovation among researchers, pharmacists, and engineers in the field of phytochemicals. By overcoming these challenges, phytochemicals can offer new opportunities and solutions for human health and well-being.

ACKNOWLEDGMENT

The authors are thankful to the parent institution for their unconditional support in the preparation of this manuscript.

REFERENCES

Aizawa, Kiyoharu, and Tadashi Inakuma. "Inhibitory Effect of Carotenoid-Rich Extract from Citrus Fruits on Helicobacter Pylori Infection in Human Gastric Mucosa." *Journal of Functional Foods* 16 (2015): 534–40.

Ayertey, Frederick, Ebenezer Ofori-Attah, Stephen Antwi, et al. "Anti-inflammatory Activity and Mechanism of Action of Ethanolic Leaf Extract of Morinda lucida Benth." *Journal of Traditional and Complementary Medicine* 11, no. 1 (2021): 21–31.

Ballard C. R., and M. R. Maróstica. *Bioactive Compounds*. Cambridge, UK: Woodhead Publishing, 2019.

Basu, T. K., N. J. Temple, and M. Garg. "Chronic Disease and Micronutrients: An Overview." In *Antioxidants in Human Health and Disease*, edited by T. K. Basu, N. J. Temple, and M. Garg, 1–14. Wallingford: CAB International, 2001.

Botelho, Ana Flávia M., Felipe Pierezan, Benito Soto-Blanco, and Marília Martins Melo. "A Review of Cardiac Glycosides: Structure, Toxicokinetics, Clinical Signs, Diagnosis and Antineoplastic Potential." *Journal of Venomous Animals and Toxins Including Tropical Diseases* 25 (2019): e20190001.

Braga, F. C., W. Kreis, R. A. Récio, and A. B. De Oliveira. "Quantitative evaluation of cardenolides in a brazilian digitalis lanata cultivar and perspectives of its industrial use." *Química Nova* 20, no. 5 (1997): 481–85.

Capilla, A. S., I. Sánchez, D. H. Caignard, P. Renard, and M. D. Pujol. "Antitumor Agents. Synthesis and Biological Evaluation of New Compounds Related to Podophyllotoxin, Containing the 2,3-Dihydro-1,4-Benzodioxin System." *European Journal of Medicinal Chemistry* 36 (2001): 11.

Cassels, B. K. "Analysis of a Maasai Arrow Poison." *Journal of Ethnopharmacology* 14, no. 2–3 (1985): 273–81.

Cui, Qinghua, Ruikun Du, Miaomiao Liu, and Lijun Rong. 2020. "Lignans and Their Derivatives from Plants as Antivirals." *Molecules* 25, no. 1: 183. https://doi.org/10.3390/molecules25010183

Cvetkovic, Dragan, Jasna Canadanovic-Brunet, Gordana Djilas, Sonja Cetkovic, Vesna Tumbas, and Jelena Savatovic. *Carotenoids from Apricots (Prunus Armeniaca L.)*, 1–26. New York: Nova Science Publishers, 2013.

D' Archivio, Massimo, Carmela Filesi, Rosaria Di Benedetto, Roberta Gargiulo, Claudio Giovannini, and Roberta Masella. "Polyphenols, Dietary Sources, and Bioavailability." *Annali dell'Istituto Superiore di Sanità* 43 (2007): 348. https://www.ncbi.nlm.nih.gov/pmc/articles/PMC2781139/.

Dai, J., and R. J. Mumper. "Plant Phenolics: Extraction, Analysis and Their Antioxidant and Anticancer Properties." *Molecules* 15, no. 10 (2010): 7313–352.

Desai, Sapna D., Dhruv G. Desai, and Harmeet Kaur. "Saponins and Their Biological Activities." *Pharmacognosy Reviews* 10, no. 19 (2016): 65–74.

Divekar, P. A., S. Narayana, B. A. Divekar, et al. "Plant Secondary Metabolites as Defense Tools against Herbivores for Sustainable Crop Protection." *International Journal of Molecular Sciences* 23, no. 5: 2690.

Eljounaidi, K., and B. R. Lichman. "Nature's Chemists: The Discovery and Engineering of Phytochemical Biosynthesis." *Frontiers in Chemistry* 8 (2020): 596479. https://doi.org/10.3389/fchem.2020.596479

Faccio, G. "Shadowing Plants: Phytochemicals for Sustainable Sun Protection, A Commentary." *Cosmetics & Toiletries* 136 (2021): 36–41.

Falsaperla, M., G. Morgia, A. Tartarone, R. Ardito, and G. Romano. "Support Ellagic Acid Therapy in Patients with Hormone Refractory Prostate Cancer (HRPC) on Standard Chemotherapy Using Vinorelbine and Estramustine: Phosphate." *European Urology* 47 (2005): 449.

Farias, F. M., C. S. Passos, M. D. Arbo, et al. "Strictosidinic Acid, Isolated from Psychotria Myriantha Mull. Arg. (Rubiaceae), Decreases Serotonin Levels in Rat Hippocampus." *Fitoterapia* 83, no. 6 (2012): 1138–143.

Guruvayoorappan, C., K. M. Sakthivel, Ganesan Padmavathi, Vaishali Bakliwal, Javadi Monisha, and Ajaikumar B. Kunnumakkara. "Anticancer Properties of Fruits and Vegetables: A Scientific Review." *International Journal of Pharmacognosy and Phytochemical* Research 6, no. 1 (2014): 1–18.

Gutiérrez-Grijalva, Erick Paul, Leticia Xochitl López-Martínez, Laura Aracely Contreras-Angulo, Cristina Alicia Elizalde-Romero, and José Basilio Heredia. *Plant Alkaloids: Structures and Bioactive Properties*, 85–117. Singapore: Springer, 2020.

Harborne, J. B. *Phytochemical Methods: A Guide to Modern Techniques of Plant Analysis*. 3rd ed. Springer, 1998.

Huang, M., Y. Y. Hu, X. Q. Dong, et al. "The Protective Role of Oxymatrine on Neuronal Cell Apoptosis in the Hemorrhagic Rat Brain." *Journal of Ethnopharmacology* 143, no. 1 (2012): 228–35.

Islam, Salman Ul, Muhammad Bilal Ahmed, Haseeb Ahsan, and Young-Sup Lee. "Recent Molecular Mechanisms and Beneficial Effects of Phytochemicals and Plant-Based Whole Foods in Reducing LDL-C and Preventing Cardiovascular Disease." *Antioxidants* 10, no. 5 (2021): 784.

Jang, Won Young, Mi-Yeon Kim, and Jae Youl Cho. 2022. "Antioxidant, Anti-Inflammatory, Anti-Menopausal, and Anti-Cancer Effects of Lignans and Their Metabolites." *International Journal of Molecular Sciences* 23, no. 24: 15482. https://doi.org/10.3390/ijms232415482

Kawazoe, K., A. Yutani, K. Tamemoto, S. Yuasa, H. Shibata, T. Higuti, and Y. Takaishi. "Phenylnaphthalene Compounds from the Subterranean Part of Vitex rotundifolia and Their Antibacterial Activity Against Methicillin-Resistant Staphylococcus Aureus." *Journal of Natural Products* 64 (2001): 588–91. https://doi.org/10.1021/np000600c.

Khandelwal, K. R. *Practical Pharmacognosy* (16th ed.). Pune: Nirali Prakashan, 2006.

Kumar, A., and A. Sharma. "Qualitative Tests for Preliminary Phytochemical Screening: An Overview." *International Journal of Chemical Studies* 8 (2020): 1–5. https://doi.org/10.22271/chemi.2020.v8.i2i.8834.

Kumar, Ashwani, Nirmal P. Mukul Kumar, Anina Jose, Vidisha Tomer, Emel Oz, Charalampos Proestos, Maomao Zeng, Tahra Elobeid, K. Sneha, et al. 2023. "Major Phytochemicals: Recent Advances in Health Benefits and Extraction Method." *Molecules* 28, no. 2: 887. https://doi.org/10.3390/molecules28020887

Kumar, M., M. R. Shylaja, P. A. Nazeem, and T. Babu. "6-Gingerol Is the Most Potent Anticancerous Compound in Ginger (Zingiber officinale Rosc.)." *Journal of Developing Drugs* 6 (2017): 1.

Lara, Naira Sahakyan, Agnieszka Bartoszek, Claus Jacob, Margarit Petrosyan, and Armen Trchounian. "Bioavailability of Tannins and Other Oligomeric Polyphenols: A Still to Be Studied Phenomenon." *Current Pharmacology Reports* 6, no. 3 (2020): 131–36.

Leitzmann, Claus. "Nutrition Ecology: The Contribution of Vegetarian Diets." *The American Journal of Clinical Nutrition* 78, no. 3 (2003): 657S–9S.

Li Shanyi, Tong Cai, Shichen Cui, Fangmin Liu, Rong Hu, and Wenji Li. "Prostate Cancer, Oxidative Stress, and Antioxidant Phytochemicals: A Brief Review." *Current Pharmacology Reports* 9, no. 6 (2023): 391–96

Li, Ya, Shuhan Xie, Jinchuan Ying, Wenjun Wei, and Kun Gao. 2018. "Chemical Structures of Lignans and Neolignans Isolated from Lauraceae" *Molecules* 23, no. 12: 3164. https://doi.org/10.3390/molecules23123164.

Liu, Rui Hai, and Jing Liu. "Health-Promoting Components of Fruits and Vegetables in the Diet." *Advances in Nutrition* 4, no. 3 (2013): 384S–92S.

Liu, Rui Hai, and Jing Liu. "Role of Phytochemicals in Cancer Prevention." *International Journal of Molecular Sciences* 20, no. 20 (2019): 4981.

Lopez-Corona, A. V., I. Valencia-Espinosa, F. A. González-Sánchez, A. L. Sánchez-López, L. E. Garcia-Amezquita, and R. Garcia-Varela. "Antioxidant, Anti-Inflammatory and Cytotoxic Activity of Phenolic Compound Family Extracted from Raspberries (*Rubus idaeus*): A General Review." *Antioxidants* 11, no. 6 (2022): 1192. https://doi.org/10.3390/antiox11061192.

Luo, Y., Y. Li, Y. Zhang, et al. "Condensed Tannins Improve Glycolipid Metabolism but Induce Liver Injury in Lethrinus maculatus." *Frontiers in Marine Science* 9 (2022): 902633.

Lu, H., and Liu, R. "Antioxidant Activity of Lignans." *Journal of Agricultural and Food Chemistry* 40 no. 12 (1992): 2347–2354.

Mahomoodally, M. F. "Traditional Medicines in Africa: An Appraisal of Ten Potent African Medicinal Plants." *Evidence-Based Complementary and Alternative Medicine* (2013): e617459.

Marcotullio, M. C., M. Curini, J. X. Becerra. "An Ethnopharmacological, Phytochemical and Pharmacological Review on Lignans from Mexican Bursera Spp." *Molecules* 23 (2018): 1976.

Masyita, Ayu, Reka Mustika Sari, Ayun Dwi Astuti, Budiman Yasir, Nur Rahma Rumata, Talha Bin Emran, Firzan Nainu, and Jesus Simal-Gandara. "Terpenes and Terpenoids as Main Bioactive Compounds of Essential Oils, Their Roles in Human Health and Potential Application as Natural Food Preservatives." *Food Chemistry* X, no. 13 (2022): 100217

Mieres-Castro, Daniel, and Freddy Mora-Poblete. "Saponins: Research Progress and Their Potential Role in the Post-COVID-19 Pandemic Era." *Pharmaceutics* 15, no. 2 (2023): 348.

Mithen, Richard. "Sulphur-Containing Compounds."In Plant Secondary Metabolites: Occurrence, Structure and Role in the Human Diet, edited by Alan Crozier, Michael N. Clifford, and Hiroshi Ashihara, 25–46. Oxford: Blackwell Publishing Ltd., 2008.

Newman, R. A., P. Yang, A. D. Pawlus, and K. I. Block. "Cardiac Glycosides as Novel Cancer Therapeutic Agents." *Molecular Interventions* 7 no. 1 (2007): 36–49. https://doi.org/10.1124/mi.7.1.6

Nyamai, D. W., W. Arika, P. E. Ogola, E. N. M. Njagi, and M. P. Ngugi. "Medicinally Important Phytochemicals: An Untapped Research Avenue." *International Journal of Pharmacognosy and Phytochemical Research* 8, no. 3 (2016): 399–413.

O'Brien, Peter, Catalina Carrasco-Pozo, and Hernán Speisky. "Boldine and Its Antioxidant or Health-Promoting Properties." *Chemico-Biological Interactions* 159, no. 1 (2006): 1–17.

Olthof, M. R., P. C. H. Hollman, and M. B. Katan. "Chlorogenic Acid and Caffeic Acid are Absorbed in Humans." *Journal of Nutrition* 131 (2001): 66.

Osman, W., H. Almukhtar, and R. Almukhtar. "Mechanistic Insight into the Cytotoxic Activity of Cardiac Glycosides and Their Novel Cancer Therapeutic Potential." *Oncotargets and Therapy* 10 (2017): 1171–1186. https://doi.org/10.2147/OTT.S140016

Pan, J. Y., S. L. Chen, M. H. Yang, J. Wu, J. Sinkkonen, and K. Zou. "An Update on Lignans: Natural Products and Synthesis." *Natural Product Reports* 26 (2009): 1251–92. 3.

Pecyna, Paulina, Joanna Wargula, Marek Murias, and Malgorzata Kucinska. "More Than Resveratrol: New Insights into Stilbene-Based Compounds." *Biomolecules* 10, no. 8 (2020): 1111.

Pedroza, H. P., M. G. Ferreira, J. G. Carvalho, J. G. Melo, K. D. Antunes, K. M. Keller, M. M. Melo, and B. Soto-Blanco. "Concentrações de Oleandrina nas Folhas de Nerium Oleander de Diferentes Cores de Floração." *Revista Brasileira de Plantas Medicinais* 17, no. 4 (2015): 1130–35.

PerezRodriguez, Maria L. "Carotenoids." In *Handbook of Food Analysis*, edited by Leo M. L. Nollet and Fidel Toldra, 3rd ed., 1: 409–32. Boca Raton: CRC Press, 2015.

Prassas, Ioannis, and Eleftherios P. Diamandis. "Novel Therapeutic Applications of Cardiac Glycosides." *Nature Reviews Drug Discovery* 7, no. 11 (2008): 926–35.

Qiu, Shi, Sun Hui, Zhang Ai-Hua, Xu Hong-Ying, Yan Guang-Li, Han Ying, and Wang Xi-Jun. "Natural Alkaloids: Basic Aspects, Biological Roles, and Future Perspectives." *Chinese Journal of Natural Medicines* 12, no. 6 (2014): 401–6.

Rodriguez-Concepcion, M., J. Avalos, M. L. Bonet, A. Boronat, L. Gomez-Gomez, D. Hornero-Mendez, et al. "A Global Perspective on Carotenoids: Metabolism, Biotechnology, and Benefits for Nutrition and Health." *Progress in Lipid Research* 70 (2018): 62–93.

Roy, Madhumita, and Amitava Datta. *Cancer Genetics and Therapeutics: Focus on Phytochemicals.* Singapore: Springer, 2020.

Ruhee, T., L. A. Roberts, S. Ma, and K. Suzuki. "Organosulfur Compounds: A Review of Their Anti-Inflammatory Effects in Human Health." Frontiers in Nutrition 7 (2020): 64. https://doi.org/10.3389/fnut.2020.00064

Sale, S., R. D. Verschoyle, D. Boocock, D. J. L. Jones, N. Wilsher, K. C. Ruparelia, G. A. Potter, P. B. Farmer, W. P. Steward, and A. J. Gescher. "Pharmacokinetics in Mice and Growth-Inhibitory Properties of the Putative Cancer Chemopreventive Agent Resveratrol and the Synthetic Analogue Trans 3,4,5,4′-Tetramethoxystilbene." *British Journal of Cancer* 90 (2004): 736–44.

Santhiravel, S., A. E.-D. A. Bekhit, E. Mendis, J. L. Jacobs, F. R. Dunshea, N. Rajapakse, and E. N. Ponnampalam. "The Impact of Plant Phytochemicals on the Gut Microbiota of Humans for a Balanced Life." *International Journal of Molecular Sciences* 23 (2022): 8124.

Saxena, M., Jyoti Saxena, Rajeev Nema, Dharmendra Singh, and Abhishek Gupta. "Phytochemistry of Medicinal Plants." *Journal of Pharmacognosy and Phytochemistry* 1, no. 6 (2013): 168–75.

Shi, Q., J. Fu, D. Ge, et al. "Huperzine A Ameliorates Cognitive Deficits and Oxidative Stress in the Hippocampus of Rats Exposed to Acute Hypobaric Hypoxia." *Neurochemical Research* 37, no. 9 (2012): 2042–52.

Singh, S., P. K. Chaurasia, and S. L. Bharati. "Hypoglycemic and Hypocholesterolemic Properties of Fenugreek: A Comprehensive Assessment." *Applied Food Research* 3, no. 2 (2023): 100311.

Singh, S., and S. Singh. "Phytochemical Screening and Extraction: A Review." *International Journal of Advanced Research in Chemical Science* 2 (2015): 25–32.

Sugawara, Tatsuya, Shigenori Aoki, and Kazuo Nagao. "Properties of Carotenoids in Human Skin and Carotenoid Supplementation." In *Carotenoids: Nutrition, Analysis and Technology*, 213–24. Chichester: John Wiley & Sons, 2016.

Sugiura, Makoto, Masahiro Nakamura, Yoko Ogawa, Toshio Nagao, and Shigeru Watanabe. "High Serum Carotenoid Concentration Associated with High Skeletal Muscle Strength in Japanese Men and Women." *The Journal of Nutrition, Health & Aging* 16, no. 9 (2012): 789–93.

Tetali, S. D. "Terpenes and Isoprenoids: A Wealth of Compounds for Global Use." *Planta* 249 (2019): 1–8.

Tijjani, H., C. Egbuna, and L. D. Carroll. "Biosynthesis of Phytochemicals." In *Phytochemistry*, edited by Chukwuebuka Egbuna, Jonathan Chinenye Ifemeje, Stanley Chidi Udedi, and Shashank Kumar, 42–62. New Jersey: Apple Academic Press, 2018. https://doi.org/10.1201/9780429426223-2

Vasmehjani, Abbas Ali, Zohreh Darabi, Azadeh Nadjarzadeh, Mohammad Mirzaei, and Mahdieh Hosseinzadeh. "The Relation between Dietary Phytochemical Index and Metabolic Syndrome and Its Components in a Large Sample of Iranian Adults: A Population-Based Study." *BMC Public Health* 21, no. 1 (2021): 1587.

Virgen-Carrillo, C. A., E. H. Valdés Miramontes, D. Fonseca Hernández, D. A. Luna-Vital, and L. Mojica. "West Mexico Berries Modulate α-Amylase, α-Glucosidase and Pancreatic Lipase Using in Vitro and in Silico Approaches." *Pharmaceuticals* 15, no. 9 (2022): 1081.

Wani, Tareq A., Irshad Ahmad Bhat, Khushboo Guleria, et al. "*Phytochemicals: Diversity, Sources and Their Roles*." In *Phytochemical Genomics*, 3–33. Singapore: Springer, 2023.

Wolfe, C., P. Pagano, C. M. Pillar, D. L. Shinabarger, and R. A. Boulos. "Comparison of the in Vitro Antibacterial Activity of Ramizol, Fidaxomicin, Vancomycin, and Metronidazole against 100 Clinical Isolates of Clostridium Difficile by Broth Microdilution." *Diagnostic Microbiology and Infectious Disease* 92 (2018): 250–252.

Xie, L.-W., S. Cai, T.-S. Zhao, M. Li, and Y. Tian. "Green Tea Derivative (–)-Epigallocatechin-3-Gallate (EGCG) Confers Protection against Ionizing Radiation-induced Intestinal Epithelial Cell Death Both In Vitro and In Vivo." Free Radical Biology and Medicine 161 (2020): 175–186.

Yang, Yan. "Phytochemicals and Health." In *Nutritional Toxicology*, edited by Rui Hai Liu, 407–23. Singapore: Springer, 2022.

Yeh, J. R. J., L. C. Liu, S. N. Chow, C. L. Ho, and T. Y. Chu. "High-Dose Digitalis Induces Apoptosis and Cell Cycle Deregulation in Human Cancer Cells: High-Potency Secondary Cancer Prevention." *Prostate* 56 no. 2 (2003): 140–146. https://doi.org/10.1002/pros.10233

Zhao, D.-D., Jiang, L.-L., Li, H.-Y., Yan, P.-F., and Zhang, Y.-L. (2016). "Chemical Components and Pharmacological Activities of Terpene Natural Products from the Genus Paeonia." *Molecules* 21 (2016): 1362.

Zhao, Ming Ming, Shi Hang Xiong, Guan Ding Zhao, and Hua Yu. *Organosulfur Compounds in Food*, 1741–1761. Singapore: Springer, 2021.

3 Significance of Millets in Gut Health

Samlesh Kumari[1], Smrutirekha Mallick[2], Deepika S Channe[3], Narendra Kumar[4], Vandana Kumari[5], and Ramesh Pothuraju[6]

[1]Scientist Dairy Microbiology, Centre of Excellence on Soybean Processing and Utilization, ICAR-Central Institute of Agricultural Engineering, Nabi Bagh Berasia Road, Bhopal, India
[2]Scientist, ICAR-National Institute on Foot and Mouth Disease, Arugul, Bhubaneswar, India
[3]Assistant Chief Technical Officer, PME Cell, ICAR-Central Institute of Agricultural Engineering, Nabi Bagh Berasia Road, Bhopal, India
[4]Assistant Professor, Food Safety and Quality Assurance, College of Dairy and Food Technology, Rajasthan University of Veterinary and Animal Sciences, State Agricultural University, Bassi, Jaipur, India
[5]Teaching Associate, Food Safety and Quality Assurance, College of Dairy and Food Technology, Rajasthan, India; University of Veterinary and Animal Sciences, State Agricultural University, Bassi, Jaipur, India
[6]Ramanujan Fellow, Scientist-C, Cancer Research Program, Rajiv Gandhi Centre for Biotechnology (RGCB), Thycaud Post, Poojappura, Thiruvananthapuram, Kerala, India

3.1 INTRODUCTION

Millets are small-seeded grains often considered a staple food in many parts of the globe. These traditional foods have gained popularity due to their exceptional nutritional and health benefits. Millets are drought-tolerant crops, can withstand extreme weather conditions, and require low chemical fertilizer/pesticide input. Millet belongs to the family Poaceae. The major types of millets are pearl millet (*Pennisetum glaucum*), foxtail millet (*Setaria italica*), proso millet (*Panicummiliaceum*), and finger millet (*Eleusine coracana*). Minor millets include barnyard millet (*Echinochloa spp.*), kodo millet (*Paspalumscrobiculatum*), little millet (*Panicum sumatrense*), and browntop millet (*Urochloa ramose/Brachiaria ramose/Panicumramosum*). With

 DOI: 10.1201/9781003514244-4

the rising global burden of metabolic disorders, there is an imperative to explore innovative and novel dietary approaches for preventing metabolic disorders. This chapter aims to provide a comprehensive review of the beneficial features of millets, their symbiotic relationship with gut microbiota, and their effect on human health and metabolic well-being.

3.2 NUTRITIONAL VALUE OF MILLETS

Millet-based diets are proven sources of energy, bioactive phytochemicals, vitamins, minerals, and fibrous materials (non-starch polysaccharides). Millet grains are rich in niacin (B3), thiamine, riboflavin, and folic acid. It is estimated that nearly 70% of grains are composed of carbohydrates, with most being soluble carbohydrates and dietary fiber including both soluble and insoluble fractions. Most of the millet polysaccharides are composed of amylopectin and amylose (70–80%). Millets are a good source of protein, contributing to the suppression of malnutrition. The protein content of millets is particularly significant in comparison to traditional staple grains like rice and wheat (Chandel et al., 2014). Millets serve as valuable sources of essential micronutrients, including iron, zinc, magnesium, calcium, and phosphorus, with more compared to wheat (1.5%) and rice (0.6%). The iron content in pearl millet and barnyard millet is abundant, and there is a good amount of zinc (4.1 mg) in foxtail millet (Amadou et al., 2013). In addition, millets are richer in polyphenols (0.2–0.5%). Millets' polyphenols, tannins, and phytates possess antioxidant activity, and these substances have a role in regulating health, the aging process, and metabolic diseases (Bravo, 1998). The presence of dietary fiber in the seed coat of millet grains is beneficial to human health due to the hypoglycemic and hypocholesterolemic effects (Onipe et al., 2022).

3.3 MILLETS FOR DIET

Diet plays an important role in human health and disease management. Millets are nutrient-dense cereals. They are rich in essential nutrients such as carbohydrates, proteins, essential fatty acids, vitamins, and minerals, which are present in comparably higher quantities in millets than in conventional cereals like wheat and rice. Due to their gluten-free nature and low glycemic index, they are suitable as diets for people with gluten sensitivity or diabetes. They significantly contribute to human diets owing to their high levels of energy, lipids, and high-quality proteins. In addition, they are also rich sources of dietary fiber and micronutrients (Hassan et al., 2021). A healthy diet constitutes good quality carbohydrates, proteins (specific amino acid sequences), lipids, dietary fibers (for digestion), and micronutrients (calcium, iron, zinc, phosphorus, potassium), and millets are full of these constituents. Phytochemicals of millets, such as phenolic compounds, exhibit antioxidant action via scavenging reactive oxygen species (ROS) (Saini et al., 2021).

Millets, when transformed into fermented products, are believed to add more benefits to human and animal health. Research shows that the presence of certain nutrients in millets leads to a double benefit of nourishment and cure (Chandrasekara and

Shahidi, 2012). The high fiber content in millet aids digestion, prevents constipation, and promotes a healthy gut microbiome. They can be easily included in our routine diet as per our suggestions as follows:

Baking: Millet flour can be used in baking to make bread, muffins, pancakes, cookies, and other baked goods. It adds a slightly nutty flavor and a tender texture to baked items.

Soups: Millets can be added to soups and stews to add texture, thickness, and nutritional value. They absorb the flavors of the other ingredients in the dish while providing a hearty base.

Millets as rice substitute: Millets can be cooked and used as a substitute for rice in various dishes like biryanis, risottos, and fried rice.

Fermented foods: Millets can also be fermented and used to make traditional fermented foods like idlis, dosas, and fermented drinks like kanji.

Millets in traditional beverages: Millets can be ground into flour and used to make beverages like millet milk or millet-based smoothies for a nutritious drink option.

Food products supplemented with millets have become increasingly popular due to their nutritional and economic advantages. Value-added products from millets have the potential to promote business and offer significant potential for future growth, as consumers understand that millets and millet-based foods contribute directly to their health. The combination of high fiber and a low glycemic index in millets promotes satiety, regulates blood sugar levels, and helps to maintain a healthy body weight. Additionally, millet has anti-inflammatory, antimicrobial, and anticancer properties; therefore, it offers a wide range of nutritional and health benefits. Incorporating millet into regular meals can contribute to improved overall health and well-being. Furthermore, by adopting appropriate processing techniques, value-added millet products can be used in varieties like puffed products, quick-cooking cereals, ready-to-eat snacks, supplementary foods, weaning foods, and health foods.

3.4 EFFECTS OF MILLETS ON GUT HEALTH

Millets are easily digestible, which makes them ideal for consumption by all age groups. They eliminate constipation, excess gas, and bloating. Millet porridge prevents constipation by increasing the water content of feces and promoting gastric emptying. Millets are antiacidic and reduce the risk of gastric ulcers or colon cancer. They act as a prebiotic. Intake of millet porridge alters intestinal microflora and improves gastrointestinal motility and the rate of gastrointestinal propulsion (Chen et al., 2022). Millet porridge changes the proportions of intestinal flora with increased levels of Bifidobacterium and Lactobacillus, and decreased levels of Escherichia coli, Enterococcus, and Bacteroides in the intestine (Chen et al., 2022). Finger millet, barnyard millet, and proso millet are rich sources of antimicrobial proteins and antioxidant peptides (Bisht et al., 2016). Due to their gluten-free nature, millets are non-allergenic. Hence, they can be consumed regularly by celiac patients. Foxtail millet's soluble dietary fiber inhibited the colony formation ability of HCT116 and HT-29 cells and significantly induced ROS and apoptosis of HCT116 and HT-29 cells. Dietary fiber present in millets interacts with gut microbiota and releases short-chain fatty acids (SCFAs) through fermentation. SCFAs are weak acids that lower the pH within the colon and thereby inhibit the growth and activity of pathogenic bacteria (Li and Komarek, 2017). Due to the water-absorbing property of dietary fibers found in millet, it increases transit time of food in the gut and acts as a detoxifying agent in the body.

3.5 EFFECTS OF MILLETS ON GUT MICROBIOTA

3.5.1 Gut Microbiota and Its Effect

The human body hosts a diverse community of microorganisms known as the gut microbiota. The "gut microbiota" is termed as the collective population of bacteria, archaea, and eukarya colonizing inside the human GI tract, which has co-evolved with the host over thousands of years to form a multifaceted and mutually beneficial relationship with the host (Thursby and Juge, 2017). It is estimated that the microbiota consists of approximately 160 species per person per fecal sample, and this ecosystem plays an important role in human health by modulating various physiological functions (Rodrı´guez et al., 2015). The human gastrointestinal (GI) tract comprises trillions of microbes. The chief bacteriological species/phyla are listed in Table 3.1 (Wu et al., 2019). The most prevailing phyla of human gut microorganisms are *Firmicutes*, *Bacteroidetes*, *Actinobacteria*, *Proteobacteria*, *Fusobacteria*, and *Verrucomicrobia*, with the preponderance (90%) of the gut microflora dominated by the two phyla, *Firmicutes* and *Bacteroidetes* (Arumugam et al., 2011; Rinninella et al., 2019). The Firmicutes phylum is composed of more than 200 different genera, such as *Lactobacillus*, *Bacillus*, *Clostridium*, *Enterococcus*, and *Ruminococcus*. Bacteroidetes consist of predominant genera such as *Bacteroides* and *Prevotella*. The

TABLE 3.1
Composition of Human Microbiota

	Relative Abundance (%)	
Phylum	**Infant (<1-year-old)**	**Adult**
Firmicutes	32.01	50.50
Bacteroidetes	20.08	48.00
Proteobacteria	46.14	0.60
Actinobacteria	1.28	0.20
Verrucomicrobia	0.44	0.60
Fusobacteria	0.06	0.08
Unclassified	NA	0.02

Adopted from Wu et al. 2019.

Actinobacteria phylum is proportionally less abundant and mainly represented by the *Bifidobacterium* genus (Arumugam et al., 2011). The composition of gut microbiota is highly variable and influenced by various factors, including genetics, diet, age, geography, and lifestyle, which contribute to the unique microbial fingerprint of each person. A rich and diverse microbial community leads to a well-balanced and healthy gut microbiota composition, resulting in the production of several cells and tissues within the gut to confer immunity in the system (Table 3.1).

In recent years, variations in diet have been significant regulators in modulating gut microbiota composition and functionality, which directly affects host homeostasis and biological processes, thus averting the onset of metabolic disorders (Rinninella et al., 2019). Among these dietary interventions, millets, which are rich in minerals and vitamins with low fat, dietary energy, and glycemic index values, have emerged as a potential dietary substitute and have been found to offer numerous health benefits (Srinivasan Nithiyanantham et al., 2019).

The numerous health benefits of millet grain consumption are well established (Table 3.2). These are mainly due to the bioactive phytochemicals and physiologically active substances found in these cereals, such as lignans, flavonoids, phenolics, beta-glucan, sterols, inulin, pigments, dietary fiber, and phytate (Nithiyanantham et al., 2019).

Millet-based diets are proven sources of energy, phytoconstituents, vitamins, minerals, and fibrous materials (non-starch polysaccharides). Among its vitamins, millet grains are rich in niacin (B3), thiamine, riboflavin, and folic acid. It is estimated that nearly 70% of grains are composed of carbohydrates, with most being soluble carbohydrates and dietary fiber including both soluble and insoluble fractions. Most of the millet polysaccharides are composed of amylopectin and amylose (70–80%). Millets are most commonly thought of as a good source of proteins, which play a crucial role in the suppression of malnutrition, and the protein content of millets is particularly significant in comparison to traditional staple grains like rice and wheat. Millets serve as valuable sources of essential micronutrients, including iron, zinc, magnesium, calcium, and phosphorus, with more compared to wheat (1.5%) and rice (0.6%). The iron content in pearl millet and barnyard millet is abundant, and there is a good amount of zinc (4.1 mg) in foxtail millet (Amadou et al., 2013). In addition, they are also richer in polyphenols (0.2–0.5%). Millets' polyphenols, tannins, and phytates provide the bulk of their antioxidant activity, and these substances have a role in regulating the aging process.

The commonly found flavonoids in millets are catechin, quercetin, luteolin, orientin, apigenin, isoorientin, vitexin, myricetin, isovitexin, daidzein, sponarin, violanthin, lucenin-1, and tricin (Shukla and Baunthiyal, 2018;). The presence of dietary fiber in the seed coat of millet grains is beneficial to human health, as it affects several metabolic and digestive processes, such as effects on glucose absorption and cholesterol levels.

TABLE 3.2
Different Types of Millet and Their Potential Features

Name of the Millet	Major Health Benefits	References
Pearl millet	Contains polyphenol oxidase, phytase, and amylase which improve the digestibility of millets and increase nutrient availability. Being gluten-free, allows a healthy digestive system to be maintained, and adverse reactions associated with gluten consumption to be avoided.	**Nkhata et al., 2018; Kumar et al., 2021**
Finger millet	Reduces the risk of digestive disorders such as diverticulitis, hemorrhoids, and irritable bowel syndrome. Promotes regular bowel movements and supports overall digestive health.	**Muthamilarasan et al. (2016)**
Barnyard millet	Exhibits prebiotic properties, stimulating the growth of beneficial bacteria and contributing to a healthy gut microbiota by changing the composition and richness of gut microbiota.	**Li et al., 2019; Sharma et al., 2021**
Proso millet	Improves the glycemic responses and plasma levels, and protects against D-galactosamine-induced liver injury.	**Park et al. (2008), Ito et al. (2008)**
Little millet	Reduces cholesterol levels in the case of cardiovascular disorder, reduces fasting blood glucose, blood glucose, and lipid parameters in diabetic subjects.	**Surekha (2004)**
Foxtail millet	Possesses prebiotic potential, promoting the growth of beneficial gut bacteria and inhibiting the growth of harmful bacteria. It improves gut dysbiosis by increasing probiotic bacteria, including bifidobacterium and lactobacillus, and reducing pathogenic bacteria, including E. Coli, enterococcus, and bacteroides. It is gut barrier friendly and promotes a healthy gut microenvironment. Foxtail millet supplementation improves glucose metabolism and gut microbiota in rats with high-fat diet/streptozotocin-induced diabetes. Fermented and germinated processing improved the protective effects of foxtail millet whole grain against dextran sulfate sodium-induced acute ulcerative colitis and gut microbiota dysbiosis in C57BL/6 mice (Zhang, 2015b). Foxtail millet supplementary diets are suitable for dietary intervention in IBD patients based on a mouse model, and for the treatment or prevention of other intestinal diseases.	**Balasubramanian et al., 2021**

3.6 MECHANISMS OF MILLETS-GUT MICROBIOTA INTERACTION

3.6.1 Millets as Prebiotic

According to the International Scientific Association of Probiotics and Prebiotics (ISAPP) in London, U.K., a prebiotic is defined as "a substrate that is selectively utilized by host microorganisms conferring a health benefit." Coarse cereals are important sources of nutrients for the gut microbiota and get converted into functional substances, facilitating the interaction between the hosts and leading to a healthy gut microbiome (Ren et al., 2021). As mentioned earlier, the seed coat of millet contains pectin, cellulose, and hemicelluloses, which serve as rich sources of dietary fiber and affect several metabolic and digestive processes. A synergistic relationship has been found between dietary fibers and flavonoids found in millets and gut microbiota, conferring various physiological functions and health benefits. The non-digestible fibers found in millets act as substrates for the growth of probiotic bacteria, thus enhancing their proliferation and metabolism. This leads to the inhibition of pathogenic bacteria and the enhancement of the immune system of the host and several other functions, as shown in Figure 3.1. The dietary fiber content in millets facilitates colonic fermentation, leading to the production of short-chain fatty acids (SCFAs), which improves gut barrier function by facilitating the formation of tight junctions between intestinal epithelial cells (Suzuki et al., 2020). In the colon, gut bacteria ferment millet fibers and resistant starches, leading to the production of SCFAs, including acetate, propionate, and butyrate. Furthermore, millet consumption may impact the secretion of gut hormones, such as glucagon-like peptide-1 (GLP-1) and peptide YY (PYY), which play roles in appetite regulation, satiety, and glucose homeostasis (Alyami et al., 2019). It has been suggested that millet porridge (foxtail millet) could change the proportions of intestinal flora and act as a potent prebiotic (Chen et al., 2022). The bioactive compounds present in millets, such as polyphenols and flavonoids, exhibit antioxidant and anti-inflammatory properties by directly scavenging free radicals, reducing oxidative stress and inflammation in the gut and throughout the body, contributing to improved gut health (Sharma et al., 2021).

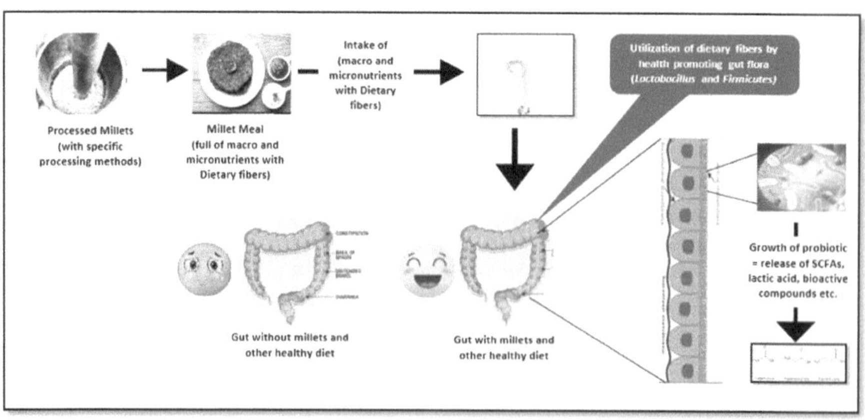

FIGURE 3.1 Changes inside gut after intake of specifically processed millets.

The phytochemicals present in millets selectively promote beneficial gut microbes alone or with dietary fiber. For instance, sorghum bran polyphenols had synergistic effects with fructooligosaccharides in promoting Bifidobacterium and Lactobacillus while independently stimulating Roseburia and Prevotella (Ashley et al., 2019). Millet shell polyphenols reduced the levels of lipopolysaccharides and inflammatory cytokines while enriching lean-associated gut bacteria Oscillospira and Ruminococcus in mice fed with a high-fat diet (Liu et al., 2021).

3.6.2 Millet-Based Probiotic Fermented Food

Fermentation of cereal is an ancient and inexpensive food preservation method, and a cultural and traditional practice in many countries, which improves the nutritional value and digestibility of raw products, enhances sensory characteristics, and improves functional qualities (Blandino et al., 2003; Borresen et al., 2012). Traditionally, fermented foods and beverages represent a major dietary component (Di Stefano et al., 2017). Several laboratories define probiotics as "live microorganisms which, when administered in adequate amounts, confer a health benefit on the host" (FAO/WHO, 2002). The common fermenting bacteria are species of *Leuconostoc*, *Lactobacillus*, *Streptococcus*, *Pediococcus*, *Micrococcus*, and *Bacillus*. The fungi genera *Aspergillus*, *Paecilomyces*, *Cladosporium*, *Fusarium*, *Penicillium*, and *Trichothecium* are the most frequently found in certain products. The common fermenting yeasts are species of Saccharomyces, which result in alcoholic fermentation. Fermentation by LAB has been shown to reduce phytic acids and tannins, thus enhancing protein availability and digestion in various cereals such as maize, sorghum, and finger millet (Stefano et al., 2017). Moreover, such fermented cereals have a higher composition of riboflavin, thiamine, niacin, and lysine (Lorri and Svanberg, 1993). In general, the natural fermentation of millet cereals leads to a decrease in the level of carbohydrates as well as some non-digestible poly- and oligosaccharides. Certain amino acids may be synthesized, and the availability of B group vitamins may be improved. Fermentation also provides optimum pH conditions for enzymatic degradation of phytate, which is present in cereals in the form of complexes with polyvalent cations such as iron, zinc, calcium, magnesium, and proteins. Such a reduction in phytate may increase the amount of soluble iron, zinc, and calcium several folds (Chavan and Kadam, 1989). Koko is made by wet milling pearl millet grains with spices.

In West Africa, ogi is a porridge made from fermented millet, sorghum, maize paste, or cake. Ogi is typically offered as a moist cake. The bacterium involved in the fermentation of ogi is *Lactobacillus plantarum*.

3.7 PROCESSING OF MILLET TO INCREASE BIOAVAILABILITY OF MICRONUTRIENTS

Processing millet in specific ways can help increase the bioavailability of micronutrients, making them more accessible for absorption by the human body. Millets, major or minor such as sorghum, pearl millet, finger millet, and others, are rich in nutrients like iron, zinc, and certain vitamins. However, they also contain anti-nutritional factors such as phytates, tannins, trypsin inhibitors, and oxalate, which can hinder the absorption of these nutrients. Anti-nutritional factors, particularly phytic acid, bind and chelate mineral ions, leading to reduced absorption and bioavailability in the gut (Samtiya et al., 2021). It's important to note that the effectiveness of processing methods may vary depending on the specific type of millet and the local varieties. Additionally, cultural preferences and dietary practices can influence the adoption of processing techniques. Combining various processing methods may also yield the best results in enhancing the nutritional quality of millets. The following sections describe some methods to enhance the bioavailability of micronutrients in millet (Figure 3.2).

3.7.1 Dehulling and Decortication

The outer layers of millet grains contain phytic acid and other anti-nutritional factors, which can bind to minerals like iron and zinc, reducing their absorption. Anti-nutrients are recognized for their capacity to influence the bioavailability and metabolic functions of minerals, proteins, and enzymes in the body by interacting with them (Samtiya et al., 2020). Consequently, decreasing the presence of anti-nutrients can enhance digestibility, minimize astringency, and result in lighter-colored products. Dehulling or decortication involves removing these outer layers, thereby reducing the phytic acid content and increasing the bioavailability of minerals (Sruthi et al., 2021).

3.7.2 Fermentation

Fermentation is a traditional processing method that can improve the nutritional quality of millets. It helps break down anti-nutritional factors and enhances the bioavailability of certain nutrients (Samtiya et al., 2021). Fermentation is a top-tier

FIGURE 3.2 Processing of millets to increase the digestibility and availability of gut microbiota.

strategy, both in traditional and industrial settings, to boost mineral accessibility in plant foods. Implementing fermentation methods that enhance mineral availability can be shared with local communities, improving household mineral intake. Studies highlight that natural and strain-specific fermentations enhance mineral bioavailability by efficiently breaking down phytic acid in plant foods. Moreover, fermentation creates favorable environmental conditions, like a low pH, fostering the growth of beneficial strains and enhancing phytase activity. This process also enhances the nutritional, physical, and sensory properties of food products, making fermentation a holistic approach for nutritional enhancement and overall product quality. Lactic acid bacteria present during fermentation can contribute to the breakdown of phytic acid, making minerals more accessible. Fermentation boosts microbial metabolism, enhancing the digestibility of macromolecules and improving the bioavailability of macro and micronutrients, as well as phytochemicals (Waters et al., 2015).

3.7.3 Germination or Sprouting

In the processing of millet, germination is employed to enhance nutritional quality by reducing the presence of anti-nutritional factors. It activates enzymes that break down anti-nutritional factors, such as phytic acid, phenols, tannins, and oxalates (Laxmi et al., 2015). These alterations are closely linked to the activation of certain endogenous enzymes, rendering germinated foods superior in nutritional quality when compared to non-germinated seeds (Zhang et al., 2015a).

3.7.4 Malting

Malting involves soaking grains in water and allowing them to germinate. This process can lead to increased enzymatic activity, breaking down anti-nutritional factors and improving nutrient availability. Combining malting with fermentation leads to a further increase in protein digestibility (Onyango et al., 2013). The processes of germination and malting play a crucial role in enzymatically breaking down carbohydrates into simple sugars. This involves the activation of endogenous enzymes like α-amylase (Oghbaei and Prakash, 2016), contributing to improved digestibility through the degradation of starch, which, in turn, provides energy for seed development (Zhang et al., 2015). Malting can be particularly beneficial for increasing the bioavailability of minerals in millets. The malting process notably augments the levels of sodium, potassium, phosphorus, calcium, and magnesium in sorghum, foxtail, and chickpea (Laxmi et al., 2015).

3.7.5 Combining with Other Foods

Pairing millets with vitamin C-rich foods can significantly boost the absorption of non-heme iron found in millets. The presence of vitamin C facilitates the conversion of iron into a more readily absorbable form, enhancing its bioavailability. This dietary combination is particularly beneficial for individuals seeking to optimize iron absorption from plant-based sources, as non-heme iron is typically less easily absorbed than heme iron from animal products. By incorporating millets into meals alongside vitamin C-rich options, one can promote better iron utilization and contribute to overall nutritional well-being.

3.7.6 Cooking Methods

Certain cooking methods, such as boiling and steaming, can help reduce anti-nutritional factors and increase the bioavailability of micronutrients. Both industrial and household cooking methods bring about significant alterations in food, leading to a widespread improvement in the bioavailability of macro- and micronutrients (Ercolini and Fogliano, 2018).

Particle size:

Extensive mechanical processing is commonly employed to enhance the palatability of foods, with the conversion of raw materials into "flours and juices" serving as a fundamental starting point for various food designs (Ercolini and Fogliano, 2018). Working with ultraprocessed, homogeneous, and flexible ingredients is advantageous for food designers. This intensive processing ensures full accessibility of nutrients to digestive enzymes, accelerating their degradation and absorption. In contrast, limiting mechanical processing can reduce nutrient bioavailability. Despite the prevailing belief that digestion initiates in the mouth, the human masticatory apparatus primarily prevents choking rather than facilitating digestion. Research indicates that food particle size inversely affects absorption, demonstrated in studies where larger particles led to decreased starch absorption and improved soybean protein digestibility. Avoiding fine grinding may substantially reduce calorie intake (Capuano et al., 2018).

Reducing millet grain particle size through milling or grinding is a vital step in processing to enhance micronutrient bioavailability. This helps break down anti-nutritional factors like phytic acid found in the outer layers, promoting the release of essential minerals during digestion. Fine milling disrupts protective layers, decreasing phytic acid concentration and improving iron and zinc availability. This enhances millets' nutritional profile, making micronutrients more accessible. Optimizing particle size through careful milling is crucial for maximizing millets' nutritional benefits and promoting their role as a nutritious staple in diverse diets, potentially increasing acceptability in various culinary applications.

Food matrix:

The food matrix refers to the complex structure of a food product, encompassing the arrangement and interaction of its components, such as proteins, carbohydrates, fats, and bioactive compounds. This intricate matrix influences the release, absorption, and bioavailability of nutrients, impacting overall nutritional quality and physiological effects.

Millets contain anti-nutritional factors, such as phytic acid and polyphenols, which can impede the absorption of essential minerals. Enhancing the food matrix involves strategic processes like dehulling, fermentation, germination, and fine milling. Removal of outer layers diminishes anti-nutritional factors (Sruthi et al., 2021), improving the overall matrix. Dehulling reduces substances like phytic acid, enhancing mineral bioavailability. Fermentation and germination alter the matrix by promoting enzyme activity, boosting micronutrient bioavailability (Samtiya et al., 2021). Formulating millet-based products with complementary ingredients positively influences the matrix, enhancing nutrient absorption. Fine milling exposes more surface area, aiding in the breakdown of anti-nutritional factors for improved micronutrient bioavailability (Ercolini and Fogliano, 2018). These methods ensure millets not only provide essential nutrients but also possess enhanced bioavailability, aligning with diverse dietary needs. This underscores millets' role as a nutritionally valuable food source, contributing to balanced diets.

Processing method:

Processing millet grains with specific methods is crucial for increasing the micronutrients' availability. Removing the outer layers of millet grains diminishes anti-nutritional factors like phytic acid, enhancing the bioavailability of crucial minerals such as iron and zinc. Fermentation with beneficial microorganisms, especially lactic acid bacteria, contributes to breaking down anti-nutritional compounds and improving nutrient accessibility. Germination activates enzymes that degrade anti-nutritional factors and enhances the bioavailability of nutrients like iron and certain vitamins. Malting, which involves soaking and germination, boosts enzymatic activity, improving the breakdown of anti-nutritional factors and overall nutritional quality. Consuming millets with vitamin-C-rich foods aids in non-heme iron absorption, elevating its bioavailability. Fine milling increases millet grain surface area, promoting enzymatic access to anti-nutritional factors and micronutrient release during digestion. Combining these methods synergistically optimizes millet grain micronutrient bioavailability, enhancing their nutritional value for human consumption.

Both heat and non-heat processing methods play a pivotal role in augmenting micronutrient bioavailability. Heat processes are essential for improving nutrient accessibility, while non-heat methods provide alternatives that can preserve certain heat-sensitive nutrients.

Processing methods with heat:

1. Roasting:

Roasting of millet involves exposing grains to dry heat. This process can improve the sensory qualities of millet, making them more palatable, and it may contribute to the breakdown of anti-nutritional factors (Gowda et al., 2022), thereby enhancing micronutrient bioavailability. Roasting of millet increased the bio-accessibility of total polyphenols, accompanied by an increase in bio-accessible flavonoid content (Rani et al., 2018; Hithamani and Srinivasan, 2014). Boiling of millet is a common cooking method that softens the grains and helps in the breakdown of anti-nutritional factors. It also increases the hydration of grains, making them more digestible. Boiling under pressure led to a decrease in total polyphenol content and reduced the percentage of free fatty acids (Hithamani and Srinivasan, 2014; Rani et al., 2018).

2. Baking:

Modern or secondary processing techniques, including baking, are employed to create value-added processed food products based on millet (Birania et al., 2020). Baking millet-based products can improve their texture and taste. The application of heat during baking contributes to the modification of the food matrix, positively affecting the bioavailability of micronutrients.

3. Pressure cooking:

Pressure cooking millets accelerates the cooking process and can improve the breakdown of anti-nutritional factors. The high temperature and pressure conditions may positively impact the nutritional quality of millets. Soaking and subsequent cooking of foxtail millet enhance the bioavailability of soluble zinc and ionizable iron (Sharma and Niranjan, 2018).

Processing methods without heat:

Allowing millet grains to undergo germination initiates the activation of enzymes that break down anti-nutritional factors, a process occurring without external heat application. This natural method significantly boosts the bioavailability of specific nutrients. Fermenting millets with microorganisms induces biochemical changes without direct heat exposure, facilitating the breakdown of anti-nutritional compounds and positively impacting micronutrient bioavailability. Malting, involving soaking and germination, fosters conditions conducive to enzymatic activity, contributing to the breakdown of anti-nutritional factors without heat. Milling or grinding millet grains, a mechanical process, reduces particle size, increasing surface area for enzymatic action and potentially enhancing the bioavailability of micronutrients. These non-heat processing methods underscore diverse approaches to improve millet nutritional quality, providing alternatives to traditional heat-based techniques. While heat processing methods are effective in improving certain nutrients and enhancing them, non-heat methods offer alternatives that may preserve heat-sensitive nutrients. Combining these approaches judiciously in millet processing can yield products with optimal nutritional profiles and sensory characteristics, catering to diverse dietary preferences and requirements.

Processing conditions:

Optimizing processing conditions is crucial for increasing the bioavailability of micronutrients in millet grains. Various factors, including time, temperature, and humidity, play key roles in determining the efficacy of processing methods. The following is an exploration of processing conditions that influence the bioavailability of micronutrients in millets:

a) Germination time and temperature:

Germination involves soaking millet grains and allowing them to sprout. The duration of germination and the temperature at which it occurs are critical factors. Longer germination periods and moderate temperatures activate enzymes that break down anti-nutritional factors, enhancing the bioavailability of micronutrients (Laxmi et al., 2015).

b) Fermentation duration and temperature:

Fermenting millets with specific microorganisms is a time- and temperature-sensitive process. Longer fermentation durations and controlled temperatures allow for a more comprehensive enzymatic breakdown of anti-nutritional factors, positively influencing nutrient bioavailability (Samtiya et al., 2021).

c) Malting conditions:

Malting, which involves soaking and germinating grains, requires careful control of conditions. Optimal temperature and humidity levels during malting contribute to enzymatic activity and improve the bioavailability of essential micronutrients (Onyango et al., 2013).

d) Heat processing temperature and duration:

Heat processing methods, such as roasting, baking, and boiling, are influenced by temperature and duration. Moderate heat, applied for an appropriate duration, contributes to the reduction of anti-nutritional factors while preserving the sensory qualities of millets (Gowda et al., 2022).

e) Particle size during milling:

Fine milling or grinding alters the particle size of millet grains. The degree of particle size reduction is a critical factor influenced by milling conditions. Smaller particle sizes increase the surface area for enzymatic action, enhancing the bioavailability of micronutrients (Sruthi et al., 2021).

f) pH conditions during processing:

The pH of the processing environment, especially during fermentation, is crucial. Controlled pH levels support the growth of useful microorganisms, influencing the breakdown of anti-nutritional factors and positively impacting micronutrient bioavailability (Waters et al., 2015).

g) Combination of processing methods:

Integrating multiple processing methods requires careful consideration of conditions. For example, combining germination with fermentation or malting can have synergistic effects on improving the bioavailability of micronutrients in millets.

Optimal processing conditions depend on the specific millet variety, local practices, and the desired product outcomes. It's essential to balance the need for nutrient enhancement with sensory qualities to produce millet-based products that are not only nutritious but also acceptable to consumers. Research and development efforts focused on refining processing conditions can contribute to the production of millet-based foods with enhanced nutritional profiles, addressing global nutritional challenges.

3.8 CONCLUSIONS

From improving gut barrier integrity to enhancing SCFA production, millets exert a profound impact on gut health and host metabolism. The synergy between millets and gut microbiota represents a novel dietary approach for promoting metabolic health and preventing metabolic disorders. Millet grain residues, being potential fiber sources, could be used in the enrichment of foods as a natural ingredient by the food industry. Processing millet grains through various methods significantly improves the bioavailability of micronutrients present in them, making them more accessible for absorption by the human body. By employing various processing techniques, the anti-nutritional factors present in millets can be mitigated, enhancing the overall nutritional quality of millets. Heat-based processing methods like roasting, baking, and boiling, as well as non-heat methods, play complementary roles in improving millets' nutritional profiles. However, careful consideration of processing conditions, such as time, temperature, and pH, is essential to optimize nutrient availability while preserving sensory qualities. Integrating multiple processing techniques can have synergistic effects, enhancing the nutritional value of millets. Millets are suitable for people of all age groups due to their nutritional and digestive advantages. More research is warranted to clinically prove their role in health and disease management. Policy support and awareness through mass media can help nutritious millet address global nutritional challenges and diverse dietary needs.

REFERENCES

Alyami, J., Whitehouse, E., Yakubov, G. E., Pritchard, S. E., Hoad, C. L., & Blackshaw, E. (2019). Glycaemic, gastrointestinal, hormonal and appetitive responses to pearl millet or oats porridge breakfasts: A randomized, crossover trial in healthy humans. British Journal of Nutrition, *122*, 1142–1154.

Amadou, I., Gounga, M. E., & Le, G. W. (2013). Millets: Nutritional composition, some health benefits and processing—A review. *Emirates Journal of Food and Agriculture*, *25*, 501–508.

Arumugam, M., Raes, J., Pelletier, E., Le Paslier, D., Yamada, T., & Mende, D. R. (2011). Enterotypes of the human gut microbiome. *Nature*, *473*, 174–180.

Ashley, D., Marasini, D., Brownmiller, C., Lee, J. A., Carbonero, F., & Lee, S. O. (2019). Impact of grain sorghum polyphenols on microbiota of normal weight and overweight/obese subjects during in vitro fecal fermentation. *Nutrients, 11*(2).

Balasubramanian, B., Soundharrajan, I., Al-Dhabi, N. A., Vijayaraghavan, P., Balasubramanian, K., & Valan, A. (2021). Probiotic characteristics of Ligilactobacillus salivarius AS22 isolated from sheep dung and its application in corn-fox tail millet silage. *Applied Sciences, 11*, 9447.

Birania, S., Rohilla, P., Kumar, R., & Kumar, N. (2020). Post harvest processing of millets: A review on value added products. *International Journal of Chemical Studies, 8*, 1824–1829.

Bisht, A., Thapliyal, M., & Singh, A. 2016. Screening and isolation of antibacterial proteins/peptides from seeds of millets. *International Journal of Current Pharmaceutical Research, 8*(3), 96–99.

Blandino, A., Al-Aseeri, M. E., Pandiella, S. S., Cantero, D., & Webb, C. (2003). Cereal-based fermented foods and beverages. *Food Research International, 36*, 527–543.

Borresen, E. C., Henderson, A. J., Kumar, A., Weir, T. L., & Ryan, E. P. (2012). Fermented foods: Patented approaches and formulations for nutritional supplementation and health promotion. *Recent Patents on Food, Nutrition & Agriculture, 4*, 134–140.

Bravo L. (1998). Polyphenols: Chemistry, dietary sources, metabolism and nutritional significance. *Nutrition Reviews, 56*, 317–333.

Capuano, E., Oliviero, T., Fogliano, V., & Pellegrini, N. (2018). (2018). Digestion and the true energy content of foods. *Nutrition Reviews, 76*, 274–289.

Chandel, G., Kumar, M., Dubey, M., & Kumar, M. (2014). Nutritional properties of minor millets: Neglected cereals with potentials to combat malnutrition. *Current Science, 107*(7), 109–111.

Chandrasekara, A., & Shahidi, F. (2012). Bioaccessibility and antioxidant potential of millet grain phenolics as afected by simulated in vitro digestion and microbial fermentation. *Journal of Functional Foods, 4*, 226–237.

Chavan, J. K., Kadam, S. S., & Beuchat, L. R. (1989). Nutritional improvement of cereals by fermentation. *Critical Reviews in Food Science & Nutrition, 28*(5), 349–400.

Chen, Y., Zhang, R., Xu, J., & Ren, Q. (2022). Alteration of intestinal microflora by the intake of millet porridge improves gastrointestinal motility. *Frontiers in Nutrition, 9*, 965687.

Di Stefano, E., White, J., Seney, S., Hekmat, S., McDowell, T., Sumarah, M., & Reid, G. (2017). A novel millet-based probiotic fermented food for the developing world. *Nutrients, 9*(5), 529.

Ercolini, D., & Fogliano, V. (2018). Food design to feed the human gut microbiota. *Journal of Agricultural and Food Chemistry, 66*, 3754–3758.

FAO/WHO. (2002). *Guidelines for the Evaluation of Probiotics in Food.* Geneva, Switzerland: Food and Agriculture Organization of the United Nations and World Health Organization Working Group.

Gowda, N. A. N., Siliveru, K., Prasad, P. V. V., Bhatt, Y., Netravati, B. P., & Gurikar, C. (2022). Modern processing of indian millets: A perspective on changes in nutritional properties. *Foods, 11*, 499.

Hassan, Z. M., Sebola N. A., & Mabelebele M. (2021). The nutritional use of millet grain for food and feed: A review. *Agriculture & Food Security, 10*, 16.

Hithamani, G., & Srinivasan, K. (2014). Effect of domestic processing on the polyphenol content and bioaccessibility in finger millet (Eleusine Coracana) and pearl millet (Pennisetum glaucum). *Food Chemistry, 164*, 55–62.

Ito, K., Ozasa, H., Noda, Y., Arii, S., & Horikawa, S. (2008). Effects of free radical scavenger on acute liver injury induced by D-galactosamine and lipopolysaccharide in rats. *Hepatology Research, 38*, 194–201.

Kumar, R. R., Bhargava, D. V., Pandit, K., Goswami, S., Shankar, S. M., Singh, S. P., et al. (2021). Lipase—The fascinating dynamics of enzyme in seed storage and germination—A real challenge to pearl millet. *Food Chemistry*, *361*, 130031.

Laxmi, G., Chaturvedi, N., & Richa, S. (2015). The impact of malting on nutritional composition of foxtail millet, wheat and chickpea. *Journal of Nutrition and Food Sciences*, *5*, 407.

Li, S., Yu, W., Guan, X., Huang, K., Liu, J., Liu, D., & Duan, R. (2019). Effects of millet whole grain supplementation on the lipid profile and gut bacteria in rats fed with high-fat diet. *Journal of Functional Foods*, *59*, 49–59,

Li, Y. O., & Komarek, A. R. (2017). Dietary fibre basics: Health, nutrition, analysis, and applications. *Food Quality and Safety*, *1*, 47–59.

Liu, F., Shan, S., Li, H., Shi, J., Hao, R., Yang, R., & Li, Z. (2021). Millet shell polyphenols prevent atherosclerosis by protecting the gut barrier and remodeling the gut microbiota in ApoE- /- mice. *Food & Function*, *12*(16), 7298–7309.

Lorri, W., & Svanberg, U. (1993). Lactic-fermented cereal gruels with improved in vitro protein digestibility. *International Journal of Food Science & Nutrition*, *44*, 29–36, 1989, *28*, 349–400.

Muthamilarasan, M., Dhaka, A., Yadav, R., & Prasad, M. (2016). Expolration of millet models for developing nutrient rich graminaceous crops. *Plant Science*, *242*, 89–97.

Nithiyanantham, S., Kalaiselvi, P., Mahomoodally, M. F., Zengin, G., Abirami, A., & Srinivasan, G. (2019). Nutritional and functional roles of millets—A review. *Journal of Food Biochemistry*, *43*(7), e12859.

Nkhata, S. G., Ayua, E., Kamau, E. H., & Shingiro, J. B. (2018). Fermentation and germination improve the nutritional value of cereals and legumes through the activation of endogenous enzymes. *Food Science & Nutrition*, *6*, 2446–2458.

Oghbaei, M., & Prakash, J. (2016). Effect of primary processing of cereals and legumes on its nutritional quality: A comprehensive review. *Cogent Food & Agriculture*, *2*, 1136015.

Onipe, O. O., & Ramashia, S. E. (2022, November 14). Finger millet seed coat-a functional nutrient-rich cereal by-product. *Molecules*, *27*(22), 7837.

Onyango, C. A., Ochanda, S. O., Mwasaru, M. A., Ochieng, J. K., Mathooko, F. M., & Kinyuru, J. N. (2013). Effects of malting and fermentation on anti-nutrient reduction and protein digestibility of red sorghum, white sorghum and pearl millet. *Journal of Food Research*, *2*, 41–49.

Park, K. O., Ito, Y., Nagasawa, T., Choi, M. R., & Nishizawa, N. (2008). Effects of dietary Korean proso-millet protein on plasma adiponec- tin, HDL cholesterol, insulin levels, and gene expression in obese type 2 diabetic mice. *Bioscience Biotechnology and Biochemistry*, *72*(11), 2918–2925.

Rani, S., Singh, R., Sehrawat, R., Kaur, B. P., & Upadhyay, A. (2018). Pearl millet processing: A review. *Nutrition & Food Science*, 30–44.

Ren, G., Fan, X., Teng, C., Li, Y., Everaert, N., & Blecker, C. (2021). The beneficial effect of coarse cereals on chronic diseases through regulating gut microbiota. *Foods*, *10*(11), 2891.

Rinninella, E., Raoul, P., Cintoni, M., Franceschi, F., Miggiano, G. A. D., Gasbarrini, A., & Mele, M. C. (2019). What is the healthy gut microbiota composition? A changing ecosystem across age, environment, diet, and diseases. *Microorganisms*, *7*(1), 14.

Rodríguez, J. M., Murphy, K., Stanton, C., Ross, R. P., Kober, O. I., Juge, N., Avershina, E., Rudi, K., Narbad, A., Jenmalm, M. C., Marchesi, J. R., & Collado, M. C. (2015). The composition of the gut microbiota throughout life, with an emphasis on early life. *Microbial Ecology in Health and Disease*, *26*.

Saini, S., Saxena, S., Samtiya, M., Puniya, M., & Dhewa, T. (2021). Potential of underutilized millets as Nutri-cereal: An overview. *Journal of Food Science and Technology, 58*, 4465–4477.

Samtiya, M., Aluko, R. E., & Dhewa, T. (2020). Plant food anti-nutritional factors and their reduction strategies: An overview. *Food Production, Processing and Nutrition, 2*(1), 1–14.

Samtiya, M., Aluko, R. E., Puniya, A. K., & Dhewa, T. (2021). Enhancing micronutrients bioavailability through fermentation of plant-based foods: A concise review. *Fermentation, 7*, 63.

Sanni, A. I., Onilude, A. A., & Ibidabpo, O. T. (1999). Biochemical composition of infant weaning food fabricated from fermented blends of cereals and soybean. *Food Chemistry, 65*, 35–39.

Sharma, N., & Niranjan, K. (2018). Foxtail millet: Properties, processing, health benefits, and uses. *Food Review International, 34*, 329–363.

Sharma, R., Sharma, S., Dar, B. N., & Singh, B. (2021). Millets as potential nutri-cereals: A review of nutrient composition, phytochemical profile, and techno-functionality. *International Journal of Food Science, 56*, 3703–3718.

Sruthi, N. U., Rao, P. S., & Rao, B. D. (2021). Decortication induced changes in the physicochemical, anti-nutrient, and functional properties of sorghum. *Journal of Food Composition and Analysis, 102*(2021), 104031.

Surekha, N. (2004). *Fabrication of Barnyard Millet Health Mix: Clinical and Shelf Life Evaluation* (MHSc, Thesis). University of Agricultural Science, Dharwad.

Suzuki T. (2020). Regulation of the intestinal barrier by nutrients: The role of tight junctions. *Animal Science Journal, Nihon ChikusanGakkaiho, 91*, e13357.

Thursby, E., & Juge, N. (2017). Introduction to the human gut microbiota. *Biochemical Journal, 474*(11), 1823–1836.

Waters, D. M., Mauch, A., Coffey, A., Arendt, E. K., & Zannini, E. (2015). Lactic acid bacteria as a cell factory for the delivery of functional biomolecules and ingredients in cereal-based beverages: A review. *Critical Reviews in Food Science and Nutrition, 55*, 503–520.

Zhang, G., Xu, Z., Gao, Y., Huang, X., & Yang, T. (2015a). Effects of germination on the nutritional properties, phenolic profiles, and antioxidant activities of buckwheat. *Journal of Food Science, 80*, H1111–H1119.

Zhang, L. Z., & Liu, R. H. (2015b). Phenolic and carotenoid profiles and antiproliferative activity of foxtail millet. *Food Chemistry, 174*, 495–501.

4 Traditional Fermented Foods

Microbes, Fermentation Processes, and Impact on Health

Yuxue Li[1] *and Kang Ning*[1]

[1]Department of Bioinformatics and Systems Biology, College of Life Science and Technology, Huazhong University of Science and Technology, Wuhan 430074, China

4.1 INTRODUCTION

Traditional fermented foods have maintained a significant presence in global markets as a result of their roots in various cultural traditions and their widespread availability (Marti-Quijal et al. 2020; Fan et al. 2023). These foods are typically created through microbial fermentation, a process where microbes break down animal or plant materials into simpler molecules (Wang, Hao, and Ren 2023). This microbial action not only makes these foods more digestible for humans but also imparts unique and often appealing flavors (Fan et al. 2023; Arain et al. 2023; Belleggia and Osimani 2023).

Despite their long history, the specific microbes involved in the fermentation of different traditional foods, the details of the fermentation processes, and their effects on human health have been shrouded in mystery for thousands of years (Galimberti et al. 2021). It is only in recent times, with the advent of advanced omics technologies, that a clearer understanding of these elements has begun to emerge. Omics technologies, encompassing genomics (Gibbons and Rinker 2015), proteomics (Abrahams 2023), metabolomics (Putri et al. 2022; Gao et al. 2021), and other related fields, have provided deeper insights into the microbial communities and biochemical pathways involved in fermentation (Chen 2022). This has led to a better appreciation of how these processes contribute to the nutritional and sensory properties of fermented foods, as well as their potential health benefits.

 DOI: 10.1201/9781003514244-5

Fermented foods are known for their probiotic qualities, which are beneficial for gut health (Parvez et al. 2006; Şanlier, Gökcen, and Sezgin 2019). The fermentation process often enhances the bioavailability of nutrients, making these foods particularly nutritious (Svanberg and Lorri 1997). For instance, the fermentation of dairy products like yogurt and cheese by lactic acid bacteria can improve digestibility and increase the levels of certain vitamins and minerals (García-Burgos et al. 2020). Similarly, fermented plant-based foods such as kimchi and sauerkraut are rich in probiotics, vitamins, and antioxidants (Shahbazi et al. 2021).

Moreover, the microbial diversity in these foods has been linked to various health benefits. Studies have shown that consuming fermented foods can positively impact the gut microbiome, potentially leading to improved digestion, enhanced immune function, and even a reduction in the risk of certain diseases (Jahn, Rekdal, and Sommer 2023; Leeuwendaal et al. 2022). However, it's also important to note that the health impacts of fermented foods can vary based on individual dietary habits and health conditions.

The integration of traditional knowledge with modern scientific techniques is paving the way for innovative approaches in food science and nutrition (Kumari et al. 2022). This synergy is not only helping to preserve traditional fermentation practices but is also enhancing our understanding of their health implications. As research in this field continues to grow, it is likely that more benefits of traditional fermented foods will be uncovered, further solidifying their role in a healthy diet (Sun and Latora 2020).

In summary, traditional fermented foods represent a rich interplay between culture, microbiology, and nutrition (Xing et al. 2023). With the advancements in omics technologies, we are now beginning to unravel the complex processes and health benefits associated with these age-old dietary staples. This growing body of knowledge is not only significant for food science but also for understanding the broader impacts of diet on human health (Dai and Shen 2022).

4.2 TRADITIONAL FERMENTED FOODS

Traditional fermented foods represent a confluence of culinary tradition and microbiological innovation (Fan et al. 2023). The diversity in these foods spans continents and cultures, each with its unique microbial signature and fermentation method. For instance, dairy-based foods like yogurt and cheese are prime examples of the role of lactic acid bacteria in fermentation (Cuamatzin-García et al. 2022). These bacteria not only preserve the food but also enhance its nutritional value by increasing the availability of calcium, phosphorus, and B vitamins.

In the realm of vegetable-based products, kimchi and sauerkraut stand out as rich sources of probiotics, fiber, and vitamins. The fermentation process in these foods involves a complex interplay of various bacteria, including Lactobacillus and Leuconostoc species, contributing to their distinct taste and health benefits (Diez-Ozaeta and Astiazaran 2022; Swain et al. 2014).

Soy-based foods like tempeh and natto are fermented with a combination of bacteria and fungi, adding to their unique texture and flavor. These foods are not only protein-rich but also contain bioactive peptides and isoflavones, which have been linked to health benefits such as reduced cholesterol levels and improved gut health (Cao et al. 2019).

Traditional beverages like kefir and kombucha are fermented using a symbiotic culture of bacteria and yeast (SCOBY), which imparts a fizzy texture and a unique blend of acids, contributing to their refreshing taste and digestive benefits.

Traditional fermented foods have been a cornerstone of human diets for centuries, offering a rich diversity of flavors and health benefits. These foods are produced using age-old fermentation techniques passed down through generations (Figure 4.1).

Cultural significance: Traditional fermented foods are an integral part of the cultural heritage of many societies. They are often associated with specific rituals, festivals, and culinary traditions (Tomar et al. 2023). These foods not only provide sustenance but also connect people to their history and heritage (Table 4.1).

Diverse range of foods: The range of traditional fermented foods is vast, encompassing dairy products like yogurt and cheese, vegetable-based products like

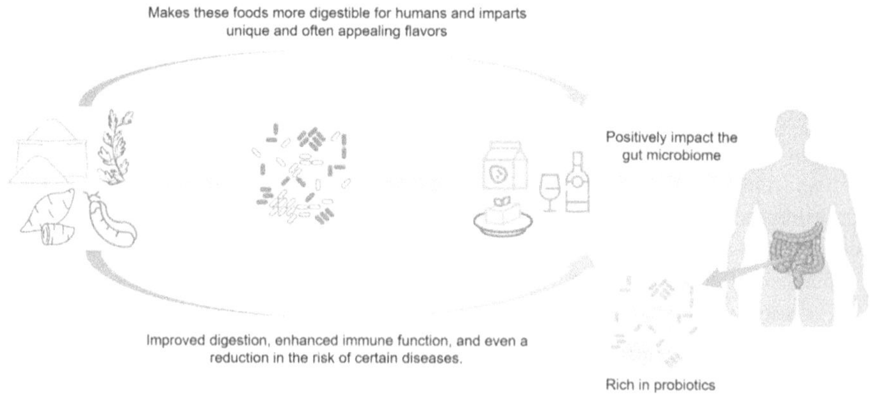

FIGURE 4.1 Traditional fermented foods represent a rich interplay between culture, microbiology, and nutrition.

TABLE 4.1
Traditional Fermented Foods Are Produced Using Age-Old Fermentation Techniques Passed Down Through Generations

Material	Fermented Foods	Function
Dairy	Yogurt, cheese	Increasing the availability of calcium, phosphorus, and B vitamins
Vegetable	Kimchi, sauerkraut	Rich sources of probiotics, fiber, and vitamins
Soy	Tempeh, natto	Protein-rich, contain bioactive peptides and isoflavones

sauerkraut and kimchi, grain-based products like sourdough bread, and fermented beverages like wine and beer. Each of these foods has its own unique flavor profile and health benefits.

Health benefits: Traditional fermented foods are known for their probiotic content, which can improve gut health and boost the immune system (Marco et al. 2017). They are also a source of important nutrients, including vitamins, minerals, and antioxidants. The fermentation process can increase the bioavailability of these nutrients, making them more accessible to the body.

Preservation of traditional methods: The preservation of traditional fermentation methods is important for maintaining the diversity of the global food supply (Paul Ross, Morgan, and Hill 2002). These methods are often more sustainable and environmentally friendly than modern industrial processes. They also contribute to the preservation of biodiversity, as they rely on locally available ingredients and microbial communities.

4.3 HEALTH BENEFITS OF FERMENTED FOODS

Fermented foods have been recognized for their significant health benefits, particularly in relation to gut health. The fermentation process enhances the bioavailability of nutrients, making these foods not only more digestible but also more nutritious (Diez-Ozaeta and Astiazaran 2022). For instance, the fermentation of dairy products like yogurt and cheese results in an increase in beneficial bacteria, which can improve gut flora and aid in digestion (Aslam et al. 2020). Similarly, plant-based fermented foods like kimchi and sauerkraut are rich in probiotics, which are known to boost immune function and may even play a role in preventing certain diseases (Wuyts et al. 2020).

In addition to their probiotic content, fermented foods can also be a source of important vitamins and minerals (Nithya et al. 2023). The fermentation process can increase levels of certain B vitamins, as well as minerals like magnesium and zinc. This makes fermented foods an important part of a balanced diet, contributing to overall health and well-being (Figure 4.2).

Moreover, the health benefits of fermented foods extend beyond their nutritional value. There is growing evidence to suggest that these foods can play a role in reducing the risk of certain chronic diseases, including heart disease and diabetes (Marco et al. 2017). The beneficial bacteria found in fermented foods can help to regulate cholesterol levels and blood sugar, contributing to heart health and helping to prevent type 2 diabetes (Mukherjee et al. 2023).

4.4 MICROBES FOR FOOD FERMENTATION

The role of microbes in food fermentation is central to the production of traditional fermented foods. Different types of microbes, including lactic acid bacteria, yeasts, and molds, contribute to the fermentation process, each playing a unique role in shaping the flavor, texture, and health benefits of the food.

Lactic acid bacteria are perhaps the most well-known group of fermentation microbes, responsible for the sour taste of many fermented foods. These bacteria

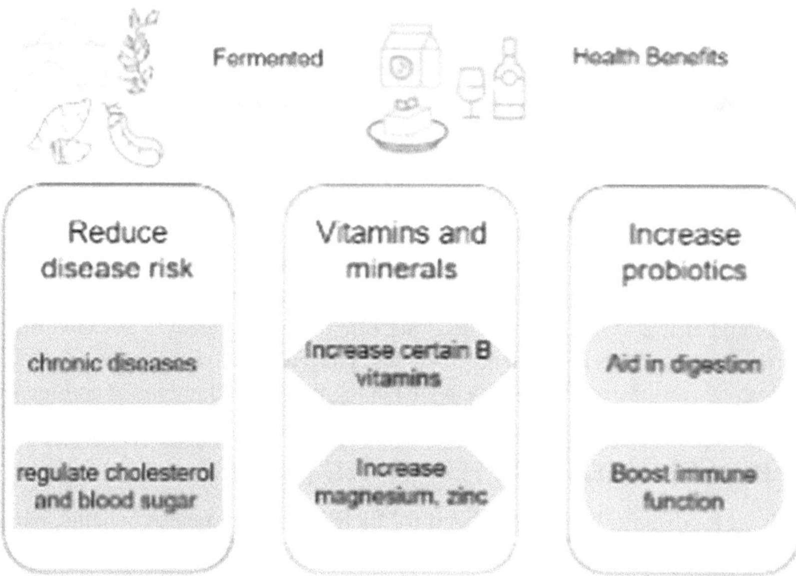

FIGURE 4.2 Health benefits of fermented foods.

ferment sugars into lactic acid, which not only preserves the food but also contributes to its flavor profile (Gänzle 2015). Yeasts, on the other hand, are responsible for the alcoholic fermentation of certain foods and beverages, converting sugars into alcohol and carbon dioxide (Maicas 2020). Molds play a role in the fermentation of foods like cheese and soy sauce, contributing to their distinctive flavors and textures.

The specific types of microbes involved in fermentation can vary widely depending on the food and the fermentation process (Voidarou et al. 2020). In some cases, a single strain of bacteria or yeast may be used, while in others, a complex community of different microbes may be involved. The interaction of these microbes with each other and with the food substrate is a complex and fascinating process, leading to a wide variety of flavors and textures in fermented foods (Figure 4.3).

Furthermore, the microbes involved in fermentation also contribute to the health benefits of the food. These beneficial bacteria can help improve gut health, boost the immune system, and even reduce the risk of certain diseases (Marco et al. 2017). The specific health benefits depend on the types of microbes involved and their activity during fermentation.

4.5 FERMENTATION PROCESSES

The fermentation process is both an art and a science, involving a delicate balance of microbial activity, environmental conditions, and human intervention (Teng et al.

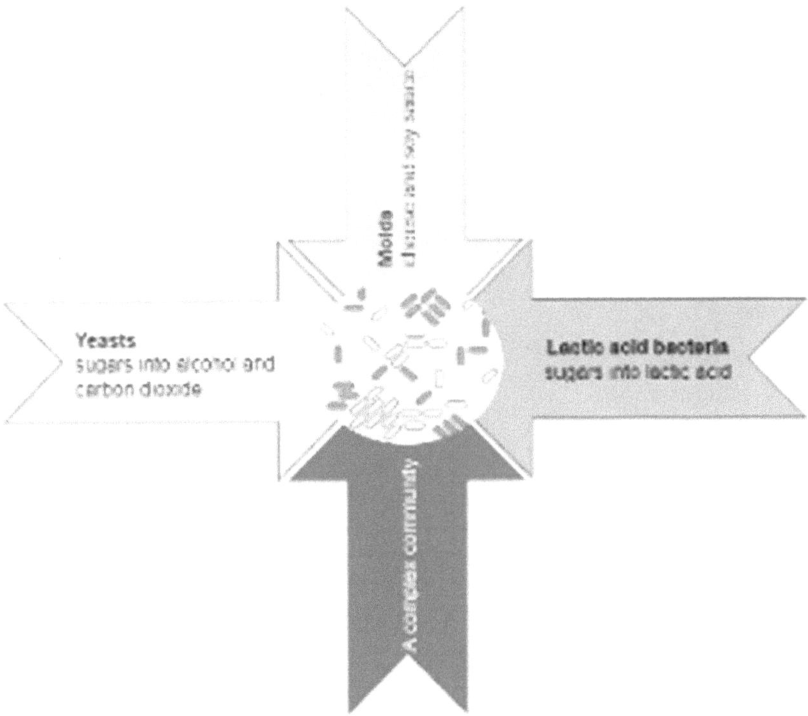

FIGURE 4.3 Microbes for food fermentation.

2021). In the production of traditional fermented foods, the fermentation process can vary widely depending on the type of food and the desired outcome.

In dairy products like cheese and yogurt, the fermentation process typically involves the addition of specific strains of lactic acid bacteria. These bacteria ferment the lactose in the milk, producing lactic acid and other compounds that give the products their characteristic flavors and textures. In grain-based products like bread and beer, the fermentation process involves the activity of yeasts, which convert sugars into alcohol and carbon dioxide.

Vegetable-based products like sauerkraut and kimchi undergo a natural fermentation process, in which the sugars in the vegetables are fermented by naturally occurring lactic acid bacteria. This process not only preserves the vegetables but also enhances their flavor and nutritional value (Figure 4.4).

The fermentation process is influenced by a variety of factors, including temperature, pH, and the presence of oxygen. Controlling these factors is essential for achieving the desired outcome, whether it be a tangy yogurt, a flavorful cheese, or a crisp, sour pickle. The fermentation process is a complex interplay of biology, chemistry, and tradition, resulting in a wide variety of delicious and healthful foods (Mengesha, Tebeje, and Tilahun 2022; Yang et al. 2021; Gupta et al. 2022).

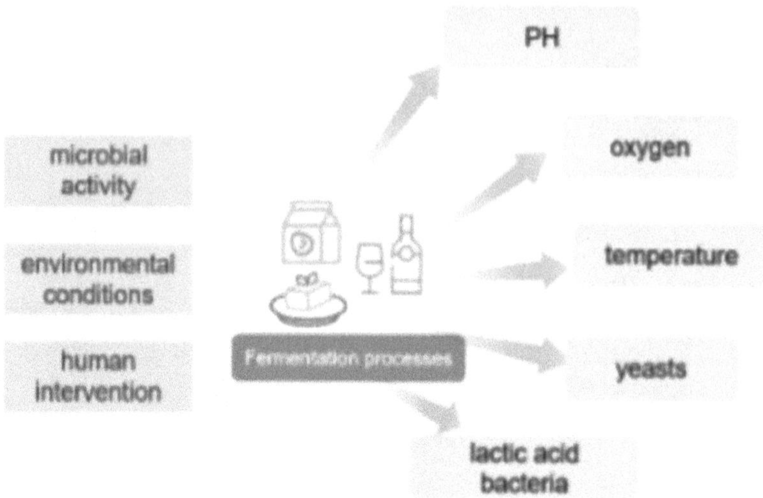

FIGURE 4.4 Different types of fermentation processes.

4.6 OMICS TECHNOLOGIES AND MICROBIAL FERMENTATION

The integration of omics technologies – genomics, proteomics, and metabolomics – in the study of microbial fermentation has been transformative. These technologies provide unparalleled insights into the complex microbial communities involved in the fermentation process and their metabolic activities.

Genomics and microbial identification: Genomic analysis has enabled the identification and characterization of the microbial species involved in fermentation (Knight et al. 2018). This includes not only the dominant species but also the less abundant ones that might play crucial roles. Understanding the genetic makeup of these microbes helps in deciphering their functional capabilities and their contributions to the fermentation process.

Proteomics and functional analysis: Proteomic studies focus on the proteins expressed by these microbes, shedding light on the enzymes and other functional molecules that drive the fermentation process (Song et al. 2020). This understanding is crucial for optimizing fermentation conditions and enhancing the nutritional and sensory properties of fermented foods.

Metabolomics and flavor profiling: Metabolomics, the study of the complete set of metabolites, plays a key role in understanding the flavor profile of fermented foods (Liu et al. 2024). It helps in identifying the specific compounds responsible for the characteristic tastes and aromas of these foods. This knowledge is invaluable for the development of new products and for the improvement of existing ones (Figure 4.5).

Integration of omics data: The integration of data from genomics, proteomics, and metabolomics provides a comprehensive picture of the fermentation process (Picard et al. 2021). This holistic approach is essential for understanding the complex

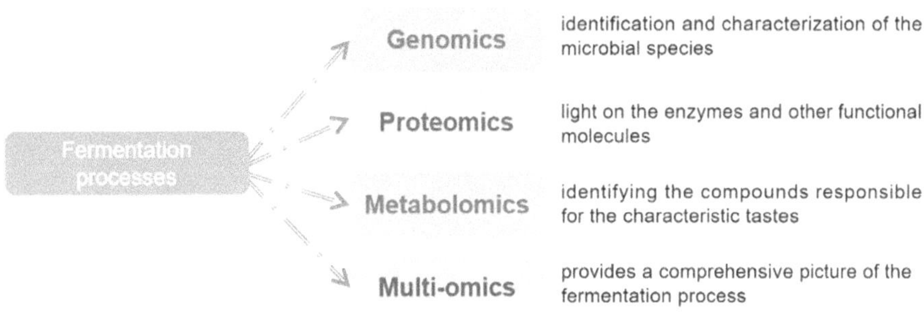

FIGURE 4.5 Omics technologies and microbial fermentation.

interactions between different microbes and how they influence the final product. It also opens up new possibilities for the manipulation of microbial communities to achieve desired outcomes.

4.7 MODERN INNOVATIONS IN FERMENTED FOODS

The field of fermented foods is at the forefront of innovation, with new technologies and approaches being developed to enhance the health benefits and sensory qualities of these products.

Tailored probiotic strains: Advances in biotechnology have enabled the development of tailored probiotic strains for use in fermented foods (Pramanik, Venkatraman, and Vaidyanathan 2023). These strains are selected for their specific health benefits, such as improving gut health, enhancing immune function, or reducing the risk of certain diseases. The incorporation of these strains into fermented foods creates functional products that deliver targeted health benefits.

Plant-based fermented foods: The rise of plant-based diets has led to the development of fermented plant-based alternatives to traditional animal-based products (Goksen et al. 2023). These products mimic the taste and texture of animal-based foods while offering the health benefits of plant-based ingredients. Fermentation is used to enhance the nutritional value of these products and to develop unique flavors.

Fermentation technology innovations: Technological innovations in fermentation processes are enabling more precise control over the fermentation environment (Terefe 2022). This allows for the optimization of conditions to maximize the growth of beneficial microbes and the production of desired compounds. These innovations are leading to the production of higher quality fermented foods with enhanced health benefits.

Personalized fermented foods: The concept of personalized nutrition is being applied to fermented foods, with products being developed to cater to the specific dietary needs and preferences of individuals (Song and Shin 2022). This includes fermented foods designed for people with specific health conditions, dietary restrictions, or flavor preferences (Figure 4.6).

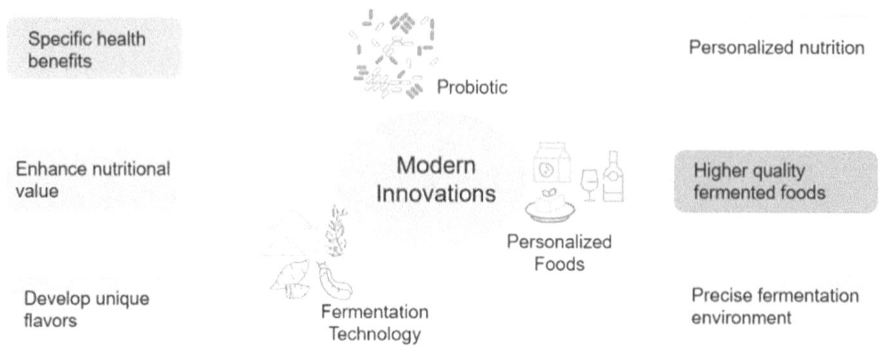

FIGURE 4.6 Modern innovations in fermented foods.

4.8 MODERN FERMENTED FOODS

Modern fermented foods represent the intersection of traditional fermentation practices and contemporary scientific knowledge. These foods are produced using advanced fermentation techniques and are designed to meet the needs of today's consumers.

Innovative products: Modern fermented foods include innovative products that combine traditional fermentation techniques with new ingredients and flavors. These products offer a modern twist on classic fermented foods, appealing to a broader range of tastes and dietary preferences (Fan et al. 2023).

Health-focused fermentation: Modern fermented foods are often developed with a focus on health and nutrition (Mukherjee et al. 2023). This includes the use of specific strains of probiotics for targeted health benefits, as well as the fortification of fermented foods with additional nutrients. These products are designed to contribute to a healthy and balanced diet.

Sustainability and environmental impact: Modern fermentation practices are increasingly focused on sustainability and reducing environmental impact. This includes the use of sustainable ingredients, energy-efficient fermentation processes, and packaging that minimizes waste.

Accessibility and convenience: Modern fermented foods are designed for convenience and accessibility. This includes ready-to-eat fermented products, as well as fermentation kits that allow consumers to ferment foods at home. These products make it easier for people to incorporate fermented foods into their diets.

This extended content adds depth and detail to the review draft, providing a comprehensive overview of traditional and modern fermented foods, the role of omics technologies in fermentation, and the latest innovations in the field.

4.9 CONCLUSIONS

The exploration of traditional and modern fermented foods, underpinned by advanced omics technologies and innovative fermentation processes, underscores a significant intersection of culinary heritage, science, and health. This review has traversed the journey of fermentation from ancient practices to cutting-edge innovations, highlighting the profound impact of fermented foods on human health, culture, and the environment.

Synthesis of tradition and innovation: One of the most striking aspects of fermented foods is the seamless blend of age-old traditions with modern scientific understanding. Traditional fermentation methods, which have been honed over centuries, are now being illuminated and enhanced by genomic, proteomic, and metabolomic analyses. These omics technologies have not only deepened our understanding of the complex microbial communities involved in fermentation but have also paved the way for the development of new fermented products tailored for health and dietary needs.

Health implications and probiotic potential: The health benefits of fermented foods, particularly their impact on gut health, have been a focal point of this review. The presence of beneficial microbes in these foods plays a crucial role in enhancing digestive health, boosting the immune system, and potentially reducing the risk of various chronic diseases. The introduction of tailored probiotic strains and the fortification of fermented foods with additional nutrients are exciting developments that could lead to more personalized nutrition and health solutions.

Cultural significance and biodiversity: Fermented foods are not just nourishing but are also imbued with cultural significance. They represent a rich tapestry of culinary traditions, reflecting the history, culture, and biodiversity of the regions from which they originate. The preservation of traditional fermentation methods is crucial for maintaining this cultural heritage and biodiversity. As modern fermentation practices evolve, it is essential that they complement rather than replace traditional methods, ensuring a diverse and sustainable food supply.

Sustainability and environmental considerations: Modern innovations in fermented foods have shown a growing emphasis on sustainability. Sustainable ingredient sourcing, energy-efficient fermentation processes, and environmentally friendly packaging are increasingly becoming the norm. These practices not only enhance the appeal of fermented foods but also align with the global imperative for sustainable food production and consumption.

Future directions in fermentation research: Looking ahead, the field of fermented food research presents vast opportunities for further exploration. The application of omics technologies is just beginning to unravel the complexities of microbial ecosystems in fermented foods. There is enormous potential for discovering new health-promoting microbes, understanding their interactions, and harnessing their benefits. Moreover, as consumer preferences evolve and awareness of health and sustainability grows, there will be a continuous demand for innovative, health-focused, and environmentally conscious fermented products.

Final reflections: In conclusion, fermented foods represent a unique confluence of flavor, health, tradition, and science. They are a testament to the ingenuity of human culinary practices and the wonders of microbial ecosystems. As we continue to explore and understand these complex fermented systems, it is imperative to balance the preservation of traditional methods with the adoption of modern scientific advancements. The future of fermented foods is not only about enhancing their health benefits or sensory qualities but also about preserving their cultural significance and promoting sustainability. By embracing both the heritage and the future of fermentation, we can ensure that these cherished foods continue to enrich our diets and our lives for generations to come.

REFERENCES

Abrahams, Adrian Mark. 2023. Chapter 23 – Proteomics and transcriptomics and their application in fermented foods. In *Indigenous Fermented Foods for the Tropics*, edited by O. A. Adebo, C. E. Chinma, A. O. Obadina, A. G. Soares, S. K. Panda, and R.-Y. Gan. Academic Press.

Arain, M. A., S. Rasheed, A. Jaweria, G. B. Khaskheli, G. S. Barham, and S. Ahmed. 2023. A review on processing opportunities for the development of camel dairy products. *Food Science of Animal Resources* 43 (3):383–401.

Aslam, H., W. Marx, T. Rocks, et al. 2020. The effects of dairy and dairy derivatives on the gut microbiota: A systematic literature review. *Gut Microbes* 12 (1):1799533.

Belleggia, L., and A. Osimani. 2023. Fermented fish and fermented fish-based products, an ever-growing source of microbial diversity: A literature review. *Food Research International* 172:113112.

Cao, Z. H., J. M. Green-Johnson, N. D. Buckley, and Q. Y. Lin. 2019. Bioactivity of soy-based fermented foods: A review. *Biotechnology Advances* 37 (1):223–238.

Chen, Wanping. 2022. Demystification of fermented foods by omics technologies. *Current Opinion in Food Science* 46:100845.

Cuamatzin-García, L., P. Rodríguez-Rugarcía, E. G. El-Kassis, et al. 2022. Traditional fermented foods and beverages from around the world and their health benefits. *Microorganisms* 10 (6):1151.

Dai, X., and L. Shen. 2022. Advances and trends in omics technology development. *Front Med (Lausanne)* 9:911861.

Diez-Ozaeta, Iñaki, and Oihana Juaristi Astiazaran. 2022. Fermented foods: An update on evidence-based health benefits and future perspectives. *Food Research International* 156:111133.

Fan, J., G. Qu, D. Wang, J. Chen, G. Du, and F. Fang. 2023. Synergistic fermentation with functional microorganisms improves safety and quality of traditional Chinese fermented foods. *Foods* 12 (15):2892.

Galimberti, A., A. Bruno, G. Agostinetto, M. Casiraghi, L. Guzzetti, and M. Labra. 2021. Fermented food products in the era of globalization: Tradition meets biotechnology innovations. *Current Opinion in Biotechnology* 70:36–41.

Gänzle, Michael G. 2015. Lactic metabolism revisited: Metabolism of lactic acid bacteria in food fermentations and food spoilage. *Current Opinion in Food Science* 2:106–117.

Gao, Y., L. Hou, J. Gao, et al. 2021. Metabolomics approaches for the comprehensive evaluation of fermented foods: A review. *Foods* 10 (10):2294.

García-Burgos, María, Jorge Moreno-Fernández, María J. M. Alférez, Javier Díaz-Castro, and Inmaculada López-Aliaga. 2020. New perspectives in fermented dairy products and their health relevance. *Journal of Functional Foods* 72:104059.

Gibbons, J. G., and D. C. Rinker. 2015. The genomics of microbial domestication in the fermented food environment. *Current Opinion in Genetics and Development* 35:1–8.

Goksen, Gulden, Qazi Sugra Altaf, Salma Farooq, et al. 2023. A glimpse into plant-based fermented products alternative to animal based products: Formulation, processing, health benefits. *Food Research International* 173:113344.

Gupta, Kuldeep, Pritam Bardhan, Devabrata Saikia, et al. 2022. Microbial fermentation: Basic fundamentals and its dynamic prospect in various industrial applications. In *Industrial Microbiology and Biotechnology*, edited by P. Verma. Springer Singapore.

Jahn, Leonie J., Vayu M. Rekdal, and Morten O. A. Sommer. 2023. Microbial foods for improving human and planetary health. *Cell* 186 (3):469–478.

Knight, Rob, Alison Vrbanac, Bryn C. Taylor, et al. 2018. Best practices for analysing microbiomes. *Nature Reviews Microbiology* 16 (7):410–422.

Kumari, Madhuree, Shipra Pandey, Ved Prakash Giri, et al. 2022. Chapter 24 – Integrated approach for technology transfer awareness of traditional knowledge for upliftment of circular bioeconomy. In *Biomass, Biofuels, Biochemicals*, edited by S. Varjani, A. Pandey, T. Bhaskar, S. V. Mohan, and D. C. W. Tsang. Elsevier.

Leeuwendaal, N. K., C. Stanton, P. W. O'Toole, and T. P. Beresford. 2022. Fermented foods, health and the gut microbiome. *Nutrients* 14 (7):1527.

Liu, Zhenshan, Hao Wang, Jing Zhang, et al. 2024. Comparative metabolomics profiling highlights unique color variation and bitter taste formation of Chinese cherry fruits. *Food Chemistry* 439:138072.

Maicas, S. 2020. The role of yeasts in fermentation processes. *Microorganisms* 8 (8):1142.

Marco, Maria L., Dustin Heeney, Sylvie Binda, et al. 2017. Health benefits of fermented foods: Microbiota and beyond. *Current Opinion in Biotechnology* 44:94–102.

Marti-Quijal, Francisco J., Fabienne Remize, Giuseppe Meca, Emilia Ferrer, María-José Ruiz, and Francisco J. Barba. 2020. Fermentation in fish and by-products processing: An overview of current research and future prospects. *Current Opinion in Food Science* 31:9–16.

Mengesha, Y., A. Tebeje, and B. Tilahun. 2022. A review on factors influencing the fermentation process of teff (eragrostis teff) and other cereal-based ethiopian Injera. *International Journal of Food Science* 2022:4419955.

Mukherjee, Arghya, Samuel Breselge, Eirini Dimidi, Maria L. Marco, and Paul D. Cotter. 2023. Fermented foods and gastrointestinal health: Underlying mechanisms. *Nature Reviews Gastroenterology & Hepatology* 21:248–266.

Nithya, A., Sourav Misra, Chirasmita Panigrahi, Chandrakant Genu Dalbhagat, and Hari Niwas Mishra. 2023. Probiotic potential of fermented foods and their role in non-communicable diseases management: An understanding through recent clinical evidences. *Food Chemistry Advances* 3:100381.

Parvez, S., K. A. Malik, S. Ah Kang, and H. Y. Kim. 2006. Probiotics and their fermented food products are beneficial for health. *Journal of Applied Microbiology* 100 (6):1171–1185.

Paul Ross, R., S. Morgan, and C. Hill. 2002. Preservation and fermentation: Past, present and future. *International Journal of Food Microbiology* 79 (1):3–16.

Picard, Milan, Marie-Pier Scott-Boyer, Antoine Bodein, Olivier Périn, and Arnaud Droit. 2021. Integration strategies of multi-omics data for machine learning analysis. *Computational and Structural Biotechnology Journal* 19:3735–3746.

Pramanik, Shreyasi, Swethaa Venkatraman, and Vinoth Kumar Vaidyanathan. 2023. Development of engineered probiotics with tailored functional properties and their application in food science. *Food Science and Biotechnology* 32 (4):453–470.

Putri, Sastia Prama, Muhammad Maulana Malikul Ikram, Arisa Sato, et al. 2022. Application of gas chromatography-mass spectrometry-based metabolomics in food science and technology. *Journal of Bioscience and Bioengineering* 133 (5):425–435.

Şanlier, N., B. B. Gökcen, and A. C. Sezgin. 2019. Health benefits of fermented foods. *Critical Reviews in Food Science and Nutrition* 59 (3):506–527.

Shahbazi, R., F. Sharifzad, R. Bagheri, N. Alsadi, H. Yasavoli-Sharahi, and C. Matar. 2021. Anti-Inflammatory and immunomodulatory properties of fermented plant foods. *Nutrients* 13 (5):1516.

Song, E. J., and J. H. Shin. 2022. Personalized diets based on the gut microbiome as a target for health maintenance: from current evidence to future possibilities. *Journal of Microbiology and Biotechnology* 32 (12):1497–1505.

Song, Yanbo, Xiaojing Shi, Yanli Zou, et al. 2020. Proteome-wide identification and functional analysis of ubiquitinated proteins in peach leaves. *Scientific Reports* 10 (1):2447.

Sun, Ye, and Vito Latora. 2020. The evolution of knowledge within and across fields in modern physics. *Scientific Reports* 10 (1):12097.

Svanberg, Ulf, and Wilbald Lorri. 1997. Fermentation and nutrient availability. *Food Control* 8 (5):319–327.

Swain, M. R., M. Anandharaj, R. C. Ray, and R. Parveen Rani. 2014. Fermented fruits and vegetables of Asia: a potential source of probiotics. *Biotechnology Research International* 2014:250424.

Teng, Ting Shien, Yi Ling Chin, Kong Fei Chai, and Wei Ning Chen. 2021. Fermentation for future food systems. *EMBO Reports* 22 (5):e52680.

Terefe, Netsanet Shiferaw. 2022. Chapter 5 – Recent developments in fermentation technology: Toward the next revolution in food production. In *Food Engineering Innovations Across the Food Supply Chain*, edited by P. Juliano, R. Buckow, M. H. Nguyen, K. Knoerzer, and J. Sellahewa. Academic Press.

Tomar, Shikha, Kumud Pant, Promila Sharma, Somya Sinha, and Debasis Mitra. 2023. Unravelling the hidden ethnic fermented treasure of the Himalayas – A review on the traditionally fermented beverages of the Northwest Indian Himalayan Region. *Food Chemistry Advances* 2:100254.

Voidarou, C., M. Antoniadou, G. Rozos, et al. 2020. Fermentative foods: Microbiology, biochemistry, potential human health benefits and public health issues. *Foods* 10 (1):69.

Wang, J., S. Hao, and Q. Ren. 2023. Uncultured microorganisms and their functions in the fermentation systems of traditional chinese fermented foods. *Foods* 12 (14):2691.

Wuyts, Sander, Wannes Van Beeck, Camille Nina Allonsius, Marianne F. L. van den Broek, and Sarah Lebeer. 2020. Applications of plant-based fermented foods and their microbes. *Current Opinion in Biotechnology* 61:45–52.

Xing, Yanxia, Mengzhen Huang, Chinasa V. Olovo, et al. 2023. Traditional fermented foods: Challenges, sources, and health benefits of fatty acids. *Fermentation* 9 (2):110.

Yang, Shanshan, Danli Yan, Yiting Zou, et al. 2021. Fermentation temperature affects yogurt quality: A metabolomics study. *Food Bioscience* 42:101104.

Section II

Gut Physiology/Microbiome

5 Food Allergies and their Management

Nikita Basant[1], Kopal Jain[1], Nishu Mittal[2], Shivam Pandey[1], Vimarsh Srivastava[1], Kritika Pandey[1], Ayushi[1], Sapna Sharma[3], and Neeraj Tiwari[1]

[1]Faculty of Biotechnology, Institute of Biosciences and Technology, Shri RamSwaroop Memorial University, Lucknow-Deva Road, Barabanki, Uttar Pradesh, India
[2]Faculty of Biosciences, Institute of Biosciences and Technology, Shri RamSwaroop Memorial University, Lucknow-Deva Road, Barabanki, Uttar Pradesh, India
[3]School of Forensic Science, Uttar Pradesh State Institute of Forensic Science, Lucknow, Uttar Pradesh, India

5.1 INTRODUCTION

Food allergies are an unpleasant immunological response brought on by proteins found in food that are swallowed; these reactions are mediated by mast cell degranulation and food-specific immunoglobin E (IgE) (Ortolani and Pastorello 2006). The symptoms can be anything from life-threatening anaphylaxis to urticaria (Sharma, Bansil, and Uygungil 2015). Over the past 20 years, food allergies have become much more common in developed nations than could be explained by purely genetic changes (Oria et al. 2016a). A complex interaction of pre- and postnatal environmental factors related to lifestyle, diet, and other exposures is likely the cause of the pathophysiology of food allergy development in the context of a genetically susceptible background (Hong and Wang 2012; Zhang et al. 2023). Food allergies have a significant detrimental impact on stress levels and the overall quality of life for afflicted individuals and their families, in addition to their immediate health effects (Flokstra-de Blok et al. 2010; Peniamina et al. 2016). Managing a food allergy involves a variety of dietary and lifestyle modifications in order to avoid the offending food allergen or allergens (Mazzocchi et al. 2017; Leone et al. 2023). This can result in physical symptoms that recur from accidental ingestion, social discrimination, challenges in locating safe foods, and an ongoing need to verify the safety of food (Peniamina et al. 2016). People's capacity to manage the difficulties presented by food allergies is impacted by this intricate circumstance. Furthermore, knowing what makes an illness adaptation successful becomes essential for improving theoretical understanding of the interaction between psychological and physical factors in individuals with food allergies as well as for clinical practice (Conner et al. 2018).

DOI: 10.1201/9781003514244-7

Though studies on other chronic conditions have yielded valuable insights, there is a dearth of specific research on the impact of personality on the types and frequency of food allergies. In individuals with chronic conditions, the big five personality traits – extraversion, neuroticism, conscientiousness, agreeableness, and openness – have been linked to quality-of-life indicators (Conner et al. 2018). Various medical conditions, such as type 1 diabetes mellitus, asthma, and congenital heart disease, have been linked to this pattern of observation (Rassart et al. 2014; Topp, Vestbo, and Mortensen 2016; Van De Ven and Engels 2011). These studies have not consistently found a link between a lower quality of life and a decreased ability to adapt to illness and new ideas or experiences. A better understanding of these personality traits can help with strategies to improve adaptation and general well-being as well as give insight into the psychological aspects of living with a food allergy. An individual's atopic predisposition is influenced by a multitude of environmental and genetic factors, which consequently establishes their vulnerability to allergen-induced allergic immune responses (Campbell et al. 2015; Falcon and Caoili 2023). When exposed to allergens, atopic individuals who are predisposed to IgE-associated allergies activate Th2 cells specific to the allergen, which causes class switching and the production of allergen-specific IgE (Shamji et al. 2021; Valenta et al. 2015). T-cell and IgE memory are produced early in life by primary allergic sensitization, and these responses can be strengthened by repeated exposure to allergens. When non-allergic people come into contact with a primary food allergen, they produce IgG and IgA that are specific to that allergen and do not cause allergic reactions (Valenta et al. 2015). The hallmark of an IgE-associated food allergy and its diagnosis is the production of IgE that is specific to food allergens (Waserman, Bégin, and Watson 2018). Living with food allergies is complex, and there are issues with food labeling, education, and awareness. For people with food allergies, accurate and clear labeling is essential. Nevertheless, the labeling guidelines that are currently in place can be difficult to understand, making it harder to identify ingredients that are allergenic (Fiocchi et al. 2021). Families with allergic children must get in touch with manufacturers directly to ensure the safety of food until universally implemented comprehensive labeling regulations are in place. Furthermore, the establishment of an environment that promotes the well-being of individuals with food allergies is greatly aided by an informed and cognizant community. A more tolerant and understanding community can be fostered by raising awareness of the consequences of food allergies in social situations, such as parties or eating out (Aljameel et al. 2023). The complex nature of food allergies is highlighted by the interaction of these immune systems, psychological variables related to personality traits, and the difficulties in food labeling and awareness.

5.2 FOOD ALLERGIES AND FOOD INTOLERANCE

There are two types of adverse food reactions: toxic and non-toxic (Figure 5.1). Non-toxic adverse food reactions can be categorized into food intolerance and allergy (Lyra et al. 2013). A food allergy is an immune response, while a food intolerance is a non-immunological reaction that exhibits symptoms resembling those of an allergy (Yu, Freeland, and Nadeau 2016; Zopf et al. 2009). Food allergies are primarily caused by fruits, vegetables, nuts, milk protein, and other foods; in contrast, food

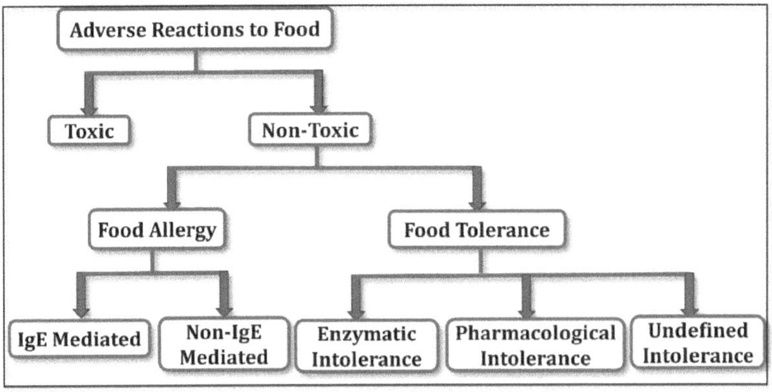

FIGURE 5.1 Adverse food reactions classification.

intolerance can be brought on by pharmacological and/or enzymatic deficits, as well as other factors (Gargano et al. 2021; Tuck et al. 2019). A food allergy is described as "an adverse health effect arising from a specific immune response that occurs reproducibly on exposure to a given food" by the National Institute of Allergy and Infectious Diseases. Nearly all immune-mediated responses, including those from the innate and adaptive immune systems, are included in this response (Valenta et al. 2015). Food sensitization is the state in which the immune system produces antigen-specific IgE (sIgE) in response to an allergenic food protein (Yu, Freeland, and Nadeau 2016). When the allergenic food is re-exposed, it binds to particular antibodies and releases mediators like histamine, which causes food allergy symptoms (Elghoudi and Narchi 2022). The prevalence of food allergies is rising worldwide, underscoring the need for better approaches to diagnosis, treatment, and prevention.

Food intolerance and allergy are distinct conditions in many ways; Table 5.1 provides a detailed comparison between food allergies and food intolerances, highlighting the many distinctions between the two conditions.

TABLE 5.1
Differences between Food Allergy and Food Intolerance

	Food Allergy	Food Intolerance
Mechanism	Immunological	Non-immunological
Onset	Rapid onset	Gradual onset
Severity	Range from mild to severe and life threatening	Less severe and not life threatening
Symptoms	Hives, eczema, wheezing, vomiting, diarrhea, anaphylaxis	Bloating, abdominal pain, headaches, migraines, itching
Diagnosis	Skin prick tests, blood tests	No standardized tests, may involve removal of suspected food items from the diet

In summary, food allergies involve an immunological response, are often rapid, and can be severe, whereas food intolerances are typically non-immune reactions that develop more gradually and are typically not life-threatening (Tordesillas, Berin, and Sampson 2017). For both conditions, a precise diagnosis and course of therapy are required, so anyone experiencing adverse food reactions ought to speak with a physician.

Food intolerance refers to an intolerant reaction to a specific food or substance, which could be immune system related or not. Some dietary intolerances, such as lactose and gluten intolerances, can also be brought on by deficiencies in digestive enzymes. A number of unpleasant symptoms, such as bloating, nausea, diarrhea, and stomach pain, are caused by food intolerance (Ozdemir et al. 2009). It can take up to 48 hours for these symptoms to manifest, although they usually do so 30 minutes after consuming the tainted meal. There may be a significant hereditary component to food intolerances. While dosage is the primary cause of most intolerance reactions to food ingredients, some people are more sensitive than others. There are two approaches to diagnosis: challenge testing and elimination (Ozdemir et al. 2009). Vaso-active pharmaceutical substances present in food or an enzymatic deficiency can cause enzymatic and pharmacological intolerance. Reactions to food that are not explained by any established mechanism are referred to as undefinable intolerance (Ortolani and Pastorello 2006). Food intolerances encompass a number of types of undesired dietary responses. The immune system plays no part in these non-allergic food reactions. One type of dietary intolerance involving an organic pathophysiological process is lactose intolerance, which is caused by a lack of an enzyme that breaks down lactose. However, some food intolerances – like many of the food intolerances associated with irritable bowel syndrome – are challenging to explain by biological processes that are currently understood (Turnbull, Adams, and Gorard 2015).

5.2.1 Types of Food Tolerances

- **Enzymatic food intolerance:** Enzymatic food intolerance refers to an unfavorable food reaction caused by abnormal levels of enzymes in the gastrointestinal tract. In the gastrointestinal tract, some foods trigger symptoms. Intolerance to lactose is the most common. This is caused by a beta-galactosidase deficiency that is common in the adult population (Malik and Panuganti 2023).
- **Pharmacological food intolerance:** Pharmacological food intolerance is caused by substances found in food, such as vasoactive amines, all of which have pharmacological activity. Vasoactive amines include dopamine, norepinephrine, histamine, phenylethylamine, tyramine, and serotonin (Shulpekova et al. 2021).
- **Undefined intolerance:** Undefined intolerance includes intolerances resulting from unidentified sources, such as reactions to food additives. However, some people who are prone to allergies might react intolerantly to some foods, especially dyes, monosodium glutamate, nitrites, sulphites,

and nitrates. Possible symptoms include urticaria, rhinitis, migraines, itching, and asthma. The underlying mechanisms of these intolerances remain unknown (Gultekin and Doguc 2013).

5.3 CLASSIFICATION OF FOOD ALLERGIES

Food allergies can be classified as mixed, non-IgE-mediated, or IgE-mediated. In order to be categorized as IgE-mediated, a food allergy needs to exhibit both the development of symptoms upon exposure and the presence of a serum-specific IgE antibody to the food allergen. This process is known as food allergen sensitization. Non-IgE-mediated food allergies are primarily driven by T-cell-mediated mechanisms, with histological signs indicating an immunological response, such as eosinophilic inflammation in the gastrointestinal tract (Turnbull, Adams, and Gorard 2015). IgE-mediated food allergies appear two to three hours after exposure to a food allergen, whereas non-IgE-mediated food allergies take longer to manifest (Lopez, Yarrarapu, and Mendez 2023).

5.3.1 IgE-Mediated Food Allergy

This type of allergy can be deadly and have a negative impact on the lives of those affected as well as their families. The mechanism underlying acute hypersensitivity is IgE-mediated food allergy. When allergenic proteins crosslink IgE on mast cells and basophils via the FcεRI receptor, histamine and other mediators are released. The symptoms of IgE-mediated food allergy include bronchospasm, vomiting, diarrhea, urticaria, angioedema, and multisystem allergy, which can result in anaphylaxis (Bird, Lack, and Perry 2015). This kind of food allergy primarily affects one or more organs and only develops a few hours after ingesting any kind of food allergen. Food allergies mediated by immunoglobulin E (IgE) are a serious public health concern that impacts 3%–10% of adults and 8% of children globally (Messina and Venter 2020). IgE-induced food allergies arise from an immune system that is unable to sustain tolerance, thereby blocking harmless food antigens from being identified as pathogens. When people acquire antibodies against the food protein that constitutes the allergen and then come into contact with it, they can experience type I hypersensitivity reactions, also referred to as IgE-mediated allergic reactions (Anvari et al. 2019). Usually, IgE antibodies are found in the peripheral blood. They therefore attach themselves to the food protein and begin to cause problems as it passes through the digestive system. Within minutes of exposure, two nearby IgE molecules bind to the food protein and cause mediator release, mast cell and basophil degranulation, or both. As a result, IgE-mediated food allergies typically cause swift and sometimes fatal signs and symptoms (Boyce et al. 2011). Eight foods – milk, eggs, wheat, fish, soy, peanuts, crustacean shellfish, and tree nuts – are responsible for 90% of IgE-mediated food allergies (Lee 2017). To identify IgE-mediated food allergies, a number of tests are performed, such as the immediate hypersensitivity skin test (IHST) and the food-specific IgE antibody test. In developed countries, 1%–3% of adults and 3%–8% of children suffer from food allergies linked to IgE.

Not only is it common, but it is also a serious and occasionally fatal medical illness that affects a person's eating and social habits and for which a proper diagnosis is necessary. Food allergies related to IgE are often triggered by fish, fruits, vegetables, wheat, eggs, peanuts, almonds, and sesame (Valenta et al. 2015). The mechanism of a IgE-mediated food allergy is shown in Figure 5.2.

5.3.2 Non-IgE-Mediated Food Allergy

Non-IgE-mediated food allergies are a group of disorders characterized by sub-acute or chronic inflammatory processes in the gastrointestinal system. While primarily impacting the gastrointestinal tract, these allergies can also extend to affect the skin or lungs. In contrast, IgE-mediated food allergies can cause multi-organ system anaphylaxis (Nowak-Węgrzyn et al. 2015). Non-IgE-mediated food allergies include food-protein-induced enterocolitis syndrome (FPIES), food-protein-induced allergic proctocolitis (FPIAP), and food-protein enteropathy (FPE). Figure 5.3 illustrates the non-IgE-mediated food allergy mechanism.

Eosinophilic inflammation of various gastrointestinal segments is a common feature of non-IgE-mediated disorders of the gastrointestinal tract. The onset of symptoms of eosinophilic esophagitis varies with age. It may show up at any age. While younger children may show signs of failure to thrive and frequent vomiting, adults and older children may experience symptoms like dysphagia, heartburn, and food impaction. Small infants who are frequently breastfed exclusively but are exposed to cow's milk or soy proteins in their mother's milk may develop eosinophilic proctocolitis, which manifests as blood and mucus in the stools (Bird, Lack, and Perry 2015). Table 5.2 provides a detailed overview of disorders associated with both IgE-mediated and non-IgE-mediated food allergies, offering a comprehensive description of these conditions.

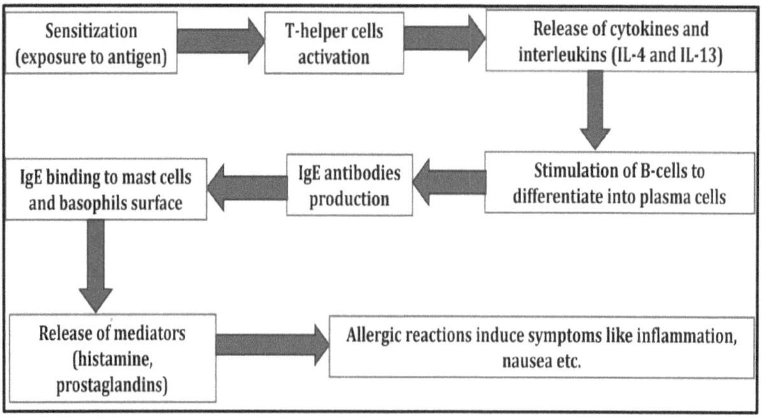

FIGURE 5.2 Mechanism of IgE-mediated food allergy.

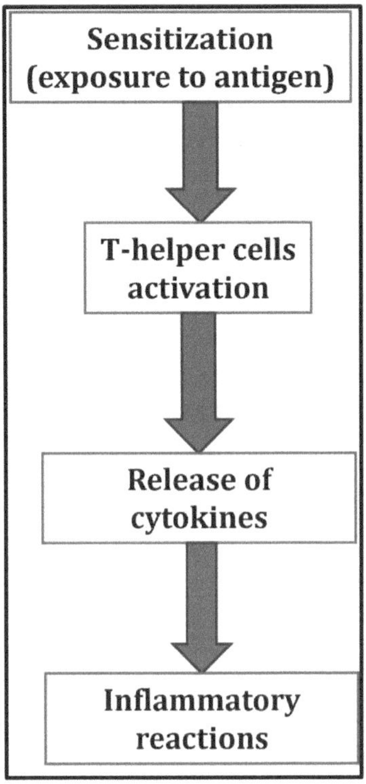

FIGURE 5.3 The non-IgE-mediated food allergy mechanism.

TABLE 5.2
IgE-Mediated and Non-IgE-Mediated Disorders

Disorders	Clinical Manifestations	Major Allergen	Target Organs
IgE-Mediated Food Reactions			
Pollen food syndrome	Mild swelling of lips, tongue, and throat	Raw fruits and vegetables	Gastrointestinal tract
Urticaria	Hives	Peanuts, tree nuts, cow milk	Skin
Anaphylaxis	Multisystemic allergy	Cow milk, eggs, shellfish	Multiple organ systems including the skin, respiratory, gastrointestinal
Non-IgE-Mediated Food Reactions			
Pulmonary hemosiderosis	Pneumonia, anemia	Cow milk	Lungs
Celiac disease	Weight loss, chronic diarrhea, abdominal pain	Gluten protein in wheat, barley	Small intestine

5.4 FOOD ALLERGENS

Environmental substances that are chemical, physical, or biological and that trigger allergic reactions in those who are sensitive to them are known as food allergens. An allergen's ability to elicit this response depends on a number of factors, including the kind, amount, duration of action, ports of entrance, and degree of sensitization of a particular organism. A variety of food allergens, including cereals, nuts, fruits, vegetables, and milk proteins, can cause allergies, and these allergies frequently cross-react with one another (Kazatsky and Wood 2016). While allergies to fish, peanuts, and tree nuts might last a lifetime, some people may eventually build tolerance against allergies to specific foods like milk, eggs, and wheat (Valenta et al. 2015). The eight main dietary allergens are known as the "Big 8" and comprise milk, tree nuts, peanuts, wheat, soybeans, fish, eggs, and crustacean shellfish (Popov-Raljić, Aleksić, and Janković 2022). The Food Allergen Labeling and Consumer Protection Act (FALCPA) covers these eight allergens (McGowan and Keet 2013). The main causes of food allergies are either glycoproteins with molecular weights between 10 and 60 kD that are stable against heat, acid, and proteases, or food allergens that are soluble in salt or water (Peters et al. 2021). The majority of food allergies are caused by allergens that remain stable against denaturation and degradation during food processing. There are more than 200 known proteinaceous allergens, and more than 100 distinct foods or food ingredients have the potential to cause negative reactions (López-Pedrouso et al. 2020). The primary allergens are those that contain gluten, such as wheat, rye, barley, oats, spelt, kamut, or their hybridized strains and their products; crustaceans; eggs; fish; peanuts; soybeans; milk and its derivatives (including lactose); and fish and eggs (Boye et al. 2012). These so-called major allergens are responsible for nearly 90% of allergy responses and intolerance. It's generally accepted that the gastrointestinal tract plays a role in sensitization to the classical food allergens, which include milk, eggs, peanuts, and fish, even though some types of food allergies are the consequence of prior sensitization to inhaled allergens like pollen (Gargano et al. 2021).

Food allergens can be divided into two main categories: heat-labile and heat-stable allergens. Most heat-stable food allergies are caused by glycoproteins with a molecular mass of 10–70 kDa. Conformational stability is provided by the molecular features of heat-stable food allergies, such as disulfide bonds, protein glycation, and lipid binding. These properties can enhance intact protein absorption in the gastrointestinal tract by protecting food allergens from heat, acid, and proteases (Francis et al. 2020). Conversely, allergens that are heat-labile exhibit high sensitivity to both heat and acid. Pollen food allergy syndrome is thought to develop as a result of exposure to homologous proteins in pollens, which sensitizes people to heat-labile allergens in plant-based foods. However, some common food allergies, such as those to cow's milk and hen's eggs, also include heat-labile allergens, which are denatured at high temperatures. Table 5.3 outlines the specific food allergens responsible for causing distinct types of food allergies.

TABLE 5.3
Common Food Allergies and Major Food Allergens Associated With Them

Food Allergy	Major Food Allergen
Cow's milk allergy	Caseins, whey protein alpha-lactalbumin, bovine serum albumin
Hen's egg allergy	Ovomucoid, ovalbumin, ovatransferrin
Legume allergy	Peanut
Tree nut allergy	Walnut, cashews
Fish and shellfish allergies	Parvalbumin, tropomysins
Grains allergy	Alpha-amylase inhibitor, lipid transfer proteins
Mammalian meat allergy	Serum albumins

5.5 GENETICS OF FOOD ALLERGY

The notion that food allergies are inherited is supported by a large body of research on heritability in families. The food allergy illness is linked to multiple genes. The profilaggrin protein is encoded by the gene filaggrin. Filaggrin monomers are produced when this protein is proteolytically broken down in the keratinocyte stratum corneum. It has been shown that mutations that cause filaggrin to lose its activity are closely associated with asthma, atopic dermatitis, and allergic rhinitis. Numerous studies indicate a direct link between filaggrin mutations and dietary allergies. However, some researchers have not found that people who have mutations that cause filaggrin to lose its activity are more likely to have food allergies. Filaggrin gene variations are linked to food sensitivity but not food allergy, according to a large population-based study on newborns with challenge-proven food allergies (Venkataraman et al. 2014).

Furthermore, genome-wide association studies have been carried out to detect common susceptibility loci among food allergy-affected populations. An extensive investigation into candidate genes has also been conducted in an effort to learn more about the genetics of food allergies. Genetic pathways investigated include HLA linkages, polymorphisms in various interleukins (e.g., IL10 and IL13), and polymorphisms in STAT6, SPINK5, and FOXP3 (Martino et al. 2017). Additionally, a recent study using RNA sequencing of peripheral blood samples from a cohort of children allergic to peanuts before, during, and after oral peanut challenges identified six potential key driver genes (LTB4R, PADI4, IL1R2, PPP1R3D, KLHL2, and ECHDC3) of interest during acute peanut-allergic reactions (Watson et al. 2017). A few monogenic disorders are associated with food allergies. These disorders present unique opportunities to explore specific pathways that may shed light on the pathophysiology of atopy and food allergies in the general population, thereby identifying potential therapeutic targets. The different monogenic disorders linked to food allergies are shown in Table 5.4, along with the clinical symptoms that correspond with each disorder.

TABLE 5.4
Monogenic Food Disorders with Inheritance Patterns and Clinical Presentations

Genetic Disorders	Inheritance Pattern	Genetic Mutation	Clinical Manifestations
Phosphoglucomutase 3 (PGM3) deficiency	Autosomal recessive	Mutation in PGM3 gene	Renal failure, skin and pulmonary infections, and food and environmental allergies
STAT3-hyper-IgE syndrome	Autosomal dominant	Loss of function of STAT3 gene	Pneumonia, candidiasis, increased IgE levels, eczematous dermatitis
Dedicatory of cytokinesis 8 (DOCK8) deficiency	Autosomal recessive	Loss of function of DOCK8 gene	Increased IgE levels, severe atopy associated with food allergy
Severe dermatitis, multiple allergies, and metabolic wasting (SAM) syndrome	Autosomal recessive	Loss of function in desmoglein-1 gene	Increased IgE levels and food allergy

5.6 FOOD ALLERGY-BASED DISORDERS AND THEIR ASSOCIATED SYMPTOMS

An unfavorable immune reaction to dietary antigens results in a wide range of disorders known as food allergies. This range includes acute, sometimes lethal, reactions as well as several long-term ailments primarily affecting the skin and gastrointestinal system. Over the past 20 years, there have been few changes to the diagnostic and treatment strategies, which still include the use of emergency drugs like epinephrine, clinical histories, physical exams, tests for particular IgE antibodies against suspected foods, elimination diets, and oral food challenges. We will discuss typical food allergy symptoms in the following section, along with associated management approaches.

5.6.1 IgE-Mediated Disorders and Their Associated Symptoms

When IgE triggers a food allergy, it activates basophils in the bloodstream and mast cells in tissues that are produced during the initial food sensitization. The rapid onset of symptoms is due to the IgE cross-linking on the surface of basophils and mast cells, which causes the preformed allergic mediators to be released quickly when the food antigen is subsequently encountered. Food allergies can cause a variety of symptoms that impact various organ systems, and they are mediated by antibodies to inflammatory proteins (IgE). A few of these symptoms include urticaria and angioedema, rhinoconjunctivitis, respiratory distress, gastrointestinal issues, and impaired heart function. Key features of each are summarized in Table 5.2.

5.6.1.1 Urticaria and Angioedema

Urticaria and angioedema develop as a result of food-specific IgE on cutaneous mast cells cross-linking. Angioedema is linked to mast cells in the deep dermis and subcutaneous tissues, while urticarial infections are linked to mast cells in the superficial dermis (Ensina et al. 2022). Urticaria, also called hives, is characterized by itchy, elevated erythematous wheals. Usually distinct or combining, these wheals frequently move throughout the body. It should be emphasized that although eating is frequently the cause of acute urticaria, contact urticaria – a condition in which hives are limited to the area in contact with the food – can also result from direct contact (Kayiran and Akdeniz 2019). Asymmetrical, nonpitting edema affecting non-gravity-influenced areas is the hallmark of angioedema. The face, limbs, or upper airway may be impacted (Memon and Tiwari 2023). The following foods are frequently linked to angioedema and acute urticaria: cow's milk, eggs, peanuts, tree nuts, soy, wheat, fish, and shellfish (Ensina et al. 2022).

5.6.1.2 Oropharyngeal Symptoms: Oral Allergy Syndrome

Oropharyngeal symptoms can manifest independently or in tandem with a more severe systemic reaction related to a food allergy. This may occur, for instance, in situations where the allergen is labile or unstable, such as in oral allergy syndrome (OAS) or in situations where the dose of the allergen ingested is minimal (Kashyap and Kashyap 2015). When raw fruits or vegetables are consumed, pollen-associated food allergy syndrome or OAS, an IgE-mediated food allergy, causes tingling and mild swelling of the lips, tongue, palate, and throat, as well as oropharyngeal pruritus. Patients usually tolerate the implicated fruits and vegetables in cooked form because the antigens involved are heat-labile (Sussman, Sussman, and Sussman 2010).

5.6.1.3 Respiratory Tract Symptoms: Rhinoconjunctivitis and Bronchospasm

IgE-mediated allergic reactions to particular foods can cause symptoms in the eyes and nose, including edema, periocular pruritus, sneezing, conjunctival erythema, congestion, lacrimation, nasal itching, and rhinorrhea. Additionally, there might be upper airway symptoms like hoarseness, stridor, and tightness in the throat, in addition to lower airway symptoms like dyspnea, wheezing, tachypnea, and cough (Al-Rabia 2016). The exception is when a person experiences isolated respiratory symptoms after consuming food, especially if they have occupational asthma or rhinitis, and work in the food preparation and packaging sector. When a particular food allergen, like wheat in the case of baker's asthma, is inhaled, these people may experience isolated upper and/or lower respiratory symptoms (James 2003).

5.6.1.4 Gastrointestinal Symptoms

Two categories of GI symptoms can be distinguished in individuals with IgE-mediated food allergies: upper GI symptoms, which typically manifest minutes to two hours after ingestion and include nausea, vomiting, and/or abdominal pain; and lower GI symptoms, which manifest two to six hours later and include cramping and diarrhea (Waserman, Bégin, and Watson 2018).

5.6.1.5 Food-Induced Anaphylaxis

Anaphylaxis is the most dangerous clinical consequence of an IgE-mediated food allergy. It is a fast-acting, potentially fatal, systemic allergic reaction. Anaphylaxis affects about 2% of the population, with food allergies accounting for up to 50% of cases. Many risk factors, such as adolescence or young adulthood, coexisting asthma, reactions to peanuts or tree nuts, and delayed epinephrine administration, increase the chance of a fatal reaction from food-induced anaphylaxis (Lopez, Yarrarapu, and Mendez 2023). Based on a small number of case studies, the foods most commonly associated with fatal food-induced anaphylaxis are peanuts, tree nuts, and cow's milk. In addition to food additives, spices, wheat, sesame, soy, and eggs, a number of other foods have also been connected to anaphylactic reactions (Cianferoni and Muraro 2012).

The multiorgan compromise observed in anaphylaxis is partly due to the rapid and extensive release of cytokines and allergic mediators into the bloodstream, including histamine, tryptase, chymase, prostaglandin D2, cysteinyl leukotrienes, interleukin 6, and tumor necrosis factor alpha. Anaphylaxis can manifest as a wide range of symptoms and indicators as a systemic reaction (Nguyen et al. 2021). The respiratory, gastrointestinal, and cardiovascular systems are the most frequently impacted organ systems, and they are vital to the potentially fatal and incapacitating aspects of anaphylaxis. Compared to adults, children are at higher risk of developing respiratory symptoms over cardiovascular symptoms. The most common symptoms in newborns and young children are cutaneous, gastrointestinal, and respiratory (Tarczoń et al. 2022).

5.6.1.5.1 Respiratory Symptoms in Anaphylaxis

Anaphylactic episodes can have respiratory symptoms and signs in as many as 70% of cases. Bronchospasm and laryngeal edema are the most frequent causes of death among children who suffer from food-induced anaphylaxis (Sharma, Bansil, and Uygungil 2015). Children who also have asthma or atopic dermatitis (AD) are more likely to experience severe respiratory symptoms related to food allergies. Furthermore, inhaling fumes from cooking eggs, fish, and shellfish is one way that airborne exposure to food allergens can occasionally cause bronchospasm (Foong, du Toit, and Fox 2017).

5.6.1.5.2 Gastrointestinal Symptoms in Anaphylaxis

Gastrointestinal (GI) signs and symptoms are present in up to 45% of anaphylactic episodes and are more common in cases of food-induced anaphylaxis than in cases with other causes. Gastrointestinal symptoms linked to food-induced anaphylaxis can cause dehydration and electrolyte imbalances, especially in newborns and young children (Keet 2011). The rapid and significant fluid shifts that occur in anaphylaxis due to increased vascular permeability and the movement of intravascular fluid into the extravascular space exacerbate the hypovolemia brought on by GI symptoms.

5.6.2 Non-IgE-Mediated Disorders and Their Associated Symptoms

Unlike IgE-mediated food allergies, which develop suddenly due to an immune mechanism, non-IgE-mediated food allergies are T cell-driven and exhibit more delayed and chronic clinical manifestations. This group of conditions includes celiac disease (CD), food-induced pulmonary hemosiderosis (Heiner syndrome), food-induced allergic proctocolitis (AP), and food-induced enterocolitis syndrome (FPIES).

5.6.2.1 Food Protein-Induced Allergic Proctocolitis

In breastfed infants between the ages of two and eight weeks, allergic proctocolitis (AP) often appears. Eosinophilic inflammation mediated by T cells that affects the colon and rectum is its defining feature. Even though they have an increased frequency of bowel movements, and mucus and blood in their stool, infants with AP seem to be in normal health otherwise (Barni et al. 2023).

5.6.2.2 Food Protein-Induced Enterocolitis Syndrome

Food protein-induced enterocolitis syndrome (FPIES) is a more severe form of T cell-mediated gastrointestinal food allergy that frequently manifests as external illness, including shock. This disorder is frequently ignored and incorrectly diagnosed in settings like emergency rooms and primary care offices (Leonard et al. 2018). It occurs more frequently in infants fed formulas containing soy or cow's milk than in babies who are breastfed. Some less common triggers are: grain (rice, oat, barley); vegetable (sweet potato, squash, string beans, peas); egg; and poultry (chicken, turkey). FPIES symptoms include frequent, severe vomiting and diarrhea. These can lead to poor growth, anemia, and hypoproteinemia in the long run, as well as short-term dehydration and fatigue that can hinder an individual's ability to thrive. Severe, potentially fatal symptoms associated with food poisoning episodes usually occur when an allergen is first encountered or when it is reintroduced into the diet (Nowak-Wegrzyn et al. 2003).

5.6.2.3 Food Protein-Induced Enteropathy: Celiac Disease

A T-cell-mediated inflammatory response to gluten, which is present in grains like wheat, rye, and barley, causes celiac disease (CD), an enteropathy. Villous atrophy, crypt hyperplasia, and mucosal inflammation are the histological characteristics of CD. The symptoms include iron deficiency anemia, delayed puberty, short stature, steatorrhea, chronic diarrhea, and distention. Malabsorption complications can also cause deficiencies in folate or vitamin B12 (Sharma et al. 2020).

5.6.2.4 Food-Induced Pulmonary Hemosiderosis (Heiner Syndrome)

It is a rare pulmonary condition primarily brought on by consuming protein from cow's milk, typically appearing in babies between the ages of 4 and 29 months. Unlike IgE antibodies found in acute food allergies, it is distinguished by the presence of milk-specific IgG antibodies. Possible complications include iron deficiency anemia, pulmonary hemosiderosis, recurrent pneumonias, and failure to thrive. Radiographic evidence of pulmonary infiltrates is commonly observed in patients with Heiner syndrome. The symptoms go away when the diet is altered to exclude milk protein (Ojuawo et al. 2019).

5.6.3 Mixed IgE and Non-IgE-Mediated Disorders and Their Associated Symptoms

The disorders and their associated symptoms resulting from mixed IgE and non-IgE-mediated food allergies are listed in the sections that follow.

5.6.3.1 Eosinophilic Esophagitis and Eosinophilic Gastroenteritis

Food allergies impacting the gastrointestinal tract that are both non-IgE (mediated by T cells) and mixed IgE include eosinophilic gastroenteritis (EG) and eosinophilic esophagitis (EoE). The age at which symptoms first appear varies among people with EoE. Infants and young children may exhibit feeding dysfunction and failure to thrive, while older children and adults often experience vomiting, abdominal pain, dysphagia, and food impaction. People who have EoE frequently have other allergic conditions like IgE-mediated food allergies and allergic rhinitis. Depending on which part of the digestive system is impacted, EG symptoms can include weight loss, malabsorption, nausea, diarrhea, and abdominal pain. Food-avoidance diets, in contrast to EoE, offer little to no benefit in EG (Zhang et al. 2021).

5.6.3.2 Atopic Dermatitis

The immunopathology of mixed IgE and non-IgE-mediated food allergies is also present in atopic dermatitis (AD). A persistent, itchy rash on the wrists, ankles, neck, antecubital and popliteal fossa, and other flexor surfaces is typically seen in children with AD. Although the rash is more common in these regions, it can also appear elsewhere on the body. Introducing food allergens may exacerbate the rash in approximately 35% of children with AD, resulting in more erythema and pruritus. If the reaction is IgE-mediated, it takes place over a few hours; if it is not, it takes place over several days. Nuts, wheat, soy, milk, and eggs are a few of the most often contaminated foods (Sharma, Bansil, and Uygungil 2015).

5.7 DIAGNOSTIC TESTS FOR FOOD ALLERGIES

For safe and reasonably restrictive dietary management, a precise diagnosis of food allergies is essential. The diagnosis of food allergies is based on clinical history in order to appropriately select allergens for testing and interpret the results of allergy tests, such as skin prick test (SPT), specific IgE (sIgE) to allergen extracts, and, more recently, specific IgE to allergen components and the basophil activation test (BAT) (Kattan and Sicherer 2015). Two incredibly sensitive methods, i.e., SPT and sIgE, are available to determine IgE sensitization to a particular food and assess the likelihood of spontaneous resolution (Gomes-Belo et al. 2018).

5.7.1 Diagnosis of IgE-Mediated Food Allergy

Tests are a supporting factor in the diagnosis of food allergies, but a thorough patient history is the primary source of confirmation. It takes specialized knowledge to conduct allergy tests and interpret the results in light of the patient's medical history, which is an essential skill. *In vivo* allergy testing may also be dangerous. Therefore, visiting a specialized clinic is usually advised for the most accurate diagnosis of food allergies. Since many allergy tests are sensitive but not specific, accuracy is crucial when conducting tests.

Therefore, conducting broad tests without discrimination for a range of allergens will result in a high number of false positive results. Any personal or family history of eczema, asthma, allergic rhinitis, or food allergies should be included in the patient's medical history. Notably, there is a strong correlation between atopic eczema and food allergies. Atopic eczema before the age of six months and severe eczema during the first year of life are linked to the development of allergies to eggs, milk, and peanuts. Lastly, it is crucial to consider the individual's reaction to medication and dietary limitations (Ansotegui et al. 2020).

5.7.2 Testing for IgE-Mediated Food Allergies

The skin prick test and the measurement of IgE antibody levels specific to a food are approved diagnostic procedures for IgE-mediated food allergies (Waserman, Bégin, and Watson 2018). The double-blind, placebo-controlled food challenge is the gold standard, but it is rarely used due to the risks, inconveniences, and costs associated with it. The use of alternative tests, some of which are available to the public, is specifically discouraged. These include hair analysis, applied kinesiology, serum-specific IgG4, lymphocyte stimulation, gastric juice analysis, endoscopic allergen provocation, facial thermography, and cytotoxicity assays (Asero et al. 2009). Atopy patch tests may be useful in evaluating T cell-mediated immune responses, but they are not helpful in identifying IgE-mediated food allergies (Ramírez-Marín et al. 2022). The issue with skin prick and specific IgE (sIgE) testing is their inability to distinguish between clinical allergy and sensitization. As such, it is crucial to tailor allergy testing based on information from the patient's medical history, ideally supplemented by a conclusive food challenge to improve testing (Ansotegui et al. 2020). Skin prick tests are typically the first choice, but sIgE tests are advised in cases of severe skin disease and dermatographism, pregnancy, high risk of severe anaphylaxis, and individuals unable to cease taking antihistamines and beta-blockers.

5.7.2.1 Food Allergen-Specific Serum IgE

The radioallergosorbent test (RAST) assay was initially conducted to measure serum IgE antibodies. However, more accurate techniques are now available for this purpose, such as tests using fluorescence enzyme labeling (Gleich & Yunginger 1981). As a result, assays like HYTEC-288 systems, ImmunoCap, and Immunlite are being used. In these assays, patient serum is incubated with a surface-fixed allergen. After washing, the fixed allergen will bind to any particular IgE antibody that is specific to the antigen and stay there. Labeled anti-IgE then binds to the bound IgE, allowing for its identification. Despite their overall reliability, these approaches necessitate distinct reference ranges for clinical decision points (Shoormasti et al. 2018). The level of a specific IgE correlates with the likelihood of a clinically significant reaction to the implicated food trigger, but it does not predict its likely severity.

If the patient's history does not support an IgE-mediated allergy, specific IgEs are not helpful because there is a reasonable risk of a false positive result (Waserman, Bégin, and Watson 2018). On the other hand, because specific IgE testing is so sensitive, it cannot rule out an allergy in situations where the patient's history is suggestive; in fact, it may miss 10%–25% of serious reactions, including anaphylaxis. For those with a strong history of IgE-mediated allergy and negative specific IgE test results, oral food challenges are advised (Calvani et al. 2019). Specific IgE tests are easier to use for non-specialists and do not carry the risk of an anaphylactic reaction when compared to skin prick testing. The development of more targeted IgE tests that target particular protein components is a promising development (Bignardi et al. 2019).

5.7.2.2 Skin Prick Tests

Apart from identifying the existence of circulating IgE against specific antigens, skin prick tests function as *in vivo* diagnostics with nearly instantaneous outcomes. After applying food extracts and controls, the skin is punctured using a lancet. After 20 minutes, the diameter of any induration is measured and compared with clinically significant wheal sizes for the corresponding allergens in tables. Skin prick tests can be carried out without the need for extracts or commercial tests, even for suspected allergies (Rasool et al. 2013). In fact, since commercial extracts are often less sensitive and degrade more quickly, it is preferable, for many foods, to use the actual food that the patient brought to the clinic, especially fruits, vegetables, and milk.

The skin prick test, like some IgE tests, cannot distinguish between sensitization and a true allergy. Still, a correlation does appear to exist between the size of the wheal and the likelihood of a genuine allergy (Peters, Gurrin, and Allen 2012). Wheal size varies according to the specific allergen and the population being studied, despite efforts to standardize significant "decision points" for each allergen.

5.7.2.3 Atopy Patch Testing

Atopy patch tests can detect both immediate and delayed urticarial reactions in addition to delayed hypersensitivity reactions. A food extract is directly applied to the patient's back during the testing procedure, and 48 to 72 hours later, the skin is examined for erythema, infiltration, and papules (Wollenberg and Vogel 2013). The atopy patch test is currently not thought to play a part in IgE-mediated food allergies. However, there is evidence to back up the theory that it might represent late-stage clinical reactions. It might be helpful in addition to some skin prick and IgE tests, potentially eliminating the need for a food challenge. Nonetheless, further efforts are required to standardize these tests.

5.7.2.4 Elimination Diets

Although they are not officially recognized for the diagnosis of IgE-mediated food allergies, elimination diets are important for managing allergies. Practically speaking, though, the outcome of an elimination diet provides valuable information even in the absence of a food challenge. Elimination diets can be used as a diagnostic tool for non-IgE-mediated allergies when specific tests are not available (Venter and Fleischer 2016).

5.7.2.5 Double-Blind Placebo Controlled Food Challenge

The double-blind, placebo-controlled food challenge is considered the most accurate test for food allergies. During the course of a day, this procedure entails progressively increasing the amount of a solution containing the suspected allergen or a placebo while closely monitoring for symptoms (Bock et al. 1988). The test is retaken a few days later with the solution that was not used the first time. Importantly, neither the patient nor the physician is aware of which of the two solutions is the placebo. Anaphylaxis is a serious risk associated with food challenges, so it's critical that these tests are only conducted in settings equipped to manage and treat anaphylactic reactions.

5.7.2.6 Scratch Test

An antigen is applied to a superficial epidermal abrasion that is about 2 mm in length as part of the scratch test. Although this test is simple to administer and takes little time, significant interindividual variation in test execution limits its sensitivity and specificity. When compared to a prick test, the main disadvantages of scratch testing are increased discomfort, poor reproducibility, and a significant rate of false positive reactions (Eigenmann, Oh, and Beyer 2011).

5.7.2.7 Specific IgE to Single Components

Component resolved diagnosis (CRD), a recent development in food allergy testing, measures particular IgE to individual allergens or components. Unlike conventional specific IgE testing, which uses extracts from the whole food, CRD concentrates on individual allergenic molecules that are present in that food. Purified allergen proteins are used in CRD; these can be produced by recombinantly expressing cDNA encoding an allergen or by purification from natural allergen sources. Certain proteins are specific to a particular species, whereas other proteins, referred to as cross-reactive components, are present in several sources of allergens (Calamelli et al. 2019).

Component testing enables the identification of particular IgE against "panallergens" or minor allergens, in addition to the major allergens found in protein structures. Component testing against major allergens and panallergens can help differentiate between species-specific and cross-reactive allergens, assess the likelihood of a severe reaction, inform choices about an oral food challenge (OFC), and offer some indication of possible allergy resolution (Callery et al. 2020).

5.7.2.8 Basophil Activation Test

The basophil activation test (BAT), a highly accurate test for food allergies, is slowly but surely moving from research labs to clinical settings. It is a functional test that evaluates whether basophils degranulate in response to the allergen. As such, it may be more similar to the clinical phenotype of patients than tests that only measure the amount of IgE specific to allergens. When a suspected food extract is exposed in a test tube to basophils that have been connected to acute allergic reactions, the BAT can be compared to an *in vitro* OFC (Briceno Noriega et al. 2021; Santos, Alpan, and Hoffmann 2021). In this test, flow cytometry is used to measure the expression of activation markers on the surface of basophils stimulated by allergens.

5.7.2.9 Oral Food Challenges

Controlled OFCs are the gold standard for diagnosing food allergies. Patients with unclear results from skin prick tests, specific IgE, or the basophil activation test (if available) should be offered an OFC in order to definitively confirm or rule out the diagnosis of food allergy (Santos and Lack 2016). The patient bears inherent risks when undergoing an OFC, so it's important to evaluate the possible severity of a reaction before moving forward. When determining whether to perform an OFC, medical history and suitable *in vivo* or *in vitro* tests are crucial in determining the likelihood of an allergy and identifying the food that may be involved (Santos and Brough 2017).

An OFC may occasionally be necessary to reach a conclusive diagnosis. Children with a history of adverse food reactions should undergo OFCs for a variety of reasons, such as diagnosing or ruling out a food allergy, determining a food's sensitivity threshold values, evaluating tolerance in cases of transient food allergies like cow's milk or hen's egg allergies, and for research purposes like clinical trials (Yu, Freeland, and Nadeau 2016).

Acute infections, unstable angina pectoris, chronic, unstable atopic disease, pregnancy, a recent history of anaphylaxis to the specific food being tested for, test results exceeding the 95% positive predictive value (PPV) cut-offs, and the use of medications that may worsen the reaction, mask its symptoms, or interfere with treatment are some of the conditions that may apply to OFCs (Santos and Brough 2017).

A double-blind placebo-controlled food challenge (DBPCFC) or an OFC may be used as the format. Even though DBPCFC is the most accurate test, its expense and time-consuming nature prevent it from being used more frequently in clinical practice (Sindher et al. 2018). Open food and single-blind challenges are more commonly used, particularly when examining late reactions or chronic symptoms or studying subjective symptoms. It is advised to use DBPCFC in research protocols, when examining subjective symptoms, delayed reactions, chronic symptoms, and situations where the clinical history and diagnostic test results are contradictory. In clinical practice, open challenges are usually the first line of action, especially when there is a high chance of a negative outcome and when IgE-mediated acute reactions with objective signs are suspected (Worm et al. 2015).

5.8 MANAGEMENT STRATEGIES

When devising a comprehensive plan for handling food allergies, a multimodal approach incorporates a number of essential components. The cornerstone is cross-contamination prevention; stringent guidelines require the use of distinct utensils, cutting boards, and cooking equipment for foods free of allergens. Simultaneously, nutritional supplementation with zinc, vitamin D, and essential fatty acids is crucial for strengthening the body's defenses against oxidative stress and inflammation, as well as promoting immunologic tolerance.

One common issue with allergens during food production, handling, and retailing is cross-contamination, which causes allergens to appear in trace amounts in the finished product (Madsen et al. 2012). Tobacco and excessive alcohol use should be avoided, and a balanced diet, regular exercise, and other healthy lifestyle choices all contribute to overall well-being (Farhud 2015). Since long term stress can affect adrenal function, stress management is essential. It's also advised to limit exposure to environmental pollutants and carcinogens. Genetic counseling may be helpful in determining potential risk factors if there is a family history of Adrenal Cortical Carcinoma (ACC) or related conditions. It's also critical to monitor and treat illnesses like Cushing's syndrome, which can impact the adrenal glands. Even though these actions promote overall health and may lower the risk of some cancers, it's crucial to speak with medical professionals for tailored guidance based on unique situations and medical background. Direct contact between foods that are allergenic and non-allergenic, or between foods with distinct allergen profiles, is typically the cause of ACC. Additionally, ACCs frequently arise from indirect contact caused by particular vectors introducing allergen particles into food, altering the allergen profiles of that food (Dzwolak 2022). These allergen transmission vectors include a variety of foods (raw materials, products, reworked products, and food components), people, media (hot water, cooling water, etc.), and any surfaces (tools, utensils, infrastructure, etc.) that come into contact with food (both inanimate and animate) (Dzwolak 2022).

The increasing number of individuals suffering from food allergies has led researchers to look into new strategies for effective management. Although avoiding food allergens is still the main treatment strategy for FA, newer approaches like oral or topical food immunotherapy are receiving more attention in the field of FA management (Mayorga et al. 2021). Notably, several studies have connected the emergence of allergic diseases with low serum vitamin D levels (Bener et al. 2014).

In order to prevent food allergies, a management strategy must include education and awareness. Employers have the ability to inform staff members about cross-contamination hazards and the warning signs and symptoms of food allergies. Additionally, they can offer training on identifying and reducing possible triggers for allergies at work. Customers can be educated about the possible dangers of cross-contamination and the precautions taken to avoid it through awareness campaigns. Employers who implement education and awareness campaigns can help make the workplace a safer place for individuals with food allergies (Oria et al. 2016). Initiatives for education and awareness are multilayered, addressing communities, schools, families, and individuals. Here, the focus should be on teaching food safety, allergen identification, label reading, and the possible advantages of dietary supplements. In addition to empowering individuals and families coping with food allergies, this comprehensive educational approach ensures that larger communities are aware of the signs, prevention, and nutritional factors (Choi, Ju, and Chang 2015).

Adrenaline auto-injectors (AAI) are vital tools in the treatment and mitigation of severe allergic reactions, especially those brought on by food allergies. They work by utilizing a straightforward but effective mechanism. The gadget has an easy-to-use spring-loaded needle that allows you to quickly inject a pre-calculated amount of adrenaline (epinephrine) into the muscle (Worm et al. 2023). Even those without medical training can operate this mechanism with ease thanks to its user-friendly design. When it comes to food allergies, AAI is an essential first-aid method. Its quick epinephrine delivery relieves severe anaphylactic symptoms like breathing difficulties, swelling, and constriction of the airways. Time-sensitive mechanisms are designed to address the critical need for prompt intervention during allergic reactions, as symptoms can escalate quickly. People can always carry AAI with them and respond promptly to allergen exposures thanks to its portability, which enhances accessibility, especially in circumstances where access to emergency medical care may be restricted.

As a useful diagnostic tool, the OFC strikes a balance between risk awareness and the possibility of gradually reintroducing allergenic foods into one's diet (Emons et al. 2020). When diagnosing an IgE-mediated food allergy, this method is considered the gold standard. In addition, people following elimination diets must strictly adhere to the guidance put forth by food authorities, such as the Food Allergen Labeling and Consumer Protection Act (FALCPA). The effectiveness of this strategy is further enhanced by knowledge of voluntary labeling phrases and available resources for assistance. The process of conducting an OFC, which is employed to determine a food allergy's final diagnosis, is guided by an evaluation of risks and benefits. The objective of the challenge, the food, and age are some of the variables that affect how OFCs are conducted. For example, a typical clinical history is used to diagnose food protein-induced enterocolitis syndrome (FPIES); however, an OFC may be required to confirm the diagnosis or assess the emergence of tolerance. Food allergies are becoming more common and present a serious clinical challenge. For a precise diagnosis and successful treatment, consulting with an allergist is essential. Currently, a comprehensive evaluation of the patient's medical history is required for diagnosis, in addition to diagnostic procedures such as serum-specific IgE testing, SPT, and, if appropriate, OFCs. The main course of treatment is to avoid foods that trigger allergic reactions and to quickly administer epinephrine when necessary. Improvements in the diagnosis, treatment, and prevention of food allergies will be fueled by new insights into the pathophysiology of this illness.

5.9 SUMMARY

Food allergies are major problems that affect people and families all over the world. Despite the fact that controlling food allergies involves awareness and flexibility, it is not an impossible task. There is a lot of promise for the future management of food allergies. By adopting a comprehensive strategy that includes cautious avoidance, anticipating reactions, dietary adjustments, and continuous education, people can confidently navigate the terrain of food allergies. Future directions presented by emerging medications and research could result in better treatment choices or possibly a cure. The development of more reliable diagnostic tools and techniques might result in an earlier and more accurate diagnosis of allergies, allowing for prompt intervention and individualized treatment plans. Furthermore, continued investigation into the processes behind allergic reactions may lead to the development of novel therapeutic strategies, such as less side-effects-prone targeted immunomodulation or even allergen-specific immunotherapy.

Although basic and translational research has made great strides, the causes of food allergies are still unknown. It is now widely accepted that a number of factors, such as genetics, early life events, lifestyle choices, and dietary components, play a significant role in the development of food allergies. More research is necessary to completely understand how the various systems interact. These factors are important in determining patterns of immunological and metabolic responses. We urgently need a strong framework linking these variables in order to establish the foundation for preventative measures. The only long-term solution to the food allergy epidemic is effective prevention.

REFERENCES

Al-Rabia, M. W. 2016. Food-induced immunoglobulin E-mediated allergic rhinitis. *J Microsc Ultrastruct* 4 (2):69–75.

Aljameel, G. M., S. A. AlSedairy, M. A. Binobead, M. H. Alhussain, M. A. Bin Obaid, L. N. Al-Harbi, G. M. Alshammari, and S. Arzoo. 2023. Awareness of food allergies and allergens on the menus of restaurants and cafes among saudi female university students. *Healthcare* 11 (9):1259.

Ansotegui, I. J., G. Melioli, G. W. Canonica, L. Caraballo, E. Villa, M. Ebisawa, G. Passalacqua, E. Savi, D. Ebo, R. M. Gómez, O. Luengo Sánchez, J. J. Oppenheimer, E. Jensen-Jarolim, D. A. Fischer, T. Haahtela, M. Antila, J. J. Bousquet, V. Cardona, W. C. Chiang, P. M. Demoly, L. M. DuBuske, M. Ferrer Puga, R. Gerth van Wijk, S. N. González Díaz, A. Gonzalez-Estrada, E. Jares, A. F. Kalpaklioğlu, L. Kase Tanno, M. L. Kowalski, D. K. Ledford, O. P. Monge Ortega, M. Morais Almeida, O. Pfaar, L. K. Poulsen, R. Pawankar, H. E. Renz, A. G. Romano, N. A. Rosário Filho, L. Rosenwasser, M. A. Sánchez Borges, E. Scala, G. E. Senna, J. C. Sisul, M. L. K. Tang, B. Y. Thong, R. Valenta, R. A. Wood, and T. Zuberbier. 2020. IgE allergy diagnostics and other relevant tests in allergy, a World Allergy Organization position paper. *World Allergy Organ J* 13 (2):100080.

Anvari, Sara, Jennifer Miller, Chih-Yin Yeh, and Carla M Davis. 2019. IgE-mediated food allergy. *Clin Rev Allergy Immunol* 57:244–260.

Asero, R., M. Fernandez-Rivas, A. C. Knulst, and C. A. Bruijnzeel-Koomen. 2009. Double-blind, placebo-controlled food challenge in adults in everyday clinical practice: A reappraisal of their limitations and real indications. *Curr Opin Allergy Clin Immunol* 9 (4):379–85.

Barni, S., F. Mori, M. Giovannini, L. Liotti, C. Mastrorilli, L. Pecoraro, F. Saretta, R. Castagnoli, S. Arasi, L. Caminiti, M. Gelsomino, A. Klain, M. M. Del Giudice, and E. Novembre. 2023. Allergic proctocolitis: Literature review and proposal of a diagnostic-therapeutic algorithm. *Life* 13 (9):1824.

Bener, A., M. S. Ehlayel, H. Z. Bener, and Q. Hamid. 2014. The impact of Vitamin D deficiency on asthma, allergic rhinitis and wheezing in children: An emerging public health problem. *J Family Community Med* 21 (3):154–161.

Bignardi, D., P. Comite, I. Mori, F. Ferrero, V. Fontana, M. Bruzzone, M. Mussap, and G. Ciprandi. 2019. Allergen-specific IgE: comparison between skin prick test and serum assay in real life. *Allergol Select* 3 (1):9–14.

Bird, J. Andrew, Gideon Lack, and Tamara T. Perry. 2015. Clinical management of food allergy. *J Allergy Clin Immunol Pract* 3 (1):1–11.

Bock, S. A., H. A. Sampson, F. M. Atkins, R. S. Zeiger, S. Lehrer, M. Sachs, R. K. Bush, and D. D. Metcalfe. 1988. Double-blind, placebo-controlled food challenge (DBPCFC) as an office procedure: A manual. *J Allergy Clin Immunol* 82 (6):986–97.

Boyce, Joshua A., Amal Assa'ad, A. Wesley Burks, Stacie M. Jones, Hugh A. Sampson, Robert A. Wood, Marshall Plaut, Susan F. Cooper, Matthew J. Fenton, and S. Hasan Arshad. 2011. Guidelines for the diagnosis and management of food allergy in the United States: Summary of the NIAID-sponsored expert panel report. *Journal of the American Academy of Dermatology* 64 (1):175–192.

Boye, J. I., A. O. Danquah, Cin Lam Thang, and X. Zhao. 2012. Food allergens. In *Food Biochem Food Process*, 798–819. John Wiley Sons.

Briceno Noriega, D., M. Teodorowicz, H. Savelkoul, and J. Ruinemans-Koerts. 2021. The basophil activation test for clinical management of food allergies: Recent advances and future directions. *J Asthma Allergy* 14:1335–1348.

Calamelli, E., L. Liotti, I. Beghetti, V. Piccinno, L. Serra, and P. Bottau. 2019. Component-resolved diagnosis in food allergies. *Medicina* 55 (8):498.

Callery, Emma L., Catherine Keymer, Nicholas A. Barnes, and Anthony W. Rowbottom. 2020. Component-resolved diagnostics in the clinical and laboratory investigation of allergy. *Ann Clin Biochem* 57 (1):26–35.

Calvani, M., A. Bianchi, C. Reginelli, M. Peresso, and A. Testa. 2019. Oral food challenge. *Medicina* 55 (10).

Campbell, D. E., R. J. Boyle, C. A. Thornton, and S. L. Prescott. 2015. Mechanisms of allergic disease–environmental and genetic determinants for the development of allergy. *Clin Exp Allergy* 45 (5):844–858.

Choi, Y., S. Ju, and H. Chang. 2015. Food allergy knowledge, perception of food allergy labeling, and level of dietary practice: A comparison between children with and without food allergy experience. *Nutr Res Pract* 9 (1):92–8.

Cianferoni, A., and A. Muraro. 2012. Food-induced anaphylaxis. *Immunol Allergy Clin North Am* 32 (1):165–95.

Conner, T. S., M. Mirosa, P. Bremer, and R. Peniamina. 2018. The role of personality in daily food allergy experiences. *Front Psychol* 9:29.

Dzwolak, Waldemar. 2022. Allergen cross-contact control plan supporting the implementation of food allergen management (FAM) in small food businesses. *Food Control* 135:108777.

Eigenmann, P. A., J. W. Oh, and K. Beyer. 2011. Diagnostic testing in the evaluation of food allergy. *Pediatr Clin North Am* 58 (2):351–62.

Elghoudi, Ahmed, and Hassib Narchi. 2022. Food allergy in children—the current status and the way forward. *World J Clin Pediatr* 11 (3):253.

Emons, Joyce, Marije van Gunst, Olivia Liem, Lonneke Landzaat, and Nicolette Arends. 2020. Improving successful introduction after a negative food challenge test: How to achieve the best result? *Nutrients* 12 (9):2731.

Ensina, L. F., T. K. Min, M. M. R. Félix, C. T. de Alcântara, and C. Costa. 2022. Acute urticaria and anaphylaxis: Differences and Similaritiessimilarities in clinical management. *Front Allergy* 3:840999.

Falcon, R. M. G., and S. E. C. Caoili. 2023. Immunologic, genetic, and ecological interplay of factors involved in allergic diseases. *Front Allergy* 4:1215616.

Farhud, D. D. 2015. Impact of lifestyle on health. *Iran J Public Health* 44 (11):1442–1444.

Fiocchi, A., D. Risso, A. DunnGalvin, S. N. González Díaz, L. Monaci, V. Fierro, and I. J. Ansotegui. 2021. Food labeling issues for severe food allergic patients. *World Allergy Organ J* 14 (10).

Flokstra-de Blok, B. M. J., A. E. J. Dubois, B. J. Vlieg-Boerstra, J. N. G. Oude Elberink, Hein Raat, A. DunnGalvin, J. O'B. Hourihane, and E. J. Duiverman. 2010. Health-related quality of life of food allergic patients: comparison with the general population and other diseases. *Allergy* 65 (2):238–244.

Foong, R. X., G. du Toit, and A. T. Fox. 2017. Asthma, food allergy, and how they relate to each other. *Front Pediatr* 5 (89):89.

Francis, Olivia L., Kathleen Y. Wang, Edwin H. Kim, and Timothy P. Moran. 2020. Common food allergens and cross-reactivity. *Journal of Food Allergy (USA)* 2 (1):17–21.

Gargano, D., R. Appanna, A. Santonicola, F. De Bartolomeis, C. Stellato, A. Cianferoni, V. Casolaro, and P. Iovino. 2021. Food Allergyallergy and intolerance: A narrative review on nutritional concerns. *Nutrients* 13 (5):1638.

Gleich, G. J., and J. W. Yunginger. 1981. Variations of the radioallergosorbent test for measurement of IgE antibody levels, allergens and blocking antibody activity. Paper read at New Trends in Allergy, 1981//, at Berlin, Heidelberg.

Gomes-Belo, J., F. Hannachi, K. Swan, and A. F. Santos. 2018. Advances in food allergy diagnosis. *Curr Pediatr Rev* 14 (3):139–149.

Gultekin, Fatih, and Duygu Kumbul Doguc. 2013. Allergic and immunologic reactions to food additives. *Clin Rev Allergy Immunol* 45:6–29.

Hong, Xiumei, and Xiaobin Wang. 2012. Early life precursors, epigenetics, and the development of food allergy. Paper read at Seminars in immunopathology.

James, J. M. 2003. Respiratory manifestations of food allergy. *Pediatrics* 111 (6 Pt 3):1625–1630.

Kashyap, R. R., and R. S. Kashyap. 2015. Oral allergy syndrome: An update for stomatologists. *J Allergy* 543928 (10):8.

Kattan, J. D., and S. H. Sicherer. 2015. Optimizing the diagnosis of food allergy. *Immunol Allergy Clin North Am* 35 (1):61–76.

Kayiran, M. A., and N. Akdeniz. 2019. Diagnosis and treatment of urticaria in primary care. *North Clin Istanb* 6 (1):93–99.

Kazatsky, Ashley M., and Robert A. Wood. 2016. Classification of food allergens and cross-reactivity. *Curr Allergy Asthm Reports* 16:1–7.

Keet, C. 2011. Recognition and management of food-induced anaphylaxis. *Pediatr Clin North Am* 58 (2):377–88.

Lee, S. 2017. IgE-mediated food allergies in children: Prevalence, triggers, and management. *Korean J Pediatr* 60 (4):99–105.

Leonard, S. A., V. Pecora, A. G. Fiocchi, and A. Nowak-Wegrzyn. 2018. Food protein-induced enterocolitis syndrome: A review of the new guidelines. *World Allergy Organ J* 11 (1):017–0182.

Leone, Ludovica, Alessandra Mazzocchi, Laura Maffeis, Valentina De Cosmi, and Carlo Agostoni. 2023. Nutritional management of food allergies: Prevention and treatment. *Front Allergy* 3:1083669.

López-Pedrouso, M., J. M. Lorenzo, M. Gagaoua, and D. Franco. 2020. Current trends in proteomic advances for food allergen analysis. *Biology* 9 (9):247.

Lopez, Claudia M., Siva Naga S Yarrarapu, and Magda D Mendez. 2023. Food allergies. In *StatPearls*. StatPearls Publishing.

Lyra, N. R., M. E. Motta, L. A. Rocha, D. Solé, D. M. Peixoto, J. A. Rizzo, L. Taborda-Barata, and E. S. Sarinho. 2013. Adverse reactions to foods and food allergy: Development and reproducibility of a questionnaire for clinical diagnosis. *J Allergy* 920679 (10):1.

Madsen, Charlotte Bernhard, S. Hattersley, K. J. Allen, K. Beyer, C.-H. Chan, S. B. Godefroy, R. Hodgson, E. N. C. Mills, A. Muñoz-Furlong, and S. Schnadt. 2012. Can we define a tolerable level of risk in food allergy? Report from a Euro Prevall/UK food standards agency workshop. *Clin Exp Allergy* 42 (1):30–37.

Malik, Talia F., and Kiran K. Panuganti. 2023. Lactose intolerance. In *StatPearls [Internet]*. StatPearls Publishing.

Martino, David J., Sarah Ashley, Jennifer Koplin, Justine Ellis, Richard Saffery, Shyamali C. Dharmage, Lyle Gurrin, Melanie C. Matheson, Birgit Kalb, and Ingo Marenholz. 2017.

Genomewide association study of peanut allergy reproduces association with amino acid polymorphisms in HLA-DRB 1. *Clin Exp Allergy* 47 (2):217–223.

Mayorga, C., F. Palomares, J. A. Cañas, N. Pérez-Sánchez, R. Núñez, M. J. Torres, and F. Gómez. 2021. New insights in therapy for food allergy. *Foods* 10 (5):1037.

Mazzocchi, A., Carina Venter, Kate Maslin, and Carlo Agostoni. 2017. The role of nutritional aspects in food allergy: prevention and management. *Nutrients* 9 (8):850.

McGowan, Emily C., and Corinne A. Keet. 2013. Prevalence of self-reported food allergy in the National Health and Nutrition Examination Survey (NHANES) 2007–2010. *J Allergy Clin Immunol* 132 (5):1216–1219. e5.

Memon, Ruba J., and Vivekanand Tiwari. 2023. Angioedema. In *StatPearls*. StatPearls Publishing.

Messina, Mark, and Carina Venter. 2020. Recent surveys on food allergy prevalence. *Nutrition Today* 55 (1):22–29.

Nguyen, S. M. T., C. P. Rupprecht, A. Haque, D. Pattanaik, J. Yusin, and G. Krishnaswamy. 2021. Mechanisms governing anaphylaxis: Inflammatory cells, mediators, endothelial gap junctions and beyond. *Int J Mol Sci* 22 (15):7785.

Nowak-Węgrzyn, Anna, Yitzhak Katz, Sam Soheil Mehr, and Sibylle Koletzko. 2015. Non–IgE-mediated gastrointestinal food allergy. *J Allergy Clin Immunol* 135 (5):1114–1124.

Nowak-Wegrzyn, A., H. A. Sampson, R. A. Wood, and S. H. Sicherer. 2003. Food protein-induced enterocolitis syndrome caused by solid food proteins. *Pediatrics* 111 (4 Pt 1):829–835.

Ojuawo, A. B., O. B. Ojuawo, A. O. Aladesanmi, M. O. Adio, M. B. Abdulkadir, and O. A. Mokuolu. 2019. Heiner Syndrome: An uncommon cause of failure to thrive. *Malawi Med J* 31 (3) September:227–229. doi: 10.4314/mmj.v31i3.11.

Oria, Maria P., Virginia A. Stallings, Engineering National Academies of Sciences, and Medicine. 2016a. Managing food allergies in retail, food service, schools, higher education, and travel settings. In *Finding a Path to Safety in Food Allergy: Assessment of the Global Burden, Causes, Prevention, Management, and Public Policy*. National Academies Press.

Oria, Maria P., Virginia A. Stallings, Engineering National Academies of Sciences, and Medicine. 2016b. Potential genetic and environmental determinants of food allergy risk and possible prevention strategies. In *Finding a Path to Safety in Food Allergy: Assessment of the Global Burden, Causes, Prevention, Management, and Public Policy*. National Academies Press.

Ortolani, Claudio, and Elide A. Pastorello. 2006. Food allergies and food intolerances. *Best Pract Res Clin Gastroenterol* 20 (3):467–483.

Ozdemir, O., E. Mete, F. Catal, and D. Ozol. 2009. Food intolerances and eosinophilic esophagitis in childhood. *Dig Dis Sci* 54 (1):8–14.

Peniamina, Rana L., Miranda Mirosa, Philip Bremer, and Tamlin S. Conner. 2016. The stress of food allergy issues in daily life. *Psychol Health* 31 (6):750–767.

Peters, Rachel L., Lyle C. Gurrin, and Katrina J. Allen. 2012. The predictive value of skin prick testing for challenge-proven food allergy: A systematic review. *Pediatr Allergy Immunol* 23 (4):347–352.

Peters, R. L., M. Krawiec, J. J. Koplin, and A. F. Santos. 2021. Update on food allergy. *Pediatr Allergy Immunol* 32 (4):647–657.

Popov-Raljić, Jovanka, Milica Aleksić, and Vesna Janković. 2022. Food allergens—food safety hazard. *Sci J Meat Technol* 63 (1):11–25.

Ramírez-Marín, H. A., A. M. Singh, P. Y. Ong, and J. I. Silverberg. 2022. Food allergy testing in atopic dermatitis. *JAAD Int* 9:50–56.

Rasool, R., I. A. Shera, S. Nissar, Z. A. Shah, N. Nayak, M. A. Siddiqi, and A. S. Sameer. 2013. Role of skin prick test in allergic disorders: A prospective study in kashmiri population in light of review. *Indian J Dermatol* 58 (1):12–7.

Rassart, Jessica, Koen Luyckx, Theo A. Klimstra, Philip Moons, Chris Groven, and Ilse Weets. 2014. Personality and illness adaptation in adults with type 1 diabetes: The intervening role of illness coping and perceptions. *J Clin Psychol Med Settings* 21:41–55.

Santos, A. F., O. Alpan, and H. J. Hoffmann. 2021. Basophil activation test: Mechanisms and considerations for use in clinical trials and clinical practice. *Allergy* 76 (8):2420–2432.

Santos, A. F., and H. A. Brough. 2017. Making the most of in vitro tests to diagnose food allergy. *J Allergy Clin Immunol Pract* 5 (2):237–248.

Santos, Alexandra F., and Gideon Lack. 2016. Basophil activation test: Food challenge in a test tube or specialist research tool? *Clin Translational Allergy* 6 (1):10.

Shamji, Mohamed H., Rudolf Valenta, Theodore Jardetzky, Valerie Verhasselt, Stephen R. Durham, Peter A. Würtzen, and R. J. Joost van Neerven. 2021. The role of allergen-specific IgE, IgG and IgA in allergic disease. *Allergy* 76 (12):3627–3641.

Sharma, Hemant P., Shweta Bansil, and Burcin Uygungil. 2015. Signs and symptoms of food allergy and food-induced anaphylaxis. *Pediatric Clinics* 62 (6):1377–1392.

Sharma, N., S. Bhatia, V. Chunduri, S. Kaur, S. Sharma, P. Kapoor, A. Kumari, and M. Garg. 2020. Pathogenesis of celiac disease and other gluten related disorders in wheat and strategies for mitigating them. *Front Nutr* 7:6.

Shoormasti, R. S., M. R. Fazlollahi, A. Kazemnejad, M. Movahedi, B. Tayebi, Z. Yazdanyar, Z. Azadi, Z. Pourpak, and M. Moin. 2018. Accuracy of immunoblotting assay for detection of specific IgE compared with ImmunoCAP in allergic patients. *Electron Physician* 10 (2):6327–6332.

Shulpekova, Y. O., V. M. Nechaev, I. R. Popova, T. A. Deeva, A. T. Kopylov, K. A. Malsagova, A. L. Kaysheva, and V. T. Ivashkin. 2021. Food intolerance: The role of histamine. *Nutrients* 13 (9):3207.

Sindher, S., A. J. Long, N. Purington, M. Chollet, S. Slatkin, S. Andorf, D. Tupa, D. Kumar, M. A. Woch, K. L. O'Laughlin, A. Assaad, J. Pongracic, J. M. Spergel, J. Tam, S. Tilles, J. Wang, S. J. Galli, K. C. Nadeau, and R. S. Chinthrajah. 2018. Analysis of a large standardized food challenge data set to determine predictors of positive outcome across multiple allergens. *Front Immunol* 9:2689.

Sussman, G., A. Sussman, and D. Sussman. 2010. Oral allergy syndrome. *Cmaj* 182 (11):1210–1211.

Tarczoń, I., E. Cichocka-Jarosz, A. Knapp, and P. Kwinta. 2022. The 2020 update on anaphylaxis in paediatric population. *Postepy Dermatol Alergol* 39 (1):13–19.

Topp, Marie, Jørgen Vestbo, and Erik Lykke Mortensen. 2016. Personality traits and mental symptoms are associated with impact of chronic obstructive pulmonary disease on patients' daily life. *COPD J Chron Obstruct Pulmon Dis* 13 (6):773–778.

Tordesillas, Leticia, M. Cecilia Berin, and Hugh A. Sampson. 2017. Immunology of food allergy. *Immunity* 47 (1):32–50.

Tuck, C. J., J. R. Biesiekierski, P. Schmid-Grendelmeier, and D. Pohl. 2019. Food intolerances. *Nutrients* 11 (7):1684.

Turnbull, J. L., H. N. Adams, and David A. Gorard. 2015. The diagnosis and management of food allergy and food intolerances. *Aliment Pharmacol Ther* 41 (1):3–25.

Valenta, R., H. Hochwallner, B. Linhart, and S. Pahr. 2015. Food allergies: The basics. *Gastroenterology* 148 (6):1120–1131.

Van De Ven, Monique O. M, and Rutger CME Engels. 2011. Quality of life of adolescents with asthma: The role of personality, coping strategies, and symptom reporting. *J Psychosom Res* 71 (3):166–173.

Venkataraman, Devasmitha, Nelís Soto-Ramírez, Ramesh J. Kurukulaaratchy, John W. Holloway, Wilfried Karmaus, Susan L. Ewart, S. Hasan Arshad, and Mich Erlewyn-Lajeunesse. 2014. Filaggrin loss-of-function mutations are associated with food allergy in childhood and adolescence. *J Allergy Clin Immunol* 134 (4):876–882. e4.

Venter, C., and D. M. Fleischer. 2016. Diets for diagnosis and management of food allergy: The role of the dietitian in eosinophilic esophagitis in adults and children. *Ann Allergy Asthma Immunol* 117 (5):468–471.

Waserman, Susan, Philippe Bégin, and Wade Watson. 2018. IgE-mediated food allergy. *Allergy Asthma Clin Immunol* 14 (2):55.

Watson, C. T., A. T. Cohain, R. S. Griffin, Y. Chun, A. Grishin, H. Hacyznska, G. E. Hoffman, N. D. Beckmann, H. Shah, and P. Dawson. 2017. Integrative transcriptomic analysis reveals key drivers of acute peanut allergic reactions. *Nat Commun* 8 (1):1943.

Wollenberg, A., and S. Vogel. 2013. Patch testing for noncontact dermatitis: the atopy patch test for food and inhalants. *Curr Allergy Asthma Rep* 13 (5):539–44.

Worm, M., A. T. Fox, M. Wickman, J. Ring, M. Ebisawa, G. Pouessel, and P. Smith. 2023. Adrenaline auto injectors pharmacokinetic/pharmacodynamic studies and potential consequences for clinical practice. *Clin Transl Allergy* 13 (12):12323.

Worm, M., I. Reese, B. Ballmer-Weber, K. Beyer, S. C. Bischoff, M. Classen, P. J. Fischer, T. Fuchs, I. Huttegger, U. Jappe, L. Klimek, B. Koletzko, L. Lange, U. Lepp, V. Mahler, B. Niggemann, U. Rabe, M. Raithel, J. Saloga, C. Schäfer, S. Schnadt, J. Schreiber, Z. Szépfalusi, R. Treudler, M. Wagenmann, B. Watzl, T. Werfel, T. Zuberbier, and J. Kleine-Tebbe. 2015. Guidelines on the management of IgE-mediated food allergies: S2k-Guidelines of the German Society for Allergology and Clinical Immunology (DGAKI) in collaboration with the German Medical Association of Allergologists (AeDA), the German Professional Association of Pediatricians (BVKJ), the German Allergy and Asthma Association (DAAB), German Dermatological Society (DDG), the German Society for Nutrition (DGE), the German Society for Gastroenterology, Digestive and Metabolic Diseases (DGVS), the German Society for Oto-Rhino-Laryngology, Head and Neck Surgery, the German Society for Pediatric and Adolescent Medicine (DGKJ), the German Society for Pediatric Allergology and Environmental Medicine (GPA), the German Society for Pneumology (DGP), the German Society for Pediatric Gastroenterology and Nutrition (GPGE), German Contact Allergy Group (DKG), the Austrian Society for Allergology and Immunology (Æ-GAI), German Professional Association of Nutritional Sciences (VDOE) and the Association of the Scientific Medical Societies Germany (AWMF). *Allergo J Int* 24:256–293.

Yu, W., D. M. H. Freeland, and K. C. Nadeau. 2016. Food allergy: Immune mechanisms, diagnosis and immunotherapy. *Nat Rev Immunol* 16 (12):751–765.

Zhang, Qiaozhi, Chi Zhang, Yong Zhang, Yinghua Liu, Jin Wang, Zhongshan Gao, Jinlyu Sun, Qianqian Li, Jiachen Sun, and Xin Cui. 2023. Early-life risk factors for food allergy: Dietary and environmental factors revisited. *Compr Rev Food Sci Food Saf* 22 (6):4355–4377.

Zhang, S., S. Sicherer, M. C. Berin, and A. Agyemang. 2021. Pathophysiology of non-IgE-mediated food allergy. *Immunotargets Ther* 10:431–446.

Zopf, Y., H. W. Baenkler, A. Silbermann, E. G. Hahn, and M. Raithel. 2009. The differential diagnosis of food intolerance. *Dtsch Arztebl Int* 106 (21):359–69.

6 Leaky Gut Syndrome and Sensitivities toward Specific Food

Anand Prakash Singh[1]

[1]Institute for Heart and Brain Health, University of Michigan, Ann Arbor, MI, U.S.A.

6.1 INTRODUCTION

The human gut is the body's largest surface area exposed to the external environment, covering around 200–300 square meters (Dommett et al. 2005; Farhadi et al. 2003). The intestinal barrier is a complex system consisting of the gut's microbial barrier, mucus, gastrointestinal functions (motility and secretion), epithelial barrier, immune components (innate and adaptive), and gut vascular and liver barriers (Tripathi et al. 2018). A healthy gut selectively allows the passage of nutrients, water, and bacterial substances regulated by various factors such as neural, cellular, immune, and hormonal mechanisms (Bischoff et al. 2014). The interplay of nutrients, gut microbiota, and the intestinal barrier is crucial for overall well-being and disease management (H. Fukui 2016a). Obesity and non-alcoholic fatty liver disease (NAFLD) are common conditions associated with higher risks of severe liver diseases and mortality (Ponziani et al. 2019). Intestinal permeability disruptions are observed in obesity and NAFLD, exacerbating the conditions for affected individuals (Carter et al. 2021). Here, we will explore the importance of the intestinal barrier and its permeability in maintaining health.

6.2 INTESTINAL BARRIER AND ITS COMPOSITION

The intestinal barrier comprises multiple tiers. Its exterior comprises the mucus layer, beneficial gut bacteria, and defensive proteins like antimicrobial proteins (AMPs) and secretory immunoglobulin A (sIgA). The middle layer comprises intestinal epithelial cells (IECs), while the innermost section comprises immune cells responsible for innate and adaptive immunity (M. Vancamelbeke and Vermeire 2017a).

DOI: 10.1201/9781003514244-8

6.2.1 The Mucus Layer

The composition of the mucus layer exhibits variation across the whole length of the intestine (M. Vancamelbeke and Vermeire 2017a). Within the small intestine, the mucus forms a slender and sporadic coating that facilitates the absorption of nutrients. In contrast, the large intestine possesses two separate layers of mucus (Vereecke, Beyaert, and van Loo 2011). The inner layer is free from microorganisms, while the outer layer acts as a physical barrier between the intestinal lumen and the epithelium. The mucus layer in the colon is thicker than in the small bowel, measuring over 800 microns in depth, similar to the height of a whole villus (varying from 500 to 1600 microns). The barrier qualities vary throughout the gastrointestinal tract, with the small intestine exhibiting an increase in pore size ranging from 4 to 5 Å at the tip of the villus to more than 20 Å at the base of the crypt (Quigley 2016). The primary role of these layers is to prevent the entry of microbiota into the apical side of the epithelium, hence offering protective properties (Pelaseyed et al. 2014). The mucus mainly comprises water, which makes up more than 98% of its content. It also contains mucins and glycoproteins, with more than 18 types of mucin-type glycoproteins identified (Ohland and MacNaughton 2010). The glycocalyx, located in the inner layer, and MUC2, found in the outer layer, play a vital role in disease prevention (Maaike Vancamelbeke and Vermeire 2017b). In addition, mucus contains secretory immunoglobulin A, antimicrobial compounds, peptides trefoil (trefoil factor family, TFF), cathelicidins, and ribonucleases (Maaike Vancamelbeke and Vermeire 2017b). These substances strengthen the physical barrier between bacteria and create a gradient from the epithelium to the lumen.

6.2.2 Intercellular Junctions of Intestinal Barrier

The intestinal barrier consists of five unique cell types in its epithelium: enterocytes, goblet cells, enteroendocrine cells, Paneth cells, and microfold cells. The assistance of a stem cell reservoir regularly renews the intestinal crypts' cells (Maaike Vancamelbeke and Vermeire 2017b).

The gut epithelium ensures that hydrophilic solutes cannot pass through, allowing chemical and nutrient flow through specialized transporters. The transportation takes place through two main pathways: the transcellular route, which involves the movement of substances through aqueous pores, active carrier-mediated absorption for nutrients, and endocytosis; and the paracellular route, which is specifically suited for the transportation of ions and hydrophilic molecules. The control of the paracellular route is governed by junctional complexes, which are comprised of several proteins such as tight junctions (TJs), adherens junctions, desmosomes, and gap junctions (Keita and Söderholm 2010). TJs have a crucial function in maintaining the integrity of the intestinal barrier (Otani and Furuse 2020; Paradis et al. 2021; Aijaz, Balda, and Matter 2006)

TJ proteins are created by arranging different proteins on the uppermost layer of epithelial cells, connecting neighboring cells. The TJs consist of transmembrane proteins like occludin, tricellulin, claudins, and junctional adhesion molecules (JAMs),

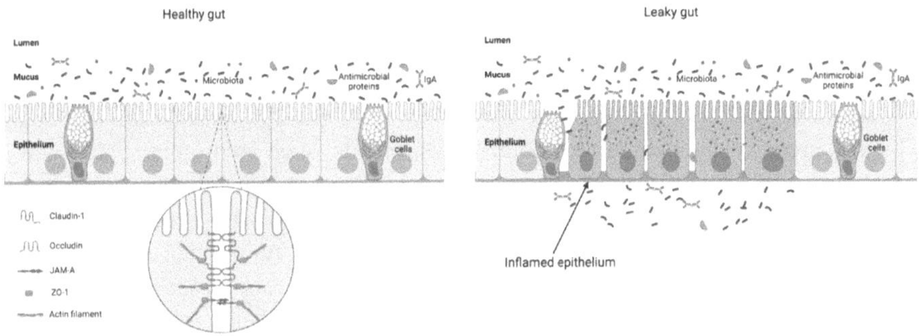

FIGURE 6.1 Components of gut barrier structure. JAM-A: Junctional Adhesion Molecule A, ZO-1: Zonula Occludens-1.

together with peripheral proteins called zonula occludens (ZO-1, ZO-2, ZO-3) which bind to actin filaments. They control the permeability of the channel for substances that travel between cells (paracellular transport), restricting the migration of ions and solutes (Otani and Furuse 2020; Paradis et al. 2021; Aijaz, Balda, and Matter 2006). Moreover, they maintain the polarity of epithelial cells by inhibiting the translocation of molecules (such as lipids and proteins) between the apical and basolateral membranes. This function plays a vital role in establishing the integrity of the intestinal barrier (Otani and Furuse 2020; Paradis et al. 2021; Aijaz, Balda, and Matter 2006).

TJ proteins can modify their function in response to the specific requirements of cells, hence regulating the closure, opening, and upkeep of paracellular transport in diverse physiological and pathological circumstances (Harhaj and Antonetti 2004). This control encompasses a variety of proteins and signaling pathways, including protein kinases C, A, and G (PKC, PKA, and PKG), phosphatase, myosin light-chain kinase (MLCK), mitogen-activated protein kinases (MAPK), and the phosphatidylinositol-3 kinase pathway (PI3K/Akt) (Rescigno 2011). Occludin phosphorylation is involved in both the initiation and termination of TJs (Rescigno 2011; Aijaz, Balda, and Matter 2006).

The versatility of TJs is crucial for the proper functioning of the gastrointestinal system, the regeneration of epithelial cells, and the development of bodily structures. Under standard physiological settings, the regulation of intestinal permeability ensures the maintenance of tolerance and equilibrium. Nevertheless, if there are any imperfections in the barrier, where TJs persistently and uncontrollably remain open, it can enable the entry of antigens from the diet or microorganisms. This disorder is generally known as "intestinal permeability" or "leaky gut" (Hollander and Kaunitz 2020) (Figure 6.1).

6.2.3 The Gut Microbiota

The gut microbiota, consisting of around 100 trillion microorganisms, plays crucial roles in protecting the intestinal tract and promoting the overall health of the host, including tasks related to protection, metabolism, and structure (Mileo, Nisticò, and

Miccadei 2019; Valdes et al. 2018). The constitution and activities of the intestinal barrier vary depending on age and health conditions, affecting its structure and functionality (Valdes et al. 2018).

The primary function of the gut microbiota is to obtain nutrients for intestinal cells and break down undigested dietary components, such as proteins and dietary fiber (Adak and Khan 2019). Its most remarkable feature is its ability to carry out anaerobic fermentation of intricate carbohydrates, producing short-chain fatty acids (SCFAs) like butyric, propionic, and acetic acids. The SCFAs provide energy for the cells that line the intestines and play a role in regulating several processes. Although SCFAs such as formate, valerate, and branched-chain fatty acids resulting from the breakdown of amino acids have limited effects on gut balance, they do promote overall gut well-being (Cunningham and Turner 2012; Miao et al. 2016).

SCFAs exert their influence via inhibiting histone deacetylases (HDACCs) and binding to G protein-coupled receptors (GPCRs), thereby promoting the proliferation of both hematopoietic and nonhematopoietic cells. They regulate the production of cytokines by neutrophils and macrophages, causing them to adopt an immunotolerogenic phenotype (Vinolo et al. 2011; Maslowski et al. 2009). In addition, SCFAs control the process of transcribing mucin genes in goblet cells (Rooks and Garrett 2016), which increases the production of the mucus layer and helps in the reconstruction of TJs. This is achieved by regulating the phosphorylation of AMP-activated protein kinase (AMPK) and myosin II regulatory light chain (MLC2), thereby strengthening the intestinal epithelial barrier (Cunningham and Turner 2012).

Furthermore, the gut microbiota has a crucial impact on the development of the host's immune system by producing metabolites, microorganism-associated molecular patterns (MAMPs), such as pathogen-associated molecular patterns (PAMPs), and antigens (Ruff, Greiling, and Kriegel 2020). Bacterial translocation, which refers to the movement of infections or their byproducts from the inside of the intestines to the lymph nodes in the mesentery, takes place. Lipopolysaccharide (LPS), which is present in the wall of Gram-negative bacteria, is a type of microbe-associated molecular pattern (MAMP) that is detected by receptors such as Toll-like receptors (TLRs) on cells of the innate immune system (Wiest, Lawson, and Geuking 2014; Kawai and Akira 2009). Activation of TLRs initiates the immune response by inducing the nuclear transcription factor-kappa B (NF-kB), which is responsible for the generation of inflammatory cytokines (Kawai and Akira 2009). TLRs, commonly found on intestinal and immune cell surfaces, can be stimulated by both pathogens and commensal bacteria. This interaction plays a crucial role in maintaining intestinal balance and harmony. Studies using experimental models have demonstrated that animals who do not have the Myd88 protein (which is triggered by the TLR pathway) or have had their microbiota reduced by antibiotics are more likely to develop severe colon inflammation when exposed to specific triggers (Rakoff-Nahoum et al. 2004). In addition, certain microbial species, such as Bacteroides fragilis, promote the growth of Foxp3+ regulatory T-cells and the production of IL-10 by expressing polysaccharide A (PSA). These cells and molecules are known for their ability to modulate the immune system and reduce inflammation in models of colitis (Mazmanian, Round, and Kasper 2008; Round and Mazmanian 2010).

6.2.4 Immune Cells

A complex network of intestinal epithelial cells, intraepithelial lymphocytes, and Paneth cells – particularly in the small intestine – contributes to intestinal immunity at the mucosal interface. In addition, the lamina propria contains both innate and adaptive immune cells (Mowat and Agace 2014). The gut-associated lymphoid tissue (GALT) appears as large clusters of lymphoid tissue called Peyer's patches in the small intestine and as solitary lymphoid follicles in both the upper intestine and colon (Mörbe et al. 2021). The gut contains cells, specifically Paneth cells, that play a crucial role in protecting the body against harmful microorganisms. These cells contribute considerably to the immune system by releasing cytokines and antimicrobial compounds, including defensins.

Paneth cells have a specific function in fighting against microorganisms and regulating the microbiota. They release lysozyme, defensins, and other proteins that help regulate the immune system (Elphick and Mahida 2005). IL-22 is particularly important among other cytokines due to its many functions in preserving the well-being of the intestines (Keir et al. 2020). IL-22 has a significant role in the regeneration of epithelial cells, maintaining mucus balance via activating mucin genes and goblet cells, and facilitating the production of antimicrobial peptides (Keir et al. 2020). The disruption of Paneth cells and IL-22 has been linked to the advancement of Crohn's disease, worsening inflammation in the intestines (Keir et al. 2020; Wehkamp and Stange 2020).

Peyer's patches play a crucial role in producing positive plasma cells for IgA in response to infections (Reboldi and Cyster 2016). They also coordinate communication between immune cells. Immunoglobulin A (IgA) is the primary immunoglobulin found in the blood and plays a vital role in protecting the body's mucosal surfaces (Papista, Berthelot, and Monteiro 2011). It helps control inflammation caused by bacteria in our bodies, preventing an imbalance in the microbial community known as dysbiosis (Okai et al. 2016). Experimental studies in mice have revealed that IgA coating plays a significant role in identifying potentially harmful bacteria in the gut microbiota, including those that can cause colitis (Palm et al. 2014).

Moreover, microfold (M) cells, a specialized epithelial cell type, have a crucial function in inflammatory reactions. They capture antigens that are present on the inner surface of the intestinal mucosa and transport them to Peyer's patches using transcytosis. Their role is crucial in regulating secretory IgA, and experimental evidence indicates their potential to reduce bacterial translocation (Ohno 2016).

6.3 DISEASES ASSOCIATED WITH LEAKY GUT

6.3.1 Inflammatory Bowel Disease

Inflammatory bowel disease (IBD) refers to a collection of chronic illnesses, specifically Crohn's disease (CD) and ulcerative colitis (UC), which primarily differ in the precise parts of the gastrointestinal tract they affect (Y.-Z. Zhang and Li 2014). Crohn's disease generally affects the ileum and cecum, whereas ulcerative colitis mainly presents in the colon and rectum. Although the exact underlying reason is unknown, heightened inflammation in the intestines is associated with a disturbance in TJs, which leads to an increase in the permeability of the intestines in affected individuals (Baumgart and Carding 2007; Edelblum and Turner 2009). Active ulcerative colitis is associated with reduced levels of specific proteins such as claudin-1, claudin-4, claudin-7, and occludin, while claudin-2 expression increases (Edelblum and Turner 2009). Conversely, CD is associated with decreased expression of claudin-3, claudin-5, and claudin-8 and increased expression of claudin-2 (Hering, Fromm, and Schulzke 2012). The changes in barrier function observed in individuals with IBD are intricately linked to inflammatory reactions and variations in TJs (B. Lee, Moon, and Kim 2018). A deeper understanding of the relationship between gut microbiota and IBD could lead to the development of new and advanced treatments and diagnostic methods.

6.3.2 Irritable Bowel Syndrome

Irritable bowel syndrome (IBS) is a condition of the digestive system characterized by persistent abdominal pain associated with changes in the frequency and consistency of stool movements (Baumgart and Carding 2007). IBS development has also been associated with intestinal permeability. IBS patients exhibit reduced amounts of ZO-1 and occludin proteins in their intestinal tissue. Furthermore, research examining the production of cytokines in peripheral blood mononuclear cells (PBMCs) has shown increased levels of TNF-α, IL-1β, and IL-6 in the bloodstream, particularly in individuals experiencing symptoms characterized by diarrhea (Hiroshi Fukui 2016b).

As stated, dysbiosis in the gut microbiota, microbial translocation, and reduced intestinal barrier function have been linked to several illnesses. Prominent among these conditions are obesity, chronic heart failure, Alzheimer's disease, cancer, diabetes, and autoimmune illnesses. Although the immune system often protects against infections and diseases, immunological disorders cause the body to react aggressively against its own cells, causing damage (Alessio Fasano 2011).

6.3.3 Obesity

Excessive fat storage is the hallmark of obesity, a medical disorder that can last for years. Obesity, as defined by the World Health Organization (WHO), occurs when an individual's body mass index (BMI) reaches or surpasses 30 kg/m^2 (WHO, 2018) There is a connection between this syndrome and increased intestinal permeability. Genetically obese mouse models exhibit an increase in intestinal permeability, plasma endotoxins, and proinflammatory cytokines such as interleukin one beta (IL-1β), interleukin-6 (IL-6), interferon-gamma (IFNγ), and tumor necrosis factor (TNF-α) as compared to non-genetically modified mice. In contrast, obesity caused by a high-fat diet (diet-induced obesity, DIO) is linked to changes in the gut microbiota, resulting in inflammation and heightened permeability of the intestines (J.Y. Lee et al. 2018). These alterations are connected to the reduced manifestation of genes related to TJs, such as ZO-1 and occludin. These findings indicate a connection between inflammation caused by obesity, modifications in the integrity of TJs, and changes in the gut microbiota (J.Y. Lee et al. 2018; Parekh, Balart, and Johnson 2015).

6.3.4 Non-Alcoholic Fatty Liver Disease

Non-alcoholic fatty liver disease (NAFLD) is the buildup of fats in liver cells that occurs without the use of alcohol. Non-alcoholic steatohepatitis (NASH) is a more severe form of NAFLD that includes inflammation and scarring of the liver (Oseini and Sanyal 2017). Alterations in the composition of gut microbiota in patients with NAFLD lead to elevated levels of lipopolysaccharides (LPS) in the bloodstream, which subsequently induce inflammation (Miele et al. 2009). The concurrent elevation of plasma LPS and proinflammatory cytokines leads to an augmentation of intestinal permeability (Miele et al. 2009). The increased permeability observed in patients with NAFLD is primarily a result of the movement of ZO-1 proteins in the crypt and the excessive development of bacteria in the small intestine (B. Lee, Moon, and Kim 2018). Both NASH and NAFLD are generally strongly linked to weakened TJ integrity.

6.3.5 Chronic Heart Failure

Chronic heart failure is defined by the heart's incapacity to efficiently circulate blood, resulting in an inadequate supply of oxygen and nutrients to the body (Morrissey, Czer, and Shah 2011). Patients suffering from chronic heart failure see a notable rise in the permeability of both the small and large intestines (Pasini et al. 2016). This increase is linked to the severity of the disease, venous blood congestion, and elevated levels of C-reactive protein (CRP) in the bloodstream. In addition, there are increased concentrations of endotoxins and inflammatory cytokines, such as TNF and STNF-R1. An examination of gut microbiota in these patients indicates a high presence of disease-causing bacteria, such as Campylobacter, Salmonella, and Candida, compared to persons in good condition. The results indicate that chronic heart failure patients with a weakened intestinal barrier experience bacterial translocation, cytokine production, and a deterioration in cardiac function (Pasini et al. 2016).

6.3.6 Celiac Disease

Celiac disease, an autoimmune condition caused by consuming gluten, leads to a reaction in individuals with celiac disease where gliadin, a type of protein found in cereals, passes through the protective layer of cells lining the intestines, causing inflammation and an increase in permeability (Thomas et al. 2006). This response could cause alterations in the structure of TJs. Elevated levels of the protein zonulin, which controls TJs and the passage of substances between cells, are linked to an increase in intestinal permeability. Gluten can stimulate zonulin release in those without health issues and those with celiac disease. However, the production of zonulin is notably more significant in patients with celiac disease. The increased permeability observed in individuals with celiac disease results in the rearrangement of the cytoskeleton via protein kinase C (PKC) and the loss of the integrity of TJs (A Fasano 2020).

6.3.7 Type 1 Diabetes Mellitus

Type 1 diabetes mellitus occurs when the body's immune system attacks the pancreatic β cells, leading to insufficient insulin generation (Kawasaki 2014). Research indicates a correlation between impaired intestinal barrier function and type 1 diabetes mellitus. People diagnosed with type 1 diabetes demonstrate compromised intestinal barrier function and heightened permeability as a result of increased zonulin synthesis, even before the development of the condition. Moreover, microbial translocation has been recognized as a causative element in the onset of type 1 diabetes. The results emphasize the crucial significance of intestinal permeability in the advancement of type 1 diabetes (Mu et al. 2017).

6.4 FOOD SENSITIVITIES IN LEAKY GUT

Diet exerts a significant influence on both the composition and function of the gut's microbial community. Dietary components have a crucial impact on the composition of the intestinal microbiota, which in turn has a substantial effect on the functioning of the gut and the integrity of its protective barrier (Brown et al. 2012). The ingestion of carbohydrates and fats can directly impact the permeability of the intestines, which may result in the development of illnesses such as irritable bowel syndrome (Binienda et al. 2020). Notably, there is a clear correlation between global variances in eating patterns and the incidence rates of intestinal disorders. Regions that have adopted Westernized cultural practices, which are typified by diets high in fats and refined carbohydrates, tend to have a higher prevalence of intestinal hyperpermeability (Steinert et al. 2017).

Carbohydrates and lipids can cause cellular inflammation, which disturbs the balance of the gut's microbial environment and affects the host's gastrointestinal metabolism and immune stability (Michielan and D'Incà 2015). Long-chain fatty acids, essential for constructing cells, have been shown to affect the intestines' permeability by altering TJs and histone acetylation. Extensive evidence indicates that fructose, glucose, and sucrose play a role in causing higher levels of intestinal permeability and malfunction in TJs (Binienda et al. 2020).

In contrast, certain carbohydrates, such as galactooligosaccharides, can promote the proliferation of advantageous bacteria (Binienda et al. 2020). At the same time, dietary fiber has been shown to benefit the permeability of the intestines (Michielan and D'Incà 2015). In conjunction with microbial fermentation, the enzymatic degradation of dietary fiber generates SCFAs, such as butyrate and propionate, which are essential for safeguarding the intestines (Binienda et al. 2020).

Moreover, there exists a correlation between food additives and leaky gut syndrome. Recent research indicates that additives disrupt TJs, promoting the entry of immunogenic antigens into the body. An optimally beneficial diet may entail avoiding fruits with high levels of fructose, dietary additives, and oils that contain ALA and GLA. On the other hand, patients suffering from leaky gut syndrome may find it advantageous to follow a diet rich in dietary fiber and low in sugar and fats (Binienda et al. 2020).

6.5 CLINICAL EVALUATION OF LEAKY GUT

There is currently no widely acknowledged standardized diagnostic technique for directly measuring intestinal permeability in patients, which prevents the formal identification of intestinal hyperpermeability as a medical diagnosis. Nevertheless, current clinical studies are examining different approaches to evaluate and identify possible indications of modified intestinal permeability.

Urine test: This evaluation entails consuming a solution containing sugars of different molecular sizes, including those that are usually not absorbed by the cells lining the intestines. The purpose of measuring the levels of these sugars in urine after consumption is to determine their capacity to cross the intestinal barrier, which indicates the permeability condition (Khoshbin et al. 2021).

Blood test: The analysis of blood samples aims to detect biomarkers indicating bacterial movement from the stomach. Identifying particular antibodies and endotoxins acts as a reliable indicator for possible disruptions in the integrity of the intestinal barrier (Hoshiko et al. 2021).

Tissue biopsy: This invasive procedure involves using an Ussing chamber to examine removed intestinal tissue. Analysts evaluate ion transport over the epithelial barrier by applying electrical currents to the tissue, which allows them to gain insights into the simultaneous dynamics of water transport (Camilleri 2019).

Confocal endomicroscopy: This utilizes advanced endoscopic techniques to provide a detailed vision of the microstructure of the intestinal mucosa. Administering a contrast agent intravenously during the examination assists in detecting possible abnormalities or discontinuities in the intestinal lining (Camilleri 2019).

Although these investigative methodologies show potential in evaluating intestinal permeability, their usefulness in clinical practice and accuracy in diagnosis require additional improvement and confirmation through significant study and controlled investigations in the medical field.

6.6 PHARMACOLOGICAL TREATMENTS

The search for pharmacological substances to strengthen the integrity of the intestinal barrier has prompted the examination of many chemicals, even though no legally sanctioned medications currently exist. These compounds demonstrate fascinating ways of working, suggesting a potentially important role in controlling intestinal permeability (Odenwald and Turner 2017).

Probiotics are living bacteria naturally present in fermented foods or intentionally added to them. When ingested in sufficient amounts, they provide health advantages to the individual (Guarner and Schaafsma 1998). The exact roles of certain probiotics, such as Lactobacillus, Bifidobacterium, and Saccharomyces, still need to be fully understood; however, evidence indicates their participation in repairing the intestinal epithelium (Krishna Rao and Samak 2013). An example is Lactobacillus plantarum MB452, which has shown the ability to increase the expression of TJ genes in Caco-2 cells (Anderson et al. 2010). Meanwhile, Lactobacillus rhamnosus GG (LGG) has a crucial function in enhancing the survival of epithelial cells by activating Akt and inhibiting p38, which are essential in responding to pro-apoptotic signaling pathways (Yan and Polk 2002).

Next-generation probiotics are a rapidly growing group of medicinal substances obtained from bacteria in the gut microbiota. These probiotics include hitherto untapped species such as Akkermansia, Bacteroides, and Faecalibacterium (O'Toole, Marchesi, and Hill 2017). Recent advanced research utilizing multi-omics methodologies has identified certain commensal microorganisms with metabolic and anti-inflammatory properties, demonstrating their promise as targeted supplements in treating chronic diseases. Although the production of Akkermansia is currently difficult, it is regarded as a highly promising option that is expected to be accessible in the near future (Almeida et al. 2020).

Prebiotics are compounds that can be fermented selectively, causing specific changes in the composition and activity of the microbiota in the gastrointestinal tract. This, in turn, leads to various health advantages (Gibson et al. 2010). Fructo-oligosaccharides (FOS) and galacto-oligosaccharides (GOS) are prebiotics that are extensively metabolized by gut bacteria, resulting in the production of SCFAs (Davani-Davari et al. 2019). SCFAs, namely butyrate, which are generated by bacteria such as Eubacterium rectale, Clostridium coccoides, and Roseburia, are vital in nourishing colonocytes (Looijer–Van Langen and Dieleman 2008). They are linked to increased mucus formation on the surface of the intestines, maintaining a balanced immune system, and boosting T-reg cells in mice (Furusawa et al. 2013). In addition, inulin, a specific type of prebiotic, demonstrates scavenging characteristics by reducing the harm caused by LPS exposure to human colonic cells (Guarino et al. 2017).

Glutamine, an amino acid that is not necessary for survival, plays a crucial function in maintaining the integrity of the intestinal barrier. It does this by activating specific protein kinases, which help with the survival and growth of intestinal cells, especially during stress (Rhoads et al. 1997). Research suggests that it is beneficial in protecting against acetaldehyde, a cancer-causing substance formed by alcohol, by blocking the breakdown of essential proteins in the cells of the human colon (Basuroy et al. 2005).

Vitamin D has demonstrated potential in improving the production of TJ proteins such as claudin and occludin, controlling the body's response to microbial infections on the surface of the intestines, and mitigating imbalances in gut bacteria (Y.-g. Zhang, Wu, and Sun 2013). Supplementation has been suggested as an additional treatment for people with IBD, spondyloarthritis (SpA), and type 1 diabetes mellitus (T1DM) (Battistini et al. 2021; Guillot, Prati, and Wendling 2014; Treiber et al. 2015).

Metformin, a frequently prescribed medicine for the management of type 2 diabetes, has recently been identified as a possible treatment for intestinal permeability. Metformin has shown in mice models that it can reduce cytokine expression and increase intestinal mucus production, thereby mitigating hepatic steatosis and adipocyte inflammation caused by a high-fat diet (Ahmadi et al. 2020).

Obeticholic acid (OCA) is a semi-synthetic compound that mimics hydrophobic bile acids and activates the farnesoid X receptor (FXR), which controls bile acid metabolism (Y. Zhang et al. 2017). OCA has been approved for the treatment of primary biliary cholangitis (Y. Zhang et al. 2017). A recent study demonstrates the potential of OCA in restoring the integrity of the intestinal barrier and limiting the movement of bacteria in experimental animals (Sorribas et al. 2019).

Divertin, a new compound, specifically interacts with the peri-junctional actomyosin ring (PAMR) found in intestinal cells. The process works by inhibiting the function of mitochondrial hydroxylase (MCLK1) on PAMR, decreasing actomyosin ring contraction, and tightening intercellular connections. Its positive impacts make it a promising candidate for treating autoimmune illnesses and IBDs (Blander 2019).

6.7 NON-PHARMACOLOGICAL TREATMENTS

The ingestion of decaffeinated coffee has shown the capacity to reduce the expression of Toll-like receptor 4 (TLR4) and LPS-binding protein (LBP) in mice fed a high-fat diet. One suggested mechanism is that coffee may alleviate intestinal permeability (Brandt et al. 2019).

Spices with high heat levels, including cayenne pepper, chili pepper, and paprika, have been found to enhance the passage of ions through the intestines and reduce the expression of TJs. Conversely, ginger has shown promise in reducing the severity of intestinal damage in a mouse model of ulcerative colitis caused by dextran sulfate sodium (DSS). This is accomplished by regulating inflammatory cytokine routes and changing the gut flora's composition (Guo et al. 2021).

Flavonoids, natural phenolic chemicals in plants, demonstrate various beneficial effects on human health (Ullah et al. 2020). Genistein, a prominent flavonoid found in soy, effectively inhibits protein kinases and has demonstrated potential in experimental models by reducing TJ damage caused by acetaldehyde (Suzuki and Hara 2011).

Finally, the transformation of tryptophan by bacteria in the intestines into tryptamine and indole metabolites such as IAA and indole aldehyde serves as a safeguard for barrier function (Taleb 2019). Lactobacilli create indole acetic acid during infections, stimulating the release of IL-22, a cytokine that can improve mucosal immunity in mice and repair epithelial integrity (Keir et al. 2020). The IL-22 pathway exhibits metabolic advantages through enhancements in insulin sensitivity, mitigation of endotoxemia, and prevention of metabolic syndrome (Wang et al. 2014).

These therapies, which can be pharmacological or non-pharmaceutical, emphasize the different approaches being investigated to strengthen and repair the integrity of the intestinal barrier. These interventions show potential for treating numerous gastrointestinal and systemic disorders.

REFERENCES

Adak, Atanu, and Mojibur R. Khan. 2019. "An insight into gut microbiota and its functionalities." *Cellular and Molecular Life Sciences* 76: 473–493.

Ahmadi, Shokouh, Atefeh Razazan, Ravinder Nagpal, Shalini Jain, Bo Wang, Sidharth P. Mishra, Shaohua Wang, Jamie Justice, Jingzhong Ding, Donald A. McClain, Stephen B. Kritchevsky, Dalane Kitzman, and Hariom Yadav. 2020. "Metformin reduces aging-related leaky gut and improves cognitive function by beneficially modulating gut microbiome/goblet cell/mucin axis." *The Journals of Gerontology: Series A* 75 (7): e9–e21. https://doi.org/10.1093/gerona/glaa056.

Aijaz, Saima, Maria S. Balda, and Karl Matter. 2006. "Tight junctions: Molecular architecture and function." In Kwang W. Jeon. (Ed). *International Review of Cytology*, 248: 261–298. Academic Press, Elsevier Inc.

Almeida, Diana, Daniela Machado, José Carlos Andrade, Sónia Mendo, Ana Maria Gomes, and Ana Cristina Freitas. 2020. "Evolving trends in next-generation probiotics: a 5W1H perspective." *Critical Reviews in Food Science and Nutrition* 60 (11): 1783–1796. https://doi.org/10.1080/10408398.2019.1599812.

Anderson, Rachel C., Adrian L. Cookson, Warren C. McNabb, Zaneta Park, Mark J. McCann, William J. Kelly, and Nicole C. Roy. 2010. "Lactobacillus plantarum MB452 enhances the function of the intestinal barrier by increasing the expression levels of genes involved in tight junction formation." *BMC Microbiology* 10 (1): 316. https://doi.org/10.1186/1471-2180-10-316.

Basuroy, S., P. Sheth, C. M. Mansbach, and R. K. Rao. 2005. "Acetaldehyde disrupts tight junctions and adherens junctions in human colonic mucosa: Protection by EGF and l-glutamine." *American Journal of Physiology-Gastrointestinal and Liver Physiology* 289 (2): G367–G375. https://doi.org/10.1152/ajpgi.00464.2004.

Battistini, Carolina, Rafael Ballan, Marcos Edgar Herkenhoff, Susana Marta Isay Saad, and Jun Sun. 2021. "Vitamin D modulates intestinal microbiota in inflammatory bowel diseases." *International Journal of Molecular Sciences* 22 (1): 362. https://www.mdpi.com/1422-0067/22/1/362.

Baumgart, Daniel C., and Simon R. Carding. 2007. "Inflammatory bowel disease: cause and immunobiology." *The Lancet* 369 (9573): 1627–1640.

Binienda, Agata, Agata Twardowska, Adam Makaro, and Maciej Salaga. 2020. "Dietary carbohydrates and lipids in the pathogenesis of leaky gut syndrome: An overview." *International Journal of Molecular Sciences* 21 (21): 8368. https://www.mdpi.com/1422-0067/21/21/8368.

Bischoff, S. C., G. Barbara, W. Buurman, T. Ockhuizen, J. D. Schulzke, M. Serino, H. Tilg, A. Watson, and J. M. Wells. 2014. "Intestinal permeability–a new target for disease prevention and therapy." *BMC Gastroenterol* 14: 189. https://doi.org/10.1186/s12876-014-0189-7. https://www.ncbi.nlm.nih.gov/pubmed/25407511.

Blander, J. Magarian. 2019. "A new approach for inflammatory bowel disease therapy." *Nature Medicine* 25 (4): 545–546. https://doi.org/10.1038/s41591-019-0416-4.

Brandt, Annette, Anika Nier, Cheng Jun Jin, Anja Baumann, Finn Jung, Vicent Ribas, Carmen García-Ruiz, Jose C. Fernández-Checa, and Ina Bergheim. 2019. "Consumption of decaffeinated coffee protects against the development of early non-alcoholic steatohepatitis: Role of intestinal barrier function." *Redox Biology* 21: 101092. https://doi.org/https://doi.org/10.1016/j.redox.2018.101092. https://www.sciencedirect.com/science/article/pii/S2213231718310516.

Brown, Kirsty, Daniella DeCoffe, Erin Molcan, and Deanna L. Gibson. 2012. "Diet-Induced dysbiosis of the intestinal microbiota and the effects on immunity and disease." *Nutrients* 4 (8): 1095–1119. https://www.mdpi.com/2072-6643/4/8/1095.

Camilleri, M. 2019. "Leaky gut: Mechanisms, measurement and clinical implications in humans." *Gut* 68 (8): 1516–1526. https://doi.org/10.1136/gutjnl-2019-318427. https://www.ncbi.nlm.nih.gov/pubmed/31076401.

Carter, J. K., D. Bhattacharya, J. N. Borgerding, M. I. Fiel, J. J. Faith, and S. L. Friedman. 2021. "Modeling dysbiosis of human NASH in mice: Loss of gut microbiome diversity and overgrowth of Erysipelotrichales." *PLoS One* 16 (1): e0244763. https://doi.org/10.1371/journal.pone.0244763. https://www.ncbi.nlm.nih.gov/pubmed/33395434.

Cunningham, K. E., and J. R. Turner. 2012. "Myosin light chain kinase: Pulling the strings of epithelial tight junction function." Annals of the New York Academy of Sciences 1258 (1): 34–42. https://doi.org/10.1111/j.1749-6632.2012.06526.x. https://www.ncbi.nlm.nih.gov/pubmed/22731713.

Davani-Davari, Dorna, Manica Negahdaripour, Iman Karimzadeh, Mostafa Seifan, Milad Mohkam, Seyed Jalil Masoumi, Aydin Berenjian, and Younes Ghasemi. 2019. "Prebiotics: Definition, types, sources, mechanisms, and clinical applications." *Foods* 8 (3): 92. https://www.mdpi.com/2304-8158/8/3/92.

Dommett, R., M. Zilbauer, J. T. George, and M. Bajaj-Elliott. 2005. "Innate immune defence in the human gastrointestinal tract." *Molecular Immunology* 42 (8): 903–12. https://doi.org/10.1016/j.molimm.2004.12.004. https://www.ncbi.nlm.nih.gov/pubmed/15829280.

Edelblum, Karen L., and Jerrold R. Turner. 2009. "The tight junction in inflammatory disease: communication breakdown." *Current Opinion in Pharmacology* 9 (6): 715–720.

Elphick, D. A., and Y. R. Mahida. 2005. "Paneth cells: Their role in innate immunity and inflammatory disease." *Gut* 54 (12): 1802–1809.

Farhadi, A., A. Banan, J. Fields, and A. Keshavarzian. 2003. "Intestinal barrier: An interface between health and disease." *Journal of Gastroenterology and Hepatology* 18 (5): 479–97. https://doi.org/10.1046/j.1440-1746.2003.03032.x. https://www.ncbi.nlm.nih.gov/pubmed/12702039.

Fasano, Alessio. 2011. "Zonulin and its regulation of intestinal barrier function: the biological door to inflammation, autoimmunity, and cancer." *Physiological Reviews.*

Fasano, A. 2020. "All disease begins in the (leaky) gut: Role of zonulin-mediated gut permeability in the pathogenesis of some chronic inflammatory diseases [version 1; peer review: 3 approved]." *F1000Research* 9 (69). https://doi.org/10.12688/f1000research.20510.1. https://f1000research.com/articles/9-69/v1.

Fukui, H. 2016a. "Increased intestinal permeability and decreased barrier function: Does it really influence the risk of inflammation?" *Inflammatory Intestinal Diseases* 1 (3): 135–145. https://doi.org/10.1159/000447252. https://www.ncbi.nlm.nih.gov/pubmed/29922669.

Fukui, Hiroshi. 2016b. "Increased intestinal permeability and decreased barrier function: Does it really influence the risk of inflammation?" *Inflammatory Intestinal Diseases* 1 (3): 135–145.

Furusawa, Yukihiro, Yuuki Obata, Shinji Fukuda, Takaho A. Endo, Gaku Nakato, Daisuke Takahashi, Yumiko Nakanishi, Chikako Uetake, Keiko Kato, Tamotsu Kato, Masumi Takahashi, Noriko N. Fukuda, Shinnosuke Murakami, Eiji Miyauchi, Shingo Hino, Koji Atarashi, Satoshi Onawa, Yumiko Fujimura, Trevor Lockett, Julie M. Clarke, David L. Topping, Masaru Tomita, Shohei Hori, Osamu Ohara, Tatsuya Morita, Haruhiko Koseki, Jun Kikuchi, Kenya Honda, Koji Hase, and Hiroshi Ohno. 2013. "Commensal microbe-derived butyrate induces the differentiation of colonic regulatory T cells." *Nature* 504 (7480): 446–450. https://doi.org/10.1038/nature12721.

Gibson, Glenn R., Karen P. Scott, R. A. Rastall, Kieran Michael Tuohy, Arland T. Hotchkiss, Alix Dubert-Ferrandon, Mélanie G. Gareau, Eileen F. Murphy, Delphine Saulnier, Gunnar Loh, Sandra Macfarlane, Nathalie M. Delzenne, Yehuda Ringel, Gunhild

Kozianowski, Robin S. Dickmann, Irene Lenoir-Wijnkoop, Carey Walker, and Randal K. Buddington. 2010. "Dietary prebiotics: Current status and new definition." *Food Science & Technology Bulletin: Functional Foods* 7: 1–19.

Guarino, Michele Pier Luca, Annamaria Altomare, Simone Barera, Vittoria Locato, Silvia Cocca, Cinzia Franchin, Giorgio Arrigoni, Candida Vannini, Sarah Grossi, Paola Campomenosi, Valentina Pasqualetti, Marcella Bracale, Rossana Alloni, Laura De Gara, and Michele Cicala. 2017. "Effect of inulin on proteome changes induced by pathogenic lipopolysaccharide in human colon." *PLoS One* 12 (1): e0169481. https://doi.org/10.1371/journal.pone.0169481.

Guarner, F., and G. J. Schaafsma. 1998. "Probiotics." *International Journal of Food Microbiology* 39 (3): 237–238. https://doi.org/https://doi.org/10.1016/S0168-1605(97)00136-0. https://www.sciencedirect.com/science/article/pii/S0168160597001360.

Guillot, Xavier, Clément Prati, and Daniel Wendling. 2014. "Vitamin D and spondyloarthritis." *Expert Review of Clinical Immunology* 10 (12): 1581–1589. https://doi.org/10.1586/1744666X.2014.972944.

Guo, S., W. Geng, S. Chen, L. Wang, X. Rong, S. Wang, T. Wang, L. Xiong, J. Huang, X. Pang, and Y. Lu. 2021. "Ginger alleviates DSS-induced ulcerative colitis severity by improving the diversity and function of gut microbiota." *Frontiers in Pharmacology* 12: 632569. https://doi.org/10.3389/fphar.2021.632569. https://www.ncbi.nlm.nih.gov/pubmed/33692694.

Harhaj, N. S., and D. A. Antonetti. 2004. "Regulation of tight junctions and loss of barrier function in pathophysiology." *International Journal of Biochemistry & Cell Biology* 36 (7): 1206–1237. https://doi.org/10.1016/j.biocel.2003.08.007. https://www.ncbi.nlm.nih.gov/pubmed/15109567.

Hering, Nina A., Michael Fromm, and Jörg-Dieter Schulzke. 2012. "Determinants of colonic barrier function in inflammatory bowel disease and potential therapeutics." *The Journal of Physiology* 590 (5): 1035–1044.

Hollander, D., and J. D. Kaunitz. 2020. "The 'leaky gut': Tight junctions but loose associations?" *Digestive Diseases and Sciences* 65 (5): 1277–1287. https://doi.org/10.1007/s10620-019-05777-2. https://www.ncbi.nlm.nih.gov/pubmed/31471860.

Hoshiko, H., E. J. M. Feskens, E. Oosterink, R. M. C. Ariens, J. J. Mes, and N. J. W. de Wit. 2021. "Identification of leaky gut-related markers as indicators of metabolic health in Dutch adults: The nutrition questionnaires plus (NQplus) study." *PLoS One* 16 (6): e0252936. https://doi.org/10.1371/journal.pone.0252936. https://www.ncbi.nlm.nih.gov/pubmed/34086828.

Kawai, T., and S. Akira. 2009. "The roles of TLRs, RLRs and NLRs in pathogen recognition." *International Immunology* 21 (4): 317–337. https://doi.org/10.1093/intimm/dxp017. https://www.ncbi.nlm.nih.gov/pubmed/19246554.

Kawasaki, Eiji. 2014. "Type 1 diabetes and autoimmunity." *Clinical Pediatric Endocrinology* 23 (4): 99–105. https://doi.org/10.1297/cpe.23.99.

Keir, Mary E., Tangsheng Yi, Timothy T. Lu, and Nico Ghilardi. 2020. "The role of IL-22 in intestinal health and disease." *Journal of Experimental Medicine* 217 (3). https://doi.org/10.1084/jem.20192195. https://doi.org/10.1084/jem.20192195.

Keita, A V., and Johan D. Söderholm. 2010. "The intestinal barrier and its regulation by neuroimmune factors." *Neurogastroenterology & Motility* 22 (7): 718–733.

Khoshbin, K., L. Khanna, D. Maselli, J. Atieh, M. Breen-Lyles, K. Arndt, D. Rhoten, R. B. Dyer, R. J. Singh, S. Nayar, S. Bjerkness, W. S. Harmsen, I. Busciglio, and M. Camilleri. 2021. "Development and validation of test for 'leaky gut' small intestinal and colonic permeability using sugars in healthy adults." *Gastroenterology* 161 (2): 463–475 e13. https://doi.org/10.1053/j.gastro.2021.04.020. https://www.ncbi.nlm.nih.gov/pubmed/33865841.

Krishna Rao, Radha, and Geetha Samak. 2013. "Protection and restitution of gut barrier by probiotics: Nutritional and clinical implications." *Current Nutrition & Food Science* 9 (2): 99–107. https://doi.org/http://dx.doi.org/10.2174/1573401311309020004. http://www.eurekaselect.com/article/50828.

Lee, Bonggi, Kyoung Mi Moon, and Choon Young Kim. 2018. "Tight junction in the intestinal epithelium: Its association with diseases and regulation by phytochemicals." *Journal of Immunology Research.*

Lee, Jessica Y, Valerie C. Wasinger, Yunki Y. Yau, Emil Chuang, Vijay Yajnik, and Rupert W. L. Leong. 2018. "Molecular pathophysiology of epithelial barrier dysfunction in inflammatory bowel diseases." *Proteomes* 6 (2): 17.

Looijer–Van Langen, Mirjam A. C., and Levinus A. Dieleman. 2008. "Prebiotics in chronic intestinal inflammation." *Inflammatory Bowel Diseases* 15 (3): 454–462. https://doi.org/10.1002/ibd.20737.

Maslowski, Kendle M., Angelica T. Vieira, Aylwin Ng, Jan Kranich, Frederic Sierro, Di Yu, Heidi C. Schilter, Michael S. Rolph, Fabienne Mackay, and David Artis. 2009. "Regulation of inflammatory responses by gut microbiota and chemoattractant receptor GPR43." *Nature* 461 (7268): 1282–1286.

Mazmanian, Sarkis K., June L. Round, and Dennis L. Kasper. 2008. "A microbial symbiosis factor prevents intestinal inflammatory disease." *Nature* 453 (7195): 620–625.

Miao, W., X. Wu, K. Wang, W. Wang, Y. Wang, Z. Li, J. Liu, L. Li, and L. Peng. 2016. "Sodium butyrate promotes reassembly of tight junctions in Caco-2 monolayers involving inhibition of MLCK/MLC2 pathway and phosphorylation of PKCbeta2." *International Journal of Molecular Sciences* 17 (10). https://doi.org/10.3390/ijms17101696. https://www.ncbi.nlm.nih.gov/pubmed/27735862.

Michielan, Andrea, and Renata D'Incà. 2015. "Intestinal permeability in inflammatory bowel disease: Pathogenesis, clinical evaluation, and therapy of leaky gut." *Mediators of Inflammation* 2015: 628157. https://doi.org/10.1155/2015/628157.

Miele, Luca, Venanzio Valenza, Giuseppe La Torre, Massimo Montalto, Giovanni Cammarota, Riccardo Ricci, Roberta Masciana, Alessandra Forgione, Maria L Gabrieli, and Germano Perotti. 2009. "Increased intestinal permeability and tight junction alterations in nonalcoholic fatty liver disease." *Hepatology* 49 (6): 1877–1887.

Mileo, Anna Maria, Paola Nisticò, and Stefania Miccadei. 2019. "Polyphenols: immunomodulatory and therapeutic implication in colorectal cancer." *Frontiers in Immunology* 10: 729.

Mörbe, Urs M., Peter B. Jørgensen, Thomas M. Fenton, Nicole von Burg, Lene B Riis, Jo Spencer, and William W. Agace. 2021. "Human gut-associated lymphoid tissues (GALT); diversity, structure, and function." *Mucosal Immunology* 14 (4): 793–802.

Morrissey, Ryan P., Lawrence Czer, and Prediman K. Shah. 2011. "Chronic heart failure: Current evidence, challenges to therapy, and future directions." *American Journal of Cardiovascular Drugs* 11: 153–171.

Mowat, Allan M., and William W. Agace. 2014. "Regional specialization within the intestinal immune system." *Nature Reviews Immunology* 14 (10): 667–685.

Mu, Qinghui, Jay Kirby, Christopher M. Reilly, and Xin M. Luo. 2017. "Leaky gut as a danger signal for autoimmune diseases." *Frontiers in Immunology* 8. https://www.frontiersin.org/articles/10.3389/fimmu.2017.00598.

O'Toole, Paul W., Julian R. Marchesi, and Colin Hill. 2017. "Next-generation probiotics: the spectrum from probiotics to live biotherapeutics." *Nature Microbiology* 2 (5): 17057. https://doi.org/10.1038/nmicrobiol.2017.57.

Odenwald, Matthew A., and Jerrold R. Turner. 2017. "The intestinal epithelial barrier: a therapeutic target?" *Nature Reviews Gastroenterology & Hepatology* 14 (1): 9–21. https://doi.org/10.1038/nrgastro.2016.169.

Ohland, Christina L., and Wallace K. MacNaughton. 2010. "Probiotic bacteria and intestinal epithelial barrier function." *American Journal of Physiology-Gastrointestinal and Liver Physiology* 298 (6): G807–G819.

Ohno, H. 2016. "Intestinal M cells." *Journal of Biochemistry* 159 (2): 151–160. https://doi.org/10.1093/jb/mvv121. https://www.ncbi.nlm.nih.gov/pubmed/26634447.

Okai, Shinsaku, Fumihito Usui, Shuhei Yokota, Yusaku Hori-i, Makoto Hasegawa, Toshinobu Nakamura, Manabu Kurosawa, Seiji Okada, Kazuya Yamamoto, and Eri Nishiyama. 2016. "High-affinity monoclonal IgA regulates gut microbiota and prevents colitis in mice." *Nature Microbiology* 1 (9): 1–11.

Organization, World Health. 2018. *Taking action on childhood obesity*. World Health Organization.

Oseini, Abdul M., and Arun J. Sanyal. 2017. "Therapies in non-alcoholic steatohepatitis (NASH)." *Liver International* 37: 97–103.

Otani, Tetsuhisa, and Mikio Furuse. 2020. "Tight junction structure and function revisited." *Trends in Cell Biology* 30 (10): 805–817.

Palm, N. W., M. R. de Zoete, T. W. Cullen, N. A. Barry, J. Stefanowski, L. Hao, P. H. Degnan, J. Hu, I. Peter, W. Zhang, E. Ruggiero, J. H. Cho, A. L. Goodman, and R. A. Flavell. 2014. "Immunoglobulin A coating identifies colitogenic bacteria in inflammatory bowel disease." *Cell* 158 (5): 1000–1010. https://doi.org/10.1016/j.cell.2014.08.006. https://www.ncbi.nlm.nih.gov/pubmed/25171403.

Papista, Christina, Laureline Berthelot, and Renato C. Monteiro. 2011. "Dysfunctions of the Iga system: A common link between intestinal and renal diseases." *Cellular & Molecular Immunology* 8 (2): 126–134.

Paradis, Tracy, Hervé Bègue, Louise Basmaciyan, Frédéric Dalle, and Fabienne Bon. 2021. "Tight junctions as a key for pathogens invasion in intestinal epithelial cells." *International Journal of Molecular Sciences* 22 (5): 2506.

Parekh, Parth J., Luis A. Balart, and David A. Johnson. 2015. "The influence of the gut microbiome on obesity, metabolic syndrome and gastrointestinal disease." *Clinical and Translational Gastroenterology* 6 (6): e91.

Pasini, Evasio, Roberto Aquilani, Cristian Testa, Paola Baiardi, Stefania Angioletti, Federica Boschi, Manuela Verri, and Francesco Dioguardi. 2016. "Pathogenic gut flora in patients with chronic heart failure." *JACC: Heart Failure* 4 (3): 220–227. https://www.jacc.org/doi/abs/10.1016/j.jchf.2015.10.009.

Pelaseyed, T., J. H. Bergstrom, J. K. Gustafsson, A. Ermund, G. M. Birchenough, A. Schutte, S. van der Post, F. Svensson, A. M. Rodriguez-Pineiro, E. E. Nystrom, C. Wising, M. E. Johansson, and G. C. Hansson. 2014. "The mucus and mucins of the goblet cells and enterocytes provide the first defense line of the gastrointestinal tract and interact with the immune system." *Immunological Reviews* 260 (1): 8–20. https://doi.org/10.1111/imr.12182. https://www.ncbi.nlm.nih.gov/pubmed/24942678.

Ponziani, Francesca Romana, Sherrie Bhoori, Chiara Castelli, Lorenza Putignani, Licia Rivoltini, Federica Del Chierico, Maurizio Sanguinetti, Daniele Morelli, Francesco Paroni Sterbini, and Valentina Petito. 2019. "Hepatocellular carcinoma is associated with gut microbiota profile and inflammation in nonalcoholic fatty liver disease." *Hepatology* 69 (1): 107–120.

Quigley, E. M. 2016. "Leaky gut – concept or clinical entity?" *Current Opinion in Gastroenterology* 32 (2): 74–79. https://doi.org/10.1097/MOG.0000000000000243. https://www.ncbi.nlm.nih.gov/pubmed/26760399.

Rakoff-Nahoum, Seth, Justin Paglino, Fatima Eslami-Varzaneh, Stephen Edberg, and Ruslan Medzhitov. 2004. "Recognition of commensal microflora by toll-like receptors is required for intestinal homeostasis." *Cell* 118 (2): 229–241.

Reboldi, Andrea, and Jason G. Cyster. 2016. "Peyer's patches: Organizing B-cell responses at the intestinal frontier." *Immunological Reviews* 271 (1): 230–245.

Rescigno, M. 2011. "The intestinal epithelial barrier in the control of homeostasis and immunity." *Trends Immunol* 32 (6): 256–264. https://doi.org/10.1016/j.it.2011.04.003. https://www.ncbi.nlm.nih.gov/pubmed/21565554.

Rhoads, J. M., R. A. Argenzio, W. Chen, R. A. Rippe, J. K. Westwick, A. D. Cox, H. M. Berschneider, and D. A. Brenner. 1997. "L-glutamine stimulates intestinal cell proliferation and activates mitogen-activated protein kinases." *American Journal of Physiology-Gastrointestinal and Liver Physiology* 272 (5): G943–G953. https://journals.physiology.org/doi/abs/10.1152/ajpgi.1997.272.5.G943.

Rooks, M. G., and W. S. Garrett. 2016. "Gut microbiota, metabolites and host immunity." *Nat Rev Immunol* 16 (6): 341–352. https://doi.org/10.1038/nri.2016.42. https://www.ncbi.nlm.nih.gov/pubmed/27231050.

Round, June L., and Sarkis K. Mazmanian. 2010. "Inducible Foxp3+ regulatory T-cell development by a commensal bacterium of the intestinal microbiota." *Proceedings of the National Academy of Sciences* 107 (27): 12204–12209.

Ruff, William E., Teri M. Greiling, and Martin A. Kriegel. 2020. "Host–microbiota interactions in immune-mediated diseases." *Nature Reviews Microbiology* 18 (9): 521–538.

Sorribas, M., M. O. Jakob, B. Yilmaz, H. Li, D. Stutz, Y. Noser, A. de Gottardi, S. Moghadamrad, M. Hassan, A. Albillos, R. Frances, O. Juanola, I. Spadoni, M. Rescigno, and R. Wiest. 2019. "FXR modulates the gut-vascular barrier by regulating the entry sites for bacterial translocation in experimental cirrhosis." *Journal of Hepatology* 71 (6): 1126–1140. https://doi.org/10.1016/j.jhep.2019.06.017. https://www.ncbi.nlm.nih.gov/pubmed/31295531.

Steinert, Robert E., Christine Feinle-Bisset, Lori Asarian, Michael Horowitz, Christoph Beglinger, and Nori Geary. 2017. "Ghrelin, CCK, GLP-1, and PYY(3–36): Secretory Controls and Physiological Roles in Eating and Glycemia in Health, Obesity, and After RYGB." *Physiological Reviews* 97 (1): 411–463. https://journals.physiology.org/doi/abs/10.1152/physrev.00031.2014.

Suzuki, Takuya, and Hiroshi Hara. 2011. "Role of flavonoids in intestinal tight junction regulation." *The Journal of Nutritional Biochemistry* 22 (5): 401–408. https://doi.org/https://doi.org/10.1016/j.jnutbio.2010.08.001. https://www.sciencedirect.com/science/article/pii/S0955286310001877.

Taleb, Soraya. 2019. "Tryptophan dietary impacts gut barrier and metabolic diseases." *Frontiers in Immunology* 10. https://www.frontiersin.org/articles/10.3389/fimmu.2019.02113.

Thomas, Karen E., Anna Sapone, Alessio Fasano, and Stefanie N. Vogel. 2006. "Gliadin stimulation of murine macrophage inflammatory gene expression and intestinal permeability are MyD88-dependent: Role of the innate immune response in celiac disease." *The Journal of Immunology* 176 (4): 2512–2521. https://doi.org/10.4049/jimmunol.176.4.2512.

Treiber, Gerlies, Barbara Prietl, Elke Fröhlich-Reiterer, Evelyne Lechner, Anja Ribitsch, Maria Fritsch, Birgit Rami-Merhar, Claudia Steigleder-Schweiger, Winfried Graninger, Martin Borkenstein, and Thomas R. Pieber. 2015. "Cholecalciferol supplementation improves suppressive capacity of regulatory T-cells in young patients with new-onset type 1 diabetes mellitus — A randomized clinical trial." *Clinical Immunology* 161 (2): 217–224. https://doi.org/https://doi.org/10.1016/j.clim.2015.08.002. https://www.sciencedirect.com/science/article/pii/S152166161530019X.

Tripathi, A., J. Debelius, D. A. Brenner, M. Karin, R. Loomba, B. Schnabl, and R. Knight. 2018. "The gut-liver axis and the intersection with the microbiome." *Nat Rev Gastroenterol Hepatol* 15 (7): 397–411. https://doi.org/10.1038/s41575-018-0011-z. https://www.ncbi.nlm.nih.gov/pubmed/29748586.

Ullah, Asad, Sidra Munir, Syed Lal Badshah, Noreen Khan, Lubna Ghani, Benjamin Gabriel Poulson, Abdul-Hamid Emwas, and Mariusz Jaremko. 2020. "Important flavonoids and their role as a therapeutic agent." *Molecules* 25 (22): 5243. https://www.mdpi.com/1420-3049/25/22/5243.

Valdes, A. M., J. Walter, E. Segal, and T. D. Spector. 2018. "Role of the gut microbiota in nutrition and health." *BMJ* 361: k2179. https://doi.org/10.1136/bmj.k2179. https://www.ncbi.nlm.nih.gov/pubmed/29899036.

Vancamelbeke, M., and S. Vermeire. 2017a. "The intestinal barrier: A fundamental role in health and disease." *Expert Review of Gastroenterology & Hepatology* 11 (9): 821–834. https://doi.org/10.1080/17474124.2017.1343143. https://www.ncbi.nlm.nih.gov/pubmed/28650209.

Vancamelbeke, Maaike, and Séverine Vermeire. 2017b. "The intestinal barrier: A fundamental role in health and disease." *Expert Review of Gastroenterology & Hepatology* 11 (9): 821–834.

Vereecke, Lars, Rudi Beyaert, and Geert van Loo. 2011. "Enterocyte death and intestinal barrier maintenance in homeostasis and disease." *Trends in Molecular Medicine* 17 (10): 584–593.

Vinolo, Marco A. R., Hosana G. Rodrigues, Elaine Hatanaka, Fábio T. Sato, Sandra C. Sampaio, and Rui Curi. 2011. "Suppressive effect of short-chain fatty acids on production of proinflammatory mediators by neutrophils." *The Journal of Nutritional Biochemistry* 22 (9): 849–855.

Wang, Xiaoting, Naruhisa Ota, Paolo Manzanillo, Lance Kates, Jose Zavala-Solorio, Celine Eidenschenk, Juan Zhang, Justin Lesch, Wyne P. Lee, Jed Ross, Lauri Diehl, Nicholas van Bruggen, Ganesh Kolumam, and Wenjun Ouyang. 2014. "Interleukin-22 alleviates metabolic disorders and restores mucosal immunity in diabetes." *Nature* 514 (7521): 237–241. https://doi.org/10.1038/nature13564.

Wehkamp, Jan, and Eduard F. Stange. 2020. "An update review on the Paneth cell as key to ileal Crohn's disease." *Frontiers in Immunology* 11: 646.

Wiest, Reiner, Melissa Lawson, and Markus Geuking. 2014. "Pathological bacterial translocation in liver cirrhosis." *Journal of Hepatology* 60 (1): 197–209.

Yan, F., and D. B. Polk. 2002. "Probiotic bacterium prevents cytokine-induced apoptosis in intestinal epithelial cells." *Journal of Biological Chemistry* 277 (52): 50959–50965. https://doi.org/10.1074/jbc.M207050200. https://www.ncbi.nlm.nih.gov/pubmed/12393915.

Zhang, Yi-Zhen, and Yong-Yu Li. 2014. "Inflammatory bowel disease: Pathogenesis." *World Journal of Gastroenterology: WJG* 20 (1): 91.

Zhang, Yong-guo, Shaoping Wu, and Jun Sun. 2013. "Vitamin D, vitamin D receptor and tissue barriers." *Tissue Barriers* 1 (1): e23118. https://doi.org/10.4161/tisb.23118.

Zhang, Yuanyuan, Jonathan P. Jackson, Robert L. St. Claire III, Kimberly Freeman, Kenneth R. Brouwer, and Jeffrey E. Edwards. 2017. "Obeticholic acid, a selective farnesoid X receptor agonist, regulates bile acid homeostasis in sandwich-cultured human hepatocytes." *Pharmacology Research & Perspectives* 5 (4): e00329. https://bpspubs.onlinelibrary.wiley.com/doi/abs/10.1002/prp2.329.

Section III

Molecular Aspects in Gut-Associated Diseases and Their Management

7 Gut-Associated Cancers and Traditional Foods

Francesca Cuffaro[1], Sara Bertorello[1], Elena Niccolai[2], and Amedeo Amedei[3]

[1]Department of Experimental and Clinical Medicine, University of Florence, 50134 Florence, Italy
[2]Department of Experimental and Clinical Medicine, University of Florence, 50134 Florence, Italy; Joint Laboratory MIA-LAB (Microbiome-Immunity Axis research for a Circular Health), University of Florence, Italy
[3]Department of Experimental and Clinical Medicine, University of Florence, 50134 Florence, Italy; Joint Laboratory MIA-LAB (Microbiome-Immunity Axis research for a Circular Health), University of Florence, Italy; Network of Immunity in Infection, Malignancy and Autoimmunity (NIIMA), Universal Scientific Education and Research Network (USERN), Florence, Italy

7.1 INTRODUCTION

Gastrointestinal cancers refer to a group of malignancies that specifically originate in the digestive system, encompassing various organs such as the stomach, small intestine, and colon. Gut-associated cancers (GAC), among the various forms of cancer, rank as the most commonly diagnosed. Colorectal cancer (CRC) stands out as the third most frequently detected, contributing to 9.4% of cancer-related deaths worldwide in 2020 (Arnold et al. 2020, Sung et al. 2021). Conversely, cancers affecting the small intestine, though rare, have shown an increasing trend over the past 40 years (Barsouk et al. 2019). Furthermore, gastrointestinal lymphomas can occur in gut regions, such as the small bowel (20–30%) and the colon (10–20%). The prevalent types of lymphomas affecting the small intestine include diffuse large B-cell lymphoma (DLBCL) and mucosa-associated lymphoid tissue (MALT) lymphoma, encompassing immunoproliferative small intestinal disease. Large intestine lymphomas, on the other hand, are rare, constituting only 10% to 20% of all gastrointestinal tract lymphomas and a mere 0.2% of all colorectal malignancies (Alvarez-Lesmes et al. 2021).

Considerable differences in incidence and mortality rates of CRC are evident among different countries and global regions. These variations stem from a range of factors, including diverse risk factors' exposure, demographic aspects, genetic

DOI: 10.1201/9781003514244-10

factors, and their impact on treatment outcomes and responsiveness within populations (Baidoun et al. 2021). In recent decades, a positive trend in both CRC incidence and mortality has emerged thanks in part to increased screening tests (primarily colonoscopy) as well as a decrease in some risk factors such as red meat consumption and smoking habits (Song and Chan 2019). Advances in therapy have also contributed to this favourable trend (Song and Chan 2019). However, a particularly concerning pattern is the increase in CRC cases in people under 50 (Islami et al. 2018). While the specific causes behind this rise in early-onset CRC remain unclear, lifestyle choices are believed to play a significant role (Islami et al. 2018). Consequently, addressing the CRC challenge, especially among young individuals, emphasizes the critical need to explore innovative prevention approaches that extend beyond routine screening (Siegel et al. 2020).

The CRC risk is intricately linked to the interplay between diet and the composition of gut microbiota (GM), which, in association with their metabolites (particularly short-chain fatty acids – SCFA), can contribute to the genesis and progression of some tumors (Pandey et al. 2023).

The human microbiome is a highly complex and dynamic ecosystem, consisting of various types of microorganisms, including bacteria, viruses and bacteriophages, fungi, and archaea. It plays a fundamental role in numerous physiological host functions, such as the maturation of the immune system and protection against infectious agents (Niccolai et al. 2020). Its composition in the healthy intestine varies spatially: longitudinally along the colon, and radially from the epithelium to the mucosa to the intestinal lumen (Flynn et al. 2018).

Alterations in GM composition, known as dysbiosis, have the potential to influence or alter how environmental factors impact CRC development. The risk factors associated with CRC, such as overweight and obesity, levels of physical activity, and dietary choices, can also influence the intestinal microbiome structure. Therefore, understanding and mitigating the impact of these environmental factors on the microbiota are critical steps in developing comprehensive strategies for CRC prevention (Song, Chan, and Sun 2020).

This chapter delves into the profound impact of traditional foods derived from diverse dietary patterns on the molecular mechanisms implicated in the development and progression of gut-associated cancers, with a focus on CRC due to its global impact.

Traditional foods, representing items and recipes fixed within specific geographic regions and bearing cultural significance for the local community, serve as the epicenter of our exploration. Examining the specific composition of traditional diets, encompassing the micro and macronutrients they contain, affords valuable insights into how these foods may either contribute to or mitigate the GAC risk. Moreover, traditional foods often encapsulate centuries-old culinary wisdom, rendering them rich sources of bioactive compounds with the potential to influence cellular signaling pathways relevant to cancer development. In a backdrop where dietary patterns undergo dynamic shifts driven by globalization and evolving lifestyle habits, the investigation of traditional foods becomes pivotal for preserving and promoting health. Unraveling the intricate link between these foods and GAC risk holds the potential to inform strategies for dietary interventions, shape public health policies, and empower people to make informed choices that can decrease their susceptibility to gut-associated cancers.

Finally, this chapter will specifically address the GM role in the context of traditional foods and CRC risk. As dietary choices influence its composition, understanding this interplay becomes crucial to defining the intricate relationship between traditional diets and colorectal cancer. By navigating through the complex network of traditional foods, dietary patterns, and gut microbiota, we aim to provide comprehensive insights into the multifaceted factors influencing the development of CRC and to inform strategies for maintaining gut health.

7.2 DIETETIC PATTERNS AND GUT-ASSOCIATED CANCERS: PREVENTIVE OR RISK FACTORS?

Diet and dietary patterns have a significant impact on health, disease prevention, and quality of life. The term "diet," derived from the Greek "*daita*," which means "lifestyle," reflects its broader connotation, meaning all the behaviors that we implement to improve our quality of life and achieve psycho-physical well-being. These include nutrition but also exercise, quality of sleep and rest, abstention from smoking and excessive alcohol consumption, as well as stress management. Lifestyles and dietary patterns are often affected by geography, religion, ethics, and culture (Glanz and Kolonel 1998). Identifying the various lifestyle factors and nutritional habits associated with different cultures is pivotal, as they have been implicated in the onset and progression of cancers.

The European Prospective Investigation into Cancer and Nutrition (EPIC) study, designed to study the relationship between diet and cancer, revealed a notable inverse association between dietary fiber, fish, calcium, and vitamin D intake, along with fruits and vegetables, and the likelihood of developing colon cancer (Bingham and Riboli 2004, Jenab et al. 2010, Gonzalez and Riboli 2010, Riboli et al. 2002). Conversely, increased consumption of red and processed meats showed a positive association with the risk of CRC (Riboli et al. 2002, Norat et al. 2005). The World Cancer Research Fund (WCRF) and the American Institute for Cancer Research (AICR) published their last report on CRC in 2018, identifying dietary compounds as preventive or risk factors (Clinton, Giovannucci, and Hursting 2020). Recognized risk factors for GAC include smoking, excess body weight, alcohol consumption, and the consumption of red and processed meats. Conversely, engaging in regular physical activity and incorporating dietary elements such as fibers, whole grains, dairy products, and calcium supplements into the diet may serve to decrease the risk of developing the disease (Clinton, Giovannucci, and Hursting 2020).

Diet has also been identified as a risk factor for small bowel cancer, with an elevated intake of animal fat and protein being linked to increased risk (Barsouk et al. 2019, Schatzkin et al. 2008). Conversely, it is documented that the consumption of dietary fibers (derived from grains, vegetables, and beans) decreases the probability of developing cancers in both the small and large intestines. This effect is thought to be achieved by accelerating the passage of food through the digestive tract, thereby minimizing exposure to carcinogens (Schatzkin et al. 2008, Barsouk et al. 2019).

In addition, nutritional habits are considered potential risk factors even for lymphomas, but to date, there are no studies that have specifically investigated the role of diet in gut-associated lymphomas. Prior research has shown uncertain results, often lacking the necessary statistical power for precise determinations (Naudin et al. 2020). An extensive prospective study investigating the connection between a composite score, encompassing various lifestyle factors (including diet, smoking, BMI, and alcohol consumption) and the likelihood of developing lymphoma has been recently performed. The results revealed that adopting healthy behaviors was associated with a reduced risk of Hodgkin lymphoma, with smoking identified as the primary factor influencing this relationship (Naudin et al. 2020).

Preventing cancer through incorporating dietary practices, lifestyle modifications, and complementary therapies remains a challenge to date (Burney and Al-Moundhri 2014, Mao et al. 2022) (Rock et al. 2020, Grosso et al. 2017). In low- and middle-income countries, there is often reference to traditional, complementary, and integrative medicine (TCIM) for disease prevention and management, due to its familiarity, affordability, and widespread availability (Trimble and Rajaraman 2017). In recent years, there has been growing interest in *integrative oncology*, which combines effective TCIM with standard cancer care and prevention strategies, bridging gaps in healthcare delivery (Mao et al. 2022). This approach incorporates lifestyle changes, mind-body therapies (such as acupuncture, massage, meditation, and yoga), and natural products (Mao et al. 2022). Within this framework, nutrition remains a cornerstone for cancer's prevention and management (Burney and Al-Moundhri 2014, Mao et al. 2022).

Studies investigating the link between dietary patterns and cancer risk typically employ an "*a posteriori* approach" to identify characteristic eating habits within the study population. A significant evolution in cancer prevention over time, reflecting current scientific evidence, is a transition from a reductionist, nutrient-centric approach to a broader concept of diet characterized by dietary patterns (Rock et al. 2020). Examining dietary patterns offers a more comprehensive perspective, considering the combined impact of various foods. People consume meals with diverse food compositions and intricate nutrient combinations, rather than isolated nutrients. The diet's complexity suggests that focusing on a single nutrient might be challenging due to interactions between food components, which often interact or work synergistically. Therefore, delving into the overall dietary pattern, which considers multiple dietary exposures simultaneously, becomes an essential complementary method in understanding the connections between diet and health well.

This transition emphasizes the importance of dietary patterns rather than focusing solely on individual nutrients and bioactive compounds, which better align with real-world eating habits and practices. New findings, primarily from epidemiological studies but also a few controlled intervention trials, indicate that adopting healthy dietary patterns, as opposed to unhealthy ones, is associated with a lower risk of cancer, particularly colon and breast cancer (Grosso et al. 2017).

While dietary patterns have been studied in cohorts from various geographic regions, the identification of traditional dietary patterns specific to a single country is a rare occurrence. We can, however, recognize some frequently studied dietary patterns associated with disease development, including CRC: the Mediterranean diet (MD) and the Western diet (WD).

7.3 THE MEDITERRANEAN DIET AND ITS ANTI-CANCER MECHANISMS

The MD is a dietetic pattern embraced by several countries in the Mediterranean basin (Italy, Spain, Greece, Morocco) and was defined in 2010 as an "intangible cultural heritage" by UNESCO. The traditional MD is characterized by elevated intakes of fruits, vegetables, whole grains, nuts, fish, and legumes, moderate consumption of dairy products, limited intakes of meat and meat products, along with a moderate intake of alcohol (usually red wine during meals) (Willett et al. 1995). Increased adherence to the MD correlates with notable improvement in health status, evidenced by substantial reductions in overall mortality (especially related to cardiovascular diseases and cancers), and incidences of neurodegenerative diseases such as Parkinson's and Alzheimer's (Sofi et al. 2010, Sofi et al. 2008).

In detail, the MD has been associated with a decrease in overall cancer rates and notably reduces occurrences of digestive tract cancers (Barak and Fridman 2017). The protective effects of the MD could be attributed to specific patterns, such as a high consumption of fruits and vegetables and limited intake of processed red meat (Grosso et al. 2013). To date, it is well established that the MD has a beneficial effect on CRC risk (Donovan et al. 2017).

Conversely, there is limited research evaluating the association between MD and the risk of lymphoma. Solans et al. demonstrate a modest inverse relation between higher adherence to MD and the overall risk of lymphomas within the EPIC study cohort (Solans et al. 2019). Currently, there are no studies evaluating the association between MD and small bowel cancers. However, a diet rich in fibers from grains and whole grains appears to have a protective effect (Schatzkin et al. 2008). The association with alcohol, animal fat, and animal protein remains controversial (Pan and Morrison 2011). Further studies are needed to define the association between MD and the risk of gut-associated lymphomas and small bowel cancers.

The relationship between MD and CRC risk has been evaluated in several clinical case-control and cohort studies, suggesting an inverse association (Agnoli et al. 2013, Dixon et al. 2007, Fung et al. 2010, Randi et al. 2010). The meta-analyses conducted by Schwingshackl and Hoffmann confirmed that MD decreased CRC risk by 14% and 17% (Schwingshackl and Hoffmann 2015, 2014).

In order to understand how a single compound can alter the signaling pathways within human CRC, preclinical investigations must be referred to. These studies consistently center on assessing the impact of each MD constituent on cancer cell lines, contributing to an understanding of their pathways and the specific molecular mechanisms influenced by each dietary compound.

Olive oil, a MD fundamental, serves as the primary source of dietary fats and stands as one of its defining elements. The consumption of olive oil has been linked to many health benefits, including its potential interference with carcinogenesis (Di Francesco et al. 2015, Banks et al. 2016). This property has been attributed to olive oil's complex composition of bioactive compounds (tocopherols, alcohols, squalene, phenolic compounds such as oleuropein, catechin, epicatechin, quercetin, hydroxytyrosol, and tyrosol) and monounsaturated fats, mainly oleic acid (Perez-Jimenez et al. 2007). Preclinical investigations have associated numerous chemopreventive effects with the constituents of olive oil, primarily due to their interference with pathways involved in the initiation, promotion, and progression of cancer development. These include the downregulation of cyclooxygenase 2 (COX-2) and B-cell lymphoma 2 (BCL-2) protein family, the generation of reactive oxygen species (ROS) in cancer cells, and the downregulation of epidermal growth factor receptor (EGFR) expression (Casaburi et al. 2013, Corona et al. 2007, Llor et al. 2003, Sun et al. 2014). A meta-analysis, incorporating findings from eight case-control studies, concluded that a high consumption of olive oil is associated with a noteworthy 30% reduction in the probability of gut-associated cancers compared to a lower intake of olive oil (Psaltopoulou et al. 2011).

Besides oleic acid and olive oil, MD is also rich in other types of beneficial fatty acids (Augimeri and Bonofiglio 2021) that are crucial for human metabolism and overall health; their characteristics vary based on chain length. In detail, short-chain fatty acids (SCFAs) consist of 1 to 6 carbon atoms, medium-chain fatty acids (MCFAs) have chains of 7 to 12 carbon atoms, and long-chain fatty acids (LCFAs) extend beyond 12 carbons. The dietary origins of these fatty acids differ significantly; LCFAs and MCFAs are primarily sourced from our diet, especially from dairy products and milk. In contrast, SCFAs mainly result from bacterial fermentation in the colon (Schönfeld and Wojtczak 2016), a topic we will delve into further in the subsequent chapter on bacterial metabolites.

Concerning LCFAs, unsaturated fatty acids (C16-24: ω3-9) display antiproliferative effects in vitro, inducing oxidative stress and modulating intracellular signaling pathways (Shuwen et al. 2019, Fauser et al. 2011).

Polyunsaturated fatty acids (PUFAs), LCFAs characterized by a backbone with two or more carbon–carbon double bonds, are not produced by mammals, and the main dietary sources are nuts and fish (Mohanty et al. 2016). They have the potential to moderate low-grade inflammation, consequently reducing the risk of cancer. Meta-analyses of prospective human studies have shown that nut consumption is associated with decreased risks of total (Aune et al. 2016) and colon cancer (Wu et al. 2015), while fish consumption is linked to lowered risks of CRC (Vieira et al. 2017). Among PUFA, docosahexaenoic acid (DHA) and eicosapentaenoic acid (EPA) have the ability to alter the membrane composition of inflammatory cells. This is achieved by reducing the levels of arachidonic acid, which is more abundant in the membrane phospholipids of cells involved in the inflammatory response and serves as a precursor of eicosanoids regulating inflammation and tumorigenesis (Calder 2012, Wang et al. 2018). Simultaneously, DHA and EPA increase the production of certain anti-inflammatory molecules such as resolvins and protectins (Calder 2012). The complex interplay of different fatty acids in metabolism, inflammation, and overall health represents a dynamic and evolving area of research, with ongoing investigations to uncover their multifaceted roles and potential health benefits (Schönfeld and Wojtczak 2016, den Besten et al. 2013).

Conversely, SCFAs and MCFAs, once produced in the intestine, enter the portal vein as free acids, where they can interact with plasma albumin. These various fatty acids influence metabolism actively, affecting processes such as glycolysis, lipogenesis, and gluconeogenesis (Schönfeld and Wojtczak 2016). Additionally, these compounds are believed to play a significant role in the inflammatory process as they interact with cell membranes, surface receptors, and intracellular receptors (Hecker et al. 2014). Lastly, they contribute to modulating gene expression (Calder 2012). MCFAs exhibit cytotoxic and antimicrobial properties both in vivo and in vitro (Huang et al. 2011). However, the current evidence regarding their potential as anti-neoplastic agents is limited.

SCFAs primarily result from the fermentation of dietary fiber and undigested carbohydrates by anaerobic colonic bacteria (den Besten et al. 2013). Apart from their involvement in host immunity, SCFAs, notably butyrate, play a relevant role in aerobic glycolysis, providing energy for colonocytes (Duizer and de Zoete 2023). In vitro models suggest that SCFAs, including butyrate, exhibit a propensity to enhance apoptosis, potentially through mechanisms involving epigenetic modifications, cell cycle arrest, and activation of pro-apoptotic genes (Shuwen et al. 2019, Fauser et al. 2011). Interestingly, the SCFAs plasma levels in CRC patients (compared to healthy controls) were significantly decreased, suggesting that reduced SCFAs may contribute to CRC progression (Niccolai et al. 2019). While butyrate is frequently highlighted, other SCFAs, such as acetate, can induce apoptosis by inhibiting cancer cell proliferation, causing DNA fragmentation, and activating caspase 3 (Duizer and de Zoete 2023).

Dietary fibers, abundant in the MD, including nonstarch polysaccharides, resistant starch, fructo-oligosaccharides, galacto-oligosaccharides, glucomannans, lignin, pectin, and β-glucans, are abundantly found in fruits, vegetables, cereals, and macrofungals. These fibers play a crucial role in promoting digestive health and consistently demonstrate an association with a reduced CRC risk (Costea et al. 2018).

Increased fiber consumption accelerates the passage of food through the digestive system, diminishing the exposure of the large intestine to potential carcinogens and reducing inflammation (Williams and Hord 2005, Ocvirk et al. 2019). In a meta-analysis of 185 prospective trials and 58 clinical studies, Reynolds et al. demonstrated an inverse correlation between fiber intake and CRC risk (Reynolds et al. 2019). Prospective analysis from the EPIC study confirms the preventive role of fiber against CRC, revealing that doubling the intake of fiber in individuals with low consumption resulted in approximately a 40% reduction in CRC cases (Bingham et al. 2003). Another study in the U.S.A., encompassing both men and women, found that individuals in the highest quintile of fiber consumption had a 27% reduced risk of developing adenomas, in contrast to those in the lowest quintile of intake (Peters et al. 2003).

One mechanism contributing to this protective effect involves the promotion of butyrate-producing bacteria in the gut and the suppression of secondary bile acid synthesis, another factor linked to CRC risk (Sánchez-Alcoholado et al. 2020). Additionally, unfermented fiber absorbs water content, normalizes bowel movements and discourages the interaction of the mucosal layer with potential carcinogens

(Dalal et al. 2021). Greater adherence to MD was associated with a higher prevalence of individuals showing adequate intake of micronutrients (Serra-Majem et al. 2009, Serra-Majem et al. 2001), which are defined as substances present in foods that are vital for human health and are needed in relatively small quantities. This category encompasses all known vitamins and essential trace minerals (Miller and Welch 2013). The adequate intake of numerous micronutrients has been correlated with CRC.

An omnivorous diet, particularly when including dairy products, spinach, kale, and various vegetables, can supply sufficient calcium (Ca) (Kohlenberg-Mueller and Raschka 2003). Low calcium intake was associated with poor adherence to MD in Spanish children (Rubio-López et al. 2017). The Ca absorption in the intestine is crucial, not only for maintaining bone health but also for broader physiological functions (Bielik and Kolisek 2021a). The gut microbiota play a role in enhancing Ca absorption and influencing the production of intestinal serotonin, which, in turn, regulates bone metabolism (D'Amelio and Sassi 2018). Regarding intestinal health and carcinogenesis, calcium acts in two ways: by promoting the formation of soaps that bind secondary bile acids and fatty acids, and by directly reducing cell proliferation and inducing apoptosis. In addition, calcium has been reported to counteract the cytotoxic effect of heme iron on the colon mucosa, although the precise molecular mechanisms are not yet fully understood (Vernia et al. 2021, Katona and Weiss 2020). Furthermore, recent studies have highlighted selenium (which can be found in various foods such as eggs, nuts, legumes, and meats) as a potent protective nutrient against certain forms of cancer. Low selenium levels, combined with diminished antioxidant defenses, have been demonstrated to increase cancer risk (Biesalski 2016, Zheng et al. 2022).

The MD basis is a high consumption of fruit and vegetables, which are also a source of vitamins and minerals (Willett et al. 1995). Increased consumption of fruits and vegetables, coupled with the avoidance of a Western diet, has been associated with a positive effect in reducing CRC incidence (Donaldson 2004).

Vitamins from the B group are essential micronutrients for many fundamental metabolic processes crucial to human health and life. As will become apparent from the following paragraph, it is very hard to unequivocally clarify and define their absolute beneficial or detrimental effects on inflammation and tumors, as their functions are extremely varied, making metabolic processes complex to understand (Peterson et al. 2020).

Vitamin B1, better known as thiamine, is received by the intestine from two primary sources: the diet and GM production (den Besten et al. 2013, Calder 2012). While the GM play a vital role in synthesizing various water-soluble vitamins, its overall impact on host vitamin status remains uncertain. Limited research has been conducted on the dynamics of vitamin movement between the microbiota and the gut, as well as their influence on gut barrier function and the immune system (Biesalski 2016).

Thiamine acts as an essential cofactor for enzymes involved in energy metabolism, including the citric acid cycle and the formation of adenosine triphosphate (ATP). As previously mentioned, the GM play a significant role in thiamine synthesis, with both the free (unphosphorylated) and phosphorylated forms of thiamine being absorbed by the colonic mucosa. This bacterial-derived thiamine has the potential to be absorbed by human colon cells, influencing thiamine status at both the colonic and systemic levels (Biesalski 2016). Thiamine exhibits contrasting anti-cancer properties, showing both cancer-promoting and anticancer effects, which can vary based on genetic and non-genetic factors (Bruce et al. 2003, Comín-Anduix et al. 2001). While thiamine supplementation may stimulate tumor cell survival, proliferation, and chemotherapy resistance, other studies have highlighted its beneficial role in cancer (LU'O'NG, Nguyễn, and proteomics 2013).

Similarly, riboflavin, also known as vitamin B2, acts as a crucial cofactor for numerous enzymes, particularly those involved in redox reactions (Biesalski 2016). A slight but not significant inverse association was observed between riboflavin intake and proximal colon tumors in women (de Vogel et al. 2008). However, a retrospective case-control study conducted by Ma et al. (2023) suggests that higher levels of riboflavin may play a role in facilitating colorectal carcinogenesis (Ma et al. 2023).

Niacin (vitamin B3) serves as a precursor for the nicotinamide nucleotide coenzymes NAD and NADP. NAD is crucial for energy-producing reactions, while NADP is involved in anabolic reactions. Additionally, NAD participates in non-redox transfer reactions related to protein modification, calcium mobilization, cell signaling, and DNA repair (Kirkland 2013). Due to these roles, it has been associated as a protective factor against inflammation and tumorigenesis (Mohseni, Taghinezhad-S, and Fu 2020).

Vitamin B6, in its coenzyme form, performs a wide array of functions primarily related to protein metabolism (Theodoratou et al. 2008). For this reason, its consumption has been positively associated with CRC, especially in women (de Vogel et al. 2008).

Finally, folic acid, a water-soluble B vitamin naturally found in foods like green leafy vegetables and partially synthesized by the GM (Biesalski 2016, Kim 2007), is crucial for DNA and RNA synthesis, cellular methylation reactions, and amino acid metabolism (Biesalski 2016). A deficiency in folic acid has been associated with an increased CRC risk, especially when combined with high alcohol consumption (Giovannucci et al. 1993). Furthermore, it seems that the protective effect of the Methylenetetrahydrofolate reductase (MTHFR) C677T gene, linked to a decreased CRC risk (but promoting other tumor forms), is observed only in subjects with adequate folate intake (La Vecchia et al. 2002).

To summarize, B vitamins might provide protection from cancer development but play a detrimental role in patients already experiencing carcinogenesis (Peterson et al. 2020).

Both vitamins A and D are critical for the intestinal immune system, mucosal barrier function, and overall gut health. In addition, these vitamins intricately influence the immune response and barrier function of the intestine, highlighting the interconnected roles that may explain the associations between malnutrition, gut microbiota, and health outcomes (Biesalski 2016). It has been advocated that vitamin D also reduces the risk of CRC by inhibiting neoangiogenesis and cellular proliferation, and inducing apoptosis (Vernia et al. 2021).

Finally, the MD is well known for its high content in bioactive compounds, such as carotenoids and polyphenols (Saura-Calixto and Goñi 2009).

Carotenoids are fat-soluble terpene pigments, including α-carotene, β-carotene, and lycopene, known for their anti-tumor and anti-inflammatory effects (Donaldson 2004). β-carotene serves as a provitamin A, contributing to essential cellular functions. It has been demonstrated to significantly reduce cell proliferation in vitro in CRC cell lines (Costea et al. 2018). Regarding lycopene, its impact on colorectal cancer is still uncertain. However, it is recognized as a potential chemopreventive agent due to its antioxidant capacity (Di Mascio, Kaiser, and Sies 1989). Lycopene effectively suppresses cell growth by inhibiting the phosphorylation of protein kinase B and, consequently, the phosphatidylinositol-4,5-bisphosphate-3-kinase signaling pathway, which is intricately linked to CRC development (Costea et al. 2018).

7.4 THE WESTERN DIET AND ITS IMPACT ON CANCER RISK

A significant shift in lifestyle has occurred over recent decades in Westernized societies, involving the adoption of Western-style diets (WD). This dietary pattern includes high consumption of ultra-processed foods, red and processed meat, convenient snacks, and sugary beverages, often lacking essential nutrients like fiber, vitamins, and minerals. The consumption of these food products has increasingly extended from high- to low-wealth countries, resulting in a simultaneous increase in WD-associated diseases, such as metabolic syndrome, autoimmune and neurodegenerative diseases, and cancers (Imamura et al. 2015). WD can lead to obesity, inflammation, and structural and functional GM changes, contributing to cancer onset (Newsome, Yang, and Jobin 2023).

A meta-analysis of 93 studies, including 21 focused on CRC, evaluated the association between a "healthy" and an "unhealthy" dietary pattern and cancer risk. The "unhealthy" pattern, resembling a WD, was positively associated with CRC risk (Grosso et al. 2017). Furthermore, this analysis revealed a connection between unhealthy eating habits and increased BMI, while also indicating that adopting healthy dietary patterns was linked to better lifestyle choices, such as increased physical activity and absence of smoking. Mehta et al. evaluated two combined cohorts, involving 137,217 participants for 32 years follow-up (Mehta et al. 2017). During this period, 3,260 people developed CRC. The researchers concluded that the WD pattern was associated with an increased risk of CRC, specifically in the distal colon and rectum, and with particular molecular subtypes.

WD influences carcinogenesis through the involvement of GM and their metabolites, which affect the immune, nervous, and hormonal systems (Newsome, Yang, and Jobin 2023). One major issue with the WD is the significant reduction in fiber intake. Adequate consumption of dietary fiber promotes the GM fermentation of carbohydrates and the production of SCFA, suppressing both oncogenic signaling and production of carcinogenic metabolites (O'Keefe and Hepatology 2019, Veettil et al. 2021, Gianfredi et al. 2018). Fibers bind bile acids, altering the gut-liver axis and lowering cholesterol levels involved in CRC progression via activation of ROS and signaling pathways, including the mitogen-activated protein kinase (MAPK) pathway (Wang et al. 2017).

Red and processed meat are key components of WD and are classified by the World Health Organization and the International Agency for Research on Cancer as "probably carcinogenic to humans" (group 2A) and "carcinogenic to humans" (group 1), respectively (Bouvard et al. 2015, Jeyakumar, Dissabandara, and Gopalan 2017). Several components found in red and processed meat seem to be linked with the onset of CRC: heme, N-nitroso compounds (NOCs), heterocyclic amines (HCAs), N-glycolylneuraminic acid (Neu5Gc), and polycyclic-aromatic hydrocarbons (PAHs) (Jeyakumar, Dissabandara, and Gopalan 2017).

Within the Italian EPIC cohort, increased consumption of red meat and fish was linked to a higher risk of CRC compared to lower intake (Agnoli et al. 2013). However, in the overall EPIC cohort, meat consumption showed no impact on CRC risk, while higher fish consumption proved to be protective, but only for men (Bamia et al. 2013). In a combined analysis of case-control studies in Italy, higher consumption of meat and meat products was associated with an elevated risk, while reduced risk was observed with fish consumption (Rosato et al. 2016).

Turner and Lloyd conducted an in-depth exploration of the correlation between red meat and CRC by systematically reviewing 40 studies employing animal and cell culture models (Turner, Lloyd et al. 2017). They documented that most of the studies administered meat or meat-related elements at levels notably higher than those typically found in the human diet. The presence of specific dietary components (such as fermentable fiber and calcium carbonate), along with the consumption of fruits, vegetables, and whole grains, appeared to mitigate and, in some cases, even counteract the potential cancer-inducing effects of meat and meat-related compounds. Adherence to a complete MD pattern might explain the reduced link between meat consumption and the risk of CRC in specific Mediterranean populations (Turner, Lloyd, and medicine 2017).

A comprehensive meta-analysis conducted by Qiao et al. in 2013 revealed a notable association between heme iron intake and an elevated risk of CRC (Qiao and Feng 2013). Additionally, the influence of iron on the GM composition is well-established (Bielik and Kolisek 2021a). This alteration favors *Escherichia coli* over lactic acid bacteria, contributing to intestinal inflammation. From a preventive perspective, the Bifidobacteriaceae family emerges as a noteworthy consideration (Bielik and Kolisek 2021b). These bacteria are able to bind iron, thereby reducing the likelihood of free radical formation, and are consequently suggested as a potential preventive measure (Bielik and Kolisek 2021a). Meat products are also one of the main sources of dietary proteins, which play a crucial role in human nutrition, beyond their primary function of building tissues. They also influence body composition, regulate diverse metabolic pathways, contribute to feelings of satiety, and impact immune responses (Ferrari et al. 2022).

Following the complete digestion of proteins in the intestinal mucosa, the resulting products, such as free amino acids, dipeptides, and tripeptides, are partially absorbed in the gut through specific transporters present in enterocytes. These absorbed components then enter the bloodstream, playing a crucial role in protein synthesis and various metabolic processes. However, some of these protein digestion products escape absorption and reach the colon, where the resident microbiota further metabolize them, producing a variety of metabolites (Lin et al. 2017) as discussed below.

While the human body typically maintains an amino acid balance, malignant tumors often disrupt this balance by relying on specific amino acids for their survival and rapid growth (Xu et al. 2023).

A study by Coker et al. revealed that several amino acids, particularly L-alanine, glycine, L-proline, L-aspartic acid, L-valine, L-leucine, and L-serine, were enriched in CRC patients compared to colorectal adenomas (CRA) and normal controls (NC) (Coker et al. 2022).

Another study by Wang et al. uncovered a new subgroup of B cells, defined LARS B cells, identified in CRC (Wang et al. 2022). LARS B cells express leucine-tRNA synthase-2 (LARS2) and exhibit a marked regulatory feature involving the secretion of transforming growth factor β1 (TGF-β1). Notably, LARS B cells preferentially use leucine as a nutrient source and demonstrate enhanced mitochondrial aminoacyl-tRNA biosynthesis. LARS B cells are associated with colorectal hyperplasia, and their presence correlates with reduced survival in CRC patients. Notably, intervention strategies such as Lars2 gene ablation or leucine blockade effectively suppress CRC immunoevasion by removing LARS B cells (Wang et al. 2022).

Furthermore, amino acids like valine (Val), leucine (Leu), and isoleucine (Ile), due to their molecular structure, are part of branched-chain amino acids (BCAAs). While they can be synthesized in bacteria, plants, and fungi, animals cannot produce them. In a review by Xu et al., BCAAs were linked to the onset and development of CRC, acting as regulatory factors that promote gut development and immune function, thereby improving gut health (Xu et al. 2023).

In addition, the human body maintains a balance regarding amino acids. However, malignant tumors often depend on specific amino acids for their survival and rapid growth (Wang et al. 2022).

Finally, the WD is characterized by a high intake of refined sugar. Diets with a higher glycemic index, promoting consistent increases in post-meal blood glucose levels, might contribute to increased colon cancer risk, potentially through the insulin/insulin-like growth factor 1 (IGF-1) signaling pathway (Kasprzak 2021).

Several studies propose that a diet low in carbohydrates and sugars may mitigate the risk of CRC (Jonoush et al. 2022), as there is convincing evidence to support the role of hyperinsulinemia and insulin resistance in CRC development. Indeed, high circulating levels of insulin and C-peptide, a marker indicative of insulin resistance, are consistently correlated with an increased risk of CRC (Yuan and Giovannucci 2020). However, conflicting results, such as those presented by the prospective study conducted by Kanehara et al. in the Japanese population, suggest that dietary sugar is not associated with CRC risk (Kanehara et al. 2023).

The resolution of these discrepancies will contribute to a more complete understanding of the intricate connections between dietary factors, insulin regulation, and colorectal cancer.

A recent review analyzed the association between a high-fat WD and the GM, revealing an association with dysbiosis that reduces SCFA production and SCFA receptor expression, contributing to inflammation and intestinal permeability

(Malesza et al. 2021). Dysbiosis is also linked to endocannabinoid system activity, which contributes to LPS translocation and increased gut permeability (Cani et al. 2007, Muccioli et al. 2010). Changes in the microbiota due to a high-fat WD can disrupt the expression of genes related to inflammation and metabolism. This diet exacerbates oxidative stress by increasing the production of ROS and reactive nitrogen species (RNS) (Kobayasi et al. 2010, Tapias et al. 2017) that, in turn, triggers endoplasmic reticulum (ER) stress, downregulating gut peptide signaling pathways and diminishing their secretion by enteroendocrine cells, ultimately contributing to a compromised gut barrier (Bleau et al. 2015, Sano and Reed 2013, Zhang et al. 2006). Dysbiosis and inflammation independently compromise the mucus layer, decreasing tight junction expression, resulting in a leaky gut, heightened LPS translocation, and metabolic endotoxemia (Guo et al. 2015, Velasquez and disorders 2018). Additionally, Toll-like receptor 4 (TLR4) activation can be triggered not only by LPS but also by saturated fatty acids (SFA), highlighting the convergence of both GM changes and high-fat WD in stimulating nuclear factor kappa B (NF-κB) and subsequent pro-inflammatory cytokine production (Ohashi et al. 2014, Shi et al. 2006). A noteworthy connection between high-fat WD and dysbiosis involves bile acids, whose increased secretion can impair gut barrier function and exert pro-inflammatory effects on both the gut and microbiota composition (Islam et al. 2011, Murakami, Tanabe, and Suzuki 2016).

The WD is commonly adopted in North American countries, including the U.S.A. and Canada, and has rapidly spread globally, spanning from the wealthiest to the poorest areas. In the next section, we will explore the correlation between the risk of gut-associated cancers and eating habits, particularly in relation to traditional foods of different world areas.

7.5 GEOGRAPHIC REGIONS' TRADITIONAL FOODS AND GUT-ASSOCIATED CANCERS

7.5.1 Central and South America's Traditional Foods

The Central and South American region has experienced a rapid epidemiologic and demographic transition, resulting in a change of the cancer profile. The prevalence of infection-related cancers, such as cervical and stomach cancers, is decreasing, while there is an emerging trend of cancers usually diagnosed in countries with the highest human development, including CRC (Sierra and Forman 2016).

The Mexican traditional diet is characterized by maize, maize tortillas, beans, squash, tomato, chili, avocado, nuts, seeds, chocolate drinks, and pulque, a fermented drink (Valerino-Perea et al. 2019). There are not specific data about gastrointestinal cancers, but one case-control study showed a 32% risk reduction for breast cancer development in women with high adherence to the Mexican traditional diet compared to those with lower adherence (Murtaugh et al. 2008). This diet is potentially high in fiber, as it contains grains, legumes, fruits, and vegetables, which, as previously described, have many health benefits, including being protective for colon cancer (Valerino-Perea et al. 2019).

Moving to Brazil, Nascimento et al. investigated the preventive and therapeutic potential of native Brazilian fruits on CRC (Nascimento et al. 2022). Brazilian fruits, such as açaí, cocoa, jaboticaba, and passionfruit, are rich in bioactive compounds, primarily phenolic compounds, characterized by antioxidant, anti-inflammatory, and anticancer properties, making them potential candidates for CRC management. Studies on açaí (*Euterpe oleracea*) and cocoa (*Theobroma cacao*) show their effects on CRC in in vitro and in vivo models with a nuanced understanding of molecular mechanisms (Nascimento et al. 2022). The primary signaling pathways influenced by these native Brazilian fruits, as studied in CRC, are linked to apoptosis (involving BCL-2 and caspase 3-related proteins), inflammation (inflammatory interleukins and enzymes), and oxidative stress (antioxidant enzymes). Furthermore, downregulation of adhesion molecules and NF-κB has been observed, suggesting positive interference in various cellular processes, including proliferation, angiogenesis, and migration. Despite these findings, further exploration of signaling pathways crucial for CRC initiation and progression remains imperative (Nascimento et al. 2022).

Finally, a study explored the relationship between dietary patterns and CRC risk in the Uruguayan population (De Stefani et al. 2011). Results observed four different dietary patterns in the male population, among which a "traditional" one was characterized by cooked vegetables, legumes, tubers, and desserts. The traditional dietary pattern was found to be inversely associated with rectal cancer, but not with colon cancer (De Stefani et al. 2011).

7.5.2 Pacific Countries' Traditional Foods

Australia and New Zealand exhibit the highest CRC incidence in the world, being the second leading cause of cancer-related death in Australia (Crawford-Williams et al. 2018). Currently, there is a limited amount of research examining the link between CRC and traditional dietary patterns in Pacific countries, but studies conducted on dietary habits show that a WD is associated with a higher occurrence of CRC, also in this geographical area (Kune 2010). On the other hand, in New Zealand, individuals of Polynesian descent, including Māori and those from various Pacific Islands, exhibit a higher likelihood of being overweight, along with increased fat and energy intake compared to Europeans. Surprisingly, despite having similar overall food plant intakes, Polynesians experience a significantly lower occurrence of CRC than their European counterparts (Ferguson et al. 1995). This discrepancy in colorectal cancer incidence suggests the possibility that variations in the consumption of specific food plants between Polynesians and Europeans may play a role (Ferguson et al. 1995). In addition, Māori New Zealanders show a lower incidence of CRC than non-Māori. Data from the 1997 New Zealand National Nutrition Survey revealed that, when compared to non-Māori populations, Māori individuals tended to have higher overall food intake, increased consumption of red meat, elevated alcohol consumption, higher intake of saturated fat, a greater obesity prevalence, and a lower percentage of individuals meeting recommended daily levels of fruit and vegetable consumption (Thomson and Shaw 2002). In term of specific traditional foods, Kumara, a typical sweet potato, is the main vegetable source for Māori people and may have anti-cancer properties, thanks to its content in sporamins A and B, which function as inhibitors. Additionally, the tubers contain coumarins such as scopoletin, aesculetin, and umbelliferone, which have anti-inflammatory properties (Cambie and Ferguson 2003).

7.5.3 Northern Europe's Traditional Foods

Northern European countries, including Germany, the Netherlands, the U.K., and the Scandinavian countries, have a diet that is characterised by the consumption of potatoes, processed food, margarines, animal fats, sweetened foods, red meat, and processed meat. Furthermore, in these countries, the intake of vegetables and fruit seems to be lower than in Southern European countries, while the consumption of legumes and vegetable oils is unusual in their diets (Slimani et al. 2002).

In recent decades, the Nordic diet (ND) model has been proposed in Scandinavian countries as a healthy and sustainable choice that considers the traditions and culture of the Nordic countries. The ND is based on traditional foods such as lingonberries, cabbage, apples, pears, root vegetables, oats, rye, and fermented milk. It promotes an increased seafood and legume consumption as the main sources of proteins and the use of canola oil as the main fat source (Blomhoff et al. 2023).

The higher adherence to an ND model is related to lower CRC incidence for both males and female (Kyrø et al. 2013). The level of protection observed was comparable to that identified in previous studies on the MD. These findings imply that advocating for the adoption of healthy regional diets is not only beneficial for health but also contributes to the preservation of cultural heritage and the environment (Kyrø et al. 2013).

Asli et al. investigated the relationship between potato consumption and CRC risk in Norwegian women. Results showed that high potato intake was associated with higher risk of CRC when comparing the highest consumption (>3 potatoes per day) with the lowest one (0–7 potatoes per week). However, the effect was only evident for women with a BMI < 25 kg/m^2, after stratifying by BMI (Åsli et al. 2017).

Finally, a systematic review investigated the association between ND-characteristic foods and health outcomes (Åkesson et al. 2013). The results showed low-grade evidence for whole grains as protective for CRC, but the study was mostly inconclusive, identifying the need for high-quality nutritional epidemiological and intervention studies about foods of the ND.

7.5.4 African Traditional Foods

CRC incidence has increased significantly in North African countries, such as Egypt, Morocco, and Libya (Awedew, Asefa, and Belay 2022). One of the potential causes identified is the transition to a Westernized lifestyle. In Libya especially, CRC is more common in urban areas, where people follow a WD and display more sedentary behaviors than in rural areas, where a relatively traditional way of life is maintained with farming, animal rearing, and small industries as the main occupations, and where people eat traditional cuisine focusing on whole grains and Mediterranean-style meals (Bodalal and Bendardaf 2014).

Furthermore, in Morocco, CRC stands as the third most prevalent cause of cancer-related mortality for both sexes (Sung et al. 2021). Simultaneously, there is a noticeable transition in the Moroccan population's dietary patterns, shifting from a Mediterranean diet to a more Westernized one.

In detail, the Moroccan diet is traditionally characterized by a significant intake of cereals, fruits, and vegetables while maintaining a low consumption of dairy and meat products. This transformation is marked by a rising consumption of red and processed meat over the past few decades. Consequently, also in Morocco, the increasing CRC incidence could be linked to these shifts in lifestyle and nutritional habits (S. Deoula et al. 2020).

The most common Moroccan traditional processed meats are *Kaddid* and *Khlii*. *Kaddid* is created by thinly slicing the ribs of beef, lamb, or camel, then combining them with olive oil and an assortment of spices and herbs like cumin, crushed garlic, salt, vinegar, and coriander. The mixture is hung in a sunny spot until it dries. Conversely, *Khlii* is made by cooking *Kaddid* in a blend of beef fat, olive oil, and water. When stored in the fat used for cooking, it can be preserved at room temperature for up to two years. A case-control study explored the relationship between red and white meat, traditional processed meat, industrial processed meat consumption, and CRC development in the Moroccan population (S. Deoula et al. 2020). The results showed an inverse association between traditional processed meat and CRC risk, and confirmed a positive association between industrial processed meat and CRC (S. Deoula et al. 2020).

An additional Moroccan study analyzed the association between modern and traditional dairy products (such as *Lben*, *Jben*, *Saykok*, *Raib*) and CRC. *Lben*, a traditional Moroccan beverage, is typically derived from cow's milk. Various other types of milk, such as from goat, sheep, camel, or a blend of these, are also suitable for making *Lben*. The fermentation process occurs over 24–48 hours at temperatures ranging from 18 to 24°C, adjusted according to the season. After coagulation, the curd is placed in a goatskin bag known as a *Chekoua* and vigorously stirred back and forth. The results showed an inverse association between higher intake of both types of dairy products and CRC risk (El Kinany et al. 2020).

Further, there has been an increase in cancer cases in Egypt, including colon cancer (Awedew, Asefa, and Belay 2022). Mahfouz et al. studied the relation between dietary and lifestyle factors and CRC development in patients attending El-Minia Oncology Center and compared them with healthy controls (Mahfouz et al. 2014). The results suggested that the CRC cases enrolled in this study had adopted a sedentary lifestyle and their dietary patterns were characterized by an unhealthy diet.

A different study evaluated the anti-cancerous properties of one of the most traditional Egyptian food components: Diallyl disulfide (DADS), a major organosulfur compound in garlic oil. In particular, the authors studied the effect of DADS on HT29 (a colon cancer cell line) and in vivo using male rabbits (*Oryctolagus cuniculus*) as a CRC animal model (Altonsy, Habib, and Hassanain ElSA 2015). DADS exhibited distinct effects on the expression of a set of genes. It down-regulated the expression of oncogenes such as CTNNB1, CCDN1, BIRC5, MYC, and AKT while concurrently up-regulating the expression of the tumor suppressor gene TP53 and the apoptosis regulator gene BAX. The apoptotic impact of DADS was further evident through the induction of cytochrome C expression and the activation of caspase-3 (Altonsy, Habib, and Hassanain ElSA 2015).

Moreover, an increase in the burden of CRC has also been described in sub-Saharan African countries, such as Zimbabwe, Kenya, South Africa, Mozambique, and Nigeria. The increase of this phenomenon has been linked to shifts in dietary patterns and lifestyle, but also to advancements in diagnostic approaches (Katsidzira et al. 2017).

One study conducted in Zimbabwe evaluated three main dietary patterns: traditional African, urban, and processed foods. The traditional African diet, characterized by the consumption of indigenous fruits and vegetables, wild animals and insects, nuts, seeds, and indigenous grains, seemed to be protective against CRC, which had no association with the other two patterns (Katsidzira et al. 2018).

In detail, the traditional West African diet consists of carbohydrate-based meals typically accompanied by vegetable soup. In rural and poor communities, meat consumption is rare compared to Caucasians, even if some authors have implicated the carcinogenic properties of charcoal-roasted meat, known as "suya" in Nigeria, due to the content of polycyclic aromatic hydrocarbons (PAHs) and heterocyclic amines (HCAs) formed during cooking. The carbohydrate-rich diet of Nigerians has demonstrated protective effects against colon cancer, thanks to the SCFA production (Costea et al. 2018).

Cassava, another staple in the Nigerian diet, may contribute to cancer protection through a different mechanism, as it contains a chemical called tamarin, responsible for hydrocyanide production (Irabor 2011). In vitro studies have shown that tamarin induces cancer cell death through self-toxicity involving hydrocyanide (Marandola et al. 2004).

Finally, Bigman et al. conducted a review of sub-Saharan Africa dietary habits and overall cancer risk, concluding that most of the studies reported an association between cancer risks and dietary intake, but they lacked relevant cofounders (Bigman et al. 2022). Large cohorts with extensive datasets, validated dietary tools, and advanced statistical analyses are essential to gain deeper insights into the crucial connections between the African diet and the risk of cancer.

7.5.5 Asian Traditional Foods

Up to a few decades ago, CRC was considered uncommon in Asia, with its highest prevalence observed in North America and Europe. Subsequently, there has been a consistent and swift rise in the incidence, prevalence, and mortality of CRC in the Asian region (Yee et al. 2009). The surge in CRC cases can be attributed to a Westernized dietary lifestyle, the aging of the population, smoking, physical inactivity, and other risk factors. Ethnicity has been identified as a significant etiological factor for colorectal CRC in Asia. Some ethnicities, including Chinese, Korean, and Japanese, exhibit a higher incidence of CRC (Onyoh et al. 2019).

A systematic review analyzing the relationship between dietary habits and CRC risk in Asia identified a positive association for red and processed meats, preserved foods, saturated/animal fats, cholesterol foods, high sugar foods, spicy foods, tubers, and refined carbohydrates (Azeem et al. 2015). On the other hand, calcium/dairy, vitamin D (by aiding calcium absorption), general consumption of vegetables, fruits, and fiber, cruciferous vegetables, soybeans/products, selenium, vitamin C, vitamin E, vitamin B12, lycopene, alpha-carotene, beta-carotene, and folic acid have a protective role against CRC risk (Azeem et al. 2015). No differences between modern and traditional foods were considered in the review.

Elsewhere, a Japanese cohort study identified a "Western" pattern, but also a "traditional" pattern, associated with higher CRC risk in females. The "traditional" pattern was characterized by Japanese traditional foods, such as rice and salted foods (pickled vegetables, salted fish and roe, and miso soup), which are hypothesized to be an indirect indicator of exposure to genotoxic carcinogens (Kim, Sasaki, et al. 2005).

Additionally, a further study in the Korean population identified a "prudent" and a "traditional" pattern, which were inversely associated with CRC risk (Park et al. 2016). On the other hand, a "Westernized" pattern showed a positive association, especially among women. The "traditional" pattern was characterized by vegetables, fish, seaweeds, soy products, mushrooms, and seasoning (Park et al. 2016).

Regarding the consumption of soy products, typical of Asian countries, a recent meta-analysis found that a higher intake of soy products is associated with lower overall gastrointestinal cancer and gastric cancer risk, but not CRC (Lu et al. 2017).

A Chinese study evaluated the diet of CRC patients and healthy controls (Seow et al. 2002) showing that high meat consumption and low intake of vegetables was associated with higher risk, and the combined effect appeared to be additive. There was no association observed with fruit, legumes, or soy intake (Seow et al. 2002).

The Singapore Chinese community has undergone significant industrialization and urbanization in recent decades, leading to a shift toward Westernized dietary habits, accompanied by an increased incidence of cancer, cardiovascular disease, and stroke (Earticle 2012). Butler et al. identified two primary dietary patterns: "vegetable-fruit-soy," characterized by a mix of traditional foods like noodles, white rice, soy products, and green leafy vegetables; and "meat-dim sum," composed of non-traditional foods such as red meat and dairy. Additionally, a prospective study (using nearly ten years of follow-up data) aimed at verifying the potential risk of these new dietary habits documented no association between specific foods or dietary patterns and the risk of CRC (Butler et al. 2008).

The traditional Indian diet is rich in foods, recipes, and spices renowned for their cancer-preventive properties (Singh and Yadav 2022) (Peters et al. 2003). The presence of spices is a characteristic element of the cuisine of this country; among the most commonly used are turmeric, ginger, chili, garlic, coriander, cloves, cinnamon, cumin, and cardamom (Singh and Yadav 2022) (Aggarwal, Kumar, and Bharti 2003).

Turmeric is one of the main ingredients in Indian recipes, including Indian curry, and has demonstrated potent antioxidant and anti-inflammatory properties, along with potential as a chemo-preventive agent (Aggarwal, Kumar, and Bharti 2003). The key chemical components of turmeric consist of a set of compounds known as curcuminoids, among which the main one is curcumin (diferuloylmethane) (Itokawa et al. 2008). Curcumin exhibits anti-cancer properties: in both in vitro and in vivo studies, it has been shown to contrast cell proliferation via a variety of mechanisms, including regulation of apoptosis (Jiang et al. 2015), suppression of angiogenesis (Bimonte et al. 2015), reduction of cell growth (Chen, Xu, and Johnson 2006), and autophagy activation (Mosieniak et al. 2012). Additionally, turmeric has been found to inhibit the growth of *Helicobacter pylori*, a carcinogenic bacterium associated with stomach adenocarcinoma and colorectal adenomas (Mahady et al. 2002).

Chili pepper exhibits similar activities thanks to its capsaicin content, which has antioxidant (Joung et al. 2007), anti-inflammatory (Kim et al. 2003), and anti-proliferative effects (Jin et al. 2014).

According to in vitro and animal studies, ginger and its components have the potential to act as chemopreventive agents by reducing COX-2 expression (Citronberg et al. 2013, Kim, Kundu, et al. 2005). In addition, garlic and its components have been shown to inhibit tumor biomarkers such as aberrant crypt foci (ACF), NF-κB, anti-apoptotic genes (Bcl-2, cIAP1/2, and XIAP), and inflammatory genes (iNOS and COX-2), as well as EGFR. Conversely, garlic promotes the expression of apoptotic genes (Ban et al. 2007, Ngo et al. 2007). It has been reported that cardamonin, found in cardamom, also demonstrates the ability to block the cell cycle and induce programmed cell death in malignant cells (Nawaz et al. 2020).

A characteristic of North Indian cuisine is the use of saffron. A study found that saffron, particularly its compound crocin, inhibits the growth and mobility of colon cancer cells induced by MACC1 (metastasis associated in colon cancer 1), a protein associated with metastasis. Saffron slows down cell cycle progression without triggering cell death and reduces levels of the cancer stem cell marker DCLK1, thereby suppressing cancer cell proliferation and migration. This highlights saffron's potential in targeting cancer progression through MACC1 inhibition (Güllü et al. 2020).

Traditional South Indian food is defined as "functional" thanks to its content of dietary fibers (whole grains and vegetables), antioxidants (spices, fruits, and vegetables), and probiotics (curds and fermented products), which all exhibit a synergistic effect on health (Devarajan and Mohanmarugaraja 2017). Functional foods encompass certain bacterial strains and products derived from plants and animals, containing physiologically active compounds that contribute to human health and lower the risk of chronic diseases (Grajek, Olejnik, and Sip 2005). A typical food containing one or more inherent functional ingredients can be termed traditional functional food (Devarajan and Mohanmarugaraja 2017).

The South Indian traditional food *Rasam*, a soup prepared using tamarind juice with a variety of spices (coriander, garlic, curry leaves, cumin, black pepper, mustard, turmeric, red chili, and asafoetida), is used from the resident population for the treatment of various health conditions, such as diabetes (Bolla et al. 2015), cold (Vijayalakshmi et al. 2016), and constipation (Swapna 2015). Some of *Rasam's* functional ingredients (such as tamarind, turmeric, garlic, ginger, and chili peppers) exhibit anti-cancer, antioxidant, and anti-inflammatory properties (Emmy De Caluwé 2010).

In detail, tamarind seeds show antioxidant activity, thanks to their phenolic compound content (Soong and Barlow 2004), and an anti-inflammatory effect due to the presence of protein inhibitors, such as serine proteinase, against human neutrophil elactase (HNE) (Fook et al. 2005).

However, to date, there have been no studies analyzing the specific association between traditional *Rasam* and cancer risk (Devarajan and Mohanmarugaraja 2017).

Another Indian traditional food, the fermented rice *Bhaati Jaanr* (BJ), seems to have antitumor activity and could serve as a potential therapeutic intervention against colon cancer. BJ consists of a traditionally fermented rice paste commonly consumed as a sweet beverage in the rural areas of Darjeeling Hills, Sikkim, and the northeast states of India. A study by Jaiswal et al. reveals significant anti-proliferative, antioxidant, and anti-inflammatory potential, aligning it with other fermented food products of the BJ extract on colon adenocarcinoma cell lines (Jaiswal et al. 2023). The beneficial properties are probably linked to the probiotic lactic acid bacteria (LAB) and yeast isolates from BJ, underscoring the positive attributes of this traditional food. Elevated counts of LAB species, such as *Pediococcus* and *Enterococcus*, in BJ suggest a substantial content of SCFA and bacteriocins (Jaiswal et al. 2023).

Overall, the fermentation of dairy products and their bacterial components is recommended for their health-promoting functions, potentially reducing the risk of colorectal cancer. *Kefir*, originating from the Caucasus and Tibet, is a traditional fermented milk product obtained by incubating *kefir* grains with raw milk or water. These grains consist of a symbiotic association of yeast and bacteria. *Kefir* grains possess health-promoting properties, including antibacterial, antifungal, anti-allergic, and anti-inflammatory effects (Cevikbas et al. 1994, Hong, Chen, and Chen 2010, Liu et al. 2002). Additionally, certain bioactive compounds present in *kefir*, such as polysaccharides and peptides, are believed to hold significant potential in inhibiting the proliferation and inducing apoptosis in tumor cells (Sharifi et al. 2017). A study demonstrated that the intake of *kefir* and *ayran*, a traditional Turkish fermented yogurt-based drink, significantly reduces DNA damage, with the effect of *ayran* being dose-dependent (Grishina et al. 2011). The supernatants of *kefir* and *ayran* were rich in acetic and lactic acid, exhibiting substantial antioxidant capacity. This evidence suggests that *kefir* and *ayran* may contribute to decreasing DNA damage due to their antioxidant capabilities (Grishina et al. 2011). Furthermore, the *kefir* anticancer potential on colorectal cancer cell lines (Caco-2 and HT-29) highlights its ability to induce cell cycle arrest at the G1 phase (Rizk et al. 2014). Finally, *kefir* decreases the expression of TGF-β1 and TGF-α in the HT-29 cell line. The upregulation of the Bax/Bcl-2 ratio supports the pro-apoptotic effect of *kefir*, along with an increase in p53-independent p21 expression (Rizk et al. 2014).

7.6 DIET, GUT MICROBIOME, AND CANCER RISK

As emerged in the previous paragraphs, the human microbiome is an extremely complex and dynamic ecosystem comprising diverse microorganisms. Diet stands out as a crucial determinant, exerting the most profound impact on the elaborate balance of microbial communities within the gut. The influence of the microbiota on host health is multifaceted, operating through both direct and indirect mechanisms. Some microbiota bacteria possess the capability to directly impact host health by producing toxins, which can exert harmful effects on the host's cells and tissues, contributing to various health conditions. Notably, specific bacterial species, such as *Fusobacterium nucleatum*, *Escherichia coli*, and *Bacteroides fragilis*, influence susceptibility and the progression of CRC (Zou, Fang et al. 2017, Genua, Raghunathan et al. 2021). Probiotic species such as *Lachnospiraceae*, *Bifidobacterium animalis*, and *Streptococcus thermophilus* are found to decrease in patients affected by this neoplastic pathology (Dai et al. 2018). In addition, other bacterial species may either contrast or promote tumor growth by modifying immune system surveillance mechanisms or the metabolism of chemotherapeutic agents (Cho and Blaser 2012, Levy, Kolodziejczyk et al. 2017, Niccolai, Boem et al. 2020). On the other hand, microbiota also employ indirect mechanisms to influence host health. Metabolic by-products generated by food components play a crucial role in shaping the host's physiological environment, contributing to either a balanced and symbiotic relationship or disruptions that may lead to health issues (Fang, Yan et al. 2021). Indeed, through the production of a diverse array of metabolites, microbiota can induce inflammation or modulate immunity (Gagnière, Raisch et al. 2016) (Figure 7.1). Consequently, the disturbance of the microbiota, including geospatial and functional aspects, influences the state of health and appears to play a role in CRC development, leading to alterations in immune responses and intestinal barrier function.

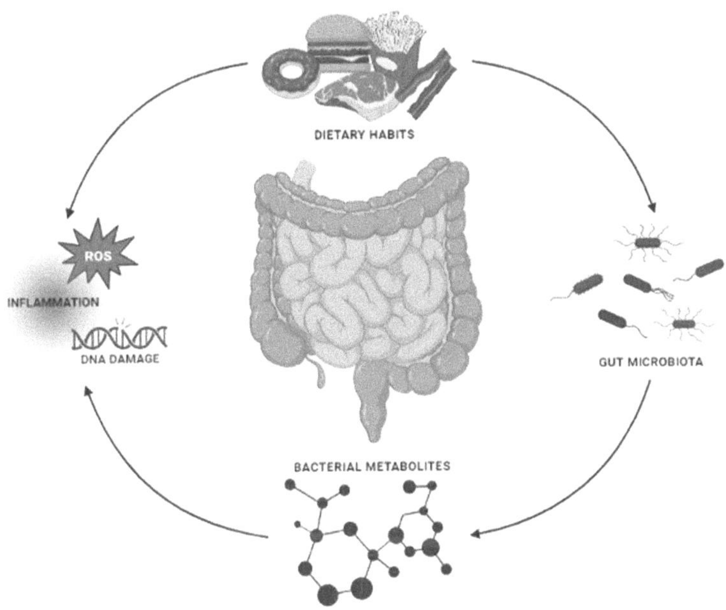

FIGURE 7.1 Microbial metabolites stimulated by food intake. Nutrients from dietary foods are used by the gut microbiota to produce bacterial metabolites. Diet can alter gut microbial diversity, consequently altering the microbial end-products. In addition, nutrients from diet can directly establish an inflammatory process, thereby involving the immune system. The sum of these events can lead to the onset and progression of colorectal cancer.

Among the potentially harmful microbial products, there is trimethylamine N-oxide (TMAO), a derivative of the gut bacteria metabolite trimethylamine (TMA), which undergoes oxidation in the liver. TMA is synthesized from specific substances, including carnitine present in poultry, red meat, and some dairy products, as well as choline (Jalandra, Dalal et al. 2021). Notably, TMAO belongs to the group of amine metabolites, and heightened levels of these substances have been linked to increased oxidative stress (Jalandra, Dalal et al. 2021). While the association between TMAO levels and cardiovascular disease is well established (Roncal et al. 2019), ongoing research is delving into its potential link with cancer development, driven by TMAO's documented ability to promote inflammation (Duizer and de Zoete 2023).

The microbiome is also instrumental in generating beneficial bioactive compounds derived from flavonoids and polyphenolic compounds found in fruits, vegetables, nuts, seeds, and plants. Some, like luteolin, exhibit antioxidant properties that may contribute to preventing inflammatory responses (Weng et al. 2015). Some studies have established a link between the consumption of flavonoids in the diet and a reduction in colorectal cancer risk (Dalal et al. 2021). Indeed, among these compounds, quercetin, abundantly found in onions, broccoli, and apples, stands out for its diverse pharmacological properties. It acts as a potent antioxidant and antitumor agent, affecting cell growth by modulating the interaction between β-catenin and Tcf-4 in colon cancer cells (Jia, Xie, and Jia 2018). Furthermore, similar to silibinin extracted from milk thistle, quercetin induces cell cycle arrest by regulating cyclin D1 and survivin gene expression in colon cancer cells (Jia, Xie, and Jia 2018). In addition, substances such as apigenin (found in parsley and celery), naringenin (found in oranges), and catechins (found in green tea) exert an inhibitory effect on carcinogenesis by inducing apoptosis in human colon cancer cells (Dalal et al. 2021).

In response to dietary fat intake, bile acids are synthesized in the liver and play a crucial role in lipid absorption in the small intestine. While most bile acids are reabsorbed during their transit through the small intestine and return to the enterohepatic circulation, a fraction is secreted into the colon, undergoing complex GM biotransformation (Cai, Sun, and Gonzalez 2022). The intricate interplay between microbiota, bile acids, and the host extends its influence on immunological functions, metabolic phenotypes, and risk factors associated with inflammatory bowel disease (IBD) and various cancers (Jia, Xie, and Jia 2018, Kim 2007). Numerous clinical studies have confirmed the association between high-fat diets, increased fecal bile acid levels, and heightened CRC risk (Ocvirk and O'Keefe 2021). Specifically, bile acids like deoxycholic acid (DCA) and, to a lesser extent, lithocholic acid (LCA) have been implicated in promoting cell proliferation, inducing DNA fragmentation, causing oxidative damage, and fostering resistance to apoptosis in human colon epithelial adenocarcinoma cell lines and colon biopsies (Ocvirk and O'Keefe 2021). In addition, some bile acids such as taurocholic acid stimulate the abundance of Desulfovibrionales, particularly *Bilophila wadsworthia*, whose growth is also stimulated by a high-fat diet. *Bilophila wadsworthia* produces hydrogen sulfide, which inhibits the oxidation of butyrate, causing hypoplasia and hyperpermeability of intestinal epithelial cells, leading to intestinal permeability and inflammation (Lam et al. 2015).

Despite its structural similarity to DCA, ursodeoxycholic acid (UDCA) is a secondary bile acid, produced by *Clostridium* species, with a documented preventive anti-CRC effect (Kim et al. 2017). UDCA inhibits CRC through various mechanisms, including enhancing the hydrophilicity of the bile pool, reducing the concentration of hydrophobic bile acids, and regulating oxidative stress in colon cancer cells and cancer stem cells (CSCs) (Kim et al. 2017). Moreover, UDCA positively regulates the major histocompatibility complex (MHC) in the colon, thereby improving immune surveillance of tumors (Fang et al. 2021). In addition, from the de-conjugation of some bile acids comes taurine, an amino sulfonic acid, which is subject to modulation by commensal bacteria. Its presence has been associated with a reduction in inflammatory bowel diseases, and it is recognized as a protective metabolite in conditions such as diabetes and colorectal cancer (Dalal et al. 2021). Its protective effect is carried out by the induction of apoptosis in CRC cells; however, the molecular pathways activated are still under investigation (Fang et al. 2021).

Regarding dietary protein intake, a diverse array of bacterial metabolites emerges, including polyamines, small hydrocarbons containing two or more amine groups, that are synthesized by both prokaryotes and eukaryotes. Among them, naturally occurring spermidine, spermine, and putrescine stand out as well-studied examples (Duizer and de Zoete 2023). Due to their positive charge, polyamines can interact with cellular molecules, including proteins, DNA, and RNA, influencing various cellular functions. Elevated levels of polyamines, especially in proliferating and tumor cells, have been implicated in CRC development (Dalal et al. 2021).

Arginine, as a precursor of polyamines, has been proposed as a potential CRC risk factor (Vernia et al. 2021) and *Fusobacterium nucleatum* has been observed to produce high levels of putrescine (Duizer and de Zoete 2023). This production is linked to the promotion of cancer cell proliferation, suggesting a potential role of polyamines, specifically putrescine, in the interaction between GM and CRC development (Duizer and de Zoete 2023).

Diets rich in processed meat, low-calorie beverages, and alcoholic drinks result in an increased abundance of hydrogen sulfide (H2S)-producing bacteria and a higher likelihood of CRC developing (Duizer and de Zoete 2023). Sulfate-reducing bacteria, stimulated especially by high-protein food residues that contain sulfates (such as protein) but also other sulfur-containing compounds such as taurine, play a key role in the intestinal ecosystem (Fang et al. 2021). Their metabolic activity leads to H_2S production. This gas, besides playing a physiological role, has been implicated in several cellular events. The consequences of high H_2S levels include cell proliferation, differentiation, apoptosis, and inflammation (Coletta et al. 2012). Notably, these events collectively contribute to the neoplastic transformation of enterocytes, underscoring the impact of dietary choices on gut health (Duizer and de Zoete 2023).

Instead, tryptophan, an essential amino acid acquired through the diet, could contribute to slowing the progression of CRC (Armand et al. 2022). In the intestine, bacteria metabolize tryptophan into various indole derivatives through tryptophanase A, expressed by a wide range of bacteria, like *E. coli* (Armand et al. 2022). These derivatives can activate the aryl hydrocarbon receptor (AHR), which in turn plays a protective role against colonic inflammation (Duizer and de Zoete 2023).

Lastly, another aspect of amino acid metabolism involves the production of nitrate, a precursor easily reduced to nitrite by the gut microbiota (Han et al. 2018). Nitrite, when combined with different nitrogen compounds such as amines, gives rise to nitroso-carcinogenic compounds. Mucin, another intermediate product of amino acid metabolism, emerges as a factor with mutagenic potential when it cooperates with bacteria (Han et al. 2018).

In association with their role in nutrient transformation, bacteria can also generate toxins. One notable example is colibactin (produced by *Escherichia coli*), which has the capacity to induce DNA damage by causing double-strand breaks (Dalal et al. 2021). Another well-known toxin, the BFT (*Bacteroides fragilis* toxin), prompts the lysis of extracellular domains of the adhesion protein E-cadherin, activating NF-κB signaling that, in turn, promotes cytokine secretion by colonocytes, contributing to inflammation. Additionally, the surface protein FadA, a key virulence factor of *Fusobacterium nucleatum*, regulates bacterial adhesion and invasion (Duizer and de Zoete 2023).

In addition to investigating the influence of individual macronutrients on the GM composition and the function of the microbiota, numerous reports have focused on the impact of broader dietary patterns, such as the Mediterranean and Western diets, in relation to the risk of CRC developing. These studies explore how the overall structure and nutritional content of these dietary patterns contribute to the complex balance of microbial communities in the gut, subsequently affecting colorectal health. A lower adherence to the MD was associated with a higher *Firmicutes-Bacteroidetes* ratio (Garcia-Mantrana et al. 2018), which may alter SCFA production and immune homeostasis, whereas increased *Bacteroidetes* presence was correlated with reduced animal protein intake (Garcia-Mantrana et al. 2018). Additionally, elevated counts of *Bifidobacteria* and higher total SCFAs were linked to greater consumption of plant-based nutrients like vegetables, proteins and polysaccharides. The study demonstrated that stronger adherence to the MD corresponded to significantly higher levels of total SCFAs (Garcia-Mantrana et al. 2018). Confirming these results, Rosés et al. established that the MD and higher consumption of fiber, legumes, vegetables, fruit, and nuts are associated with an increase in butyrate-producing taxa (Rosés et al. 2021). De Filippis et al. supported these findings, reporting that a substantial intake of plant-based foods and higher adherence to MD was linked to increased concentrations of fecal SCFA, *Prevotella*, and certain fiber-degrading *Firmicutes* (Filippis et al. 2016). Mitsou et al. also confirmed these results, finding positive correlations between adherence to the MD and the *Bifidobacteria*:*E. coli* ratio, the relative abundance of Bacteroides, and total SCFA, as well as a decrease in *E. coli* levels (Mitsou et al. 2017). Furthermore, butyrate has also been seen to be linked with the immune system, promoting immunoregulation (Roediger 1980). In contrast, individuals exhibiting lower adherence to the MD demonstrated higher levels of urinary TMAO (Filippis et al. 2016). As previously reported, decreased SCFAs levels may contribute to promoting CRC development (Niccolai et al. 2019). Furthermore, components such as TMAO may be associated with increased cancer risk due to its pro-inflammatory effects (Duizer and de Zoete 2023).

On the other hand, the WD is linked to dysbiosis in mice, characterized by increased levels of *Bacteroides* spp. and *Ruminococcus torques* (Martinez-Medina et al. 2014).

An investigation comparing European (EU) children following a WD and Burkina Faso (BF) children adopting a diet abundant in millet, sorghum, and local vegetables, characterized by minimal lipids and animal proteins, demonstrated a higher abundance of *Proteobacteria* in EU children compared to BF children (De Filippo et al. 2010). Furthermore, the microbiota of BF children exhibited enrichment in *Prevotella* and *Xylanibacter* in contrast to the GM of EU children. Yatsunenko et al. compared GM composition of participants from Venezuela, Malawi, and the United States (Yatsunenko et al. 2012). These findings indicated that, regardless of age, the GM of Malawian and Venezuelan individuals were similar, while individuals in the United States exhibited the lowest microbial diversity. Both studies highlighted that the genus *Prevotella* was less represented in WD microbiota, suggesting its potential role as a taxonomic discriminator (Yatsunenko et al. 2012, De Filippo et al. 2010). Furthermore, the WD has been shown to elevate the presence of bile-tolerant microorganisms such as *Alistipes*, *Bilophila*, and *Bacteroides* while simultaneously reducing levels of Firmicutes responsible for metabolizing dietary plant polysaccharides like *Roseburia*, *Eubacterium rectale*, and *Ruminococcus bromii* (David et al. 2014). Additionally, various studies indicate that the adverse effects of the WD on gut microbiota may also be influenced by food additives, leading to dysbiosis. In this context, Suez et al. (2014) investigated the microbiota of individuals who consumed non-caloric artificial sweeteners (NAS) compared to those who did not. Their findings revealed an increase in *Bacteroides* spp. and *Lactobacillus* spp., along with a decrease in *Clostridiales* spp. in the feces of NAS consumers (Suez et al. 2014).

The Westernization of diets can result in dysbiosis, leading to disruptions in intestinal barrier function and abnormal activation of immune cells (Rinninella et al. 2019). Furthermore, it is interesting to note that changes in the microbiota observed with a high-fat diet are followed by a decrease in antimicrobial peptides from Paneth cells, which may stimulate the development of a pro-inflammatory environment in the intestine (Guo et al. 2017). Conversely, the Mediterranean diet has been observed to positively influence GM diversity and stability. It can regulate intestinal permeability and enhance immune functions within the human body (Rinninella et al. 2019).

Applying a simple formula proposed by Wahrendorf (Wahrendorf 1987), Trichopoulou et al. hypothesized that individuals in industrialized countries could potentially decrease their CRC risk by approximately 25% by switching from a WD to the MD (Trichopoulou et al. 2000), identifying, once again, a traditional MD as a preventive dietary pattern.

7.7 CONCLUSIONS AND PERSPECTIVES

Promoting dietary guidelines can be a valuable strategy for preventing CRC and cancer more broadly. It would prove highly beneficial to educate the population, starting with young people, on decreasing the consumption of ultra-processed foods, red and processed meat, and sweet beverages, in favor of an increase of fruits, vegetables, legumes, olive oil, nuts, and fish.

Furthermore, it is essential to rekindle an appreciation for the traditional and cultural significance that food once held in various regions of the world. The literature on this subject has shown us that, in most cases, a frugal diet linked to the territory and origins of the population is associated with better health outcomes than following a Western diet, which is now becoming increasingly widespread in every corner of the world. Following a delectable and pleasurable dietary pattern, such as the Mediterranean diet, within the framework of an active lifestyle, is highly desirable. The action of preparing and enjoying meals, which is the basis of various food pyramids, holds a pivotal role in lifestyles geared towards preventing chronic diseases. It is preferable that recommendations regarding dietary choices, provided by researchers and health professionals, be viewed in the wider context of a healthy lifestyle. This encompasses engaging in enjoyable physical activities and making food choices that contribute to enhancing overall well-being.

Traditional foods often carry a wealth of cultural and regional significance, reflecting centuries-old wisdom in food choices. These foods, rich in locally sourced ingredients, have the potential to provide distinctive compounds and beneficial microbes that foster a resilient and diverse gut microbiome. Integrating traditional foods into a balanced diet not only preserves cultural heritage but also introduces a variety of bioactive compounds that have been associated with health benefits. By appreciating and incorporating traditional dietary practices, individuals can harness the synergistic effects of both modern nutritional science and time-tested traditional wisdom to optimize their gut health and reduce the risk of cancer, fostering a holistic approach to their health.

Finally, to stress the importance of a mainly plant-based and sustainable diet, including traditional foods in one's dietary habits can further enhance the positive impact on GM, overall health, and cancer risk. This holistic approach extends beyond individual health benefits to contribute to pollution reduction, thereby fostering an improvement in environmental health. Additionally, prioritizing local, low-impact foods, such as those from nearby sources or "km 0" foods, and minimizing the use of highly industrialized processes in food production further contributes to a healthier diet in terms of both individual well-being and environmental and social well-being. Furthermore, it is imperative to address future research and intervention strategies to mitigate the burden of gut-associated cancers. Culturally tailored nutrition programs and public health initiatives are essential to ensure widespread adoption and adherence among diverse populations. By acknowledging and respecting cultural contexts, these initiatives can be more effectively implemented and embraced by communities. Moreover, identifying key areas for future research is crucial in advancing our understanding of gut-associated cancers and developing targeted interventions. These efforts may include investigating the role of traditional dietary practices, exploring novel treatment modalities, and promoting early detection strategies. By prioritizing these areas of research and intervention, we can work towards reducing the prevalence and impact of gut-associated cancers on a global scale. Further investigations could offer valuable insights into the role of traditional foods in advancing the cause of environmental sustainability.

REFERENCES

Aggarwal, B. B., A. Kumar, and A. C. Bharti. 2003. "Anticancer potential of curcumin: Preclinical and clinical studies." *Anticancer Res* 23 (1a):363–398.

Agnoli, Claudia, Sara Grioni, Sabina Sieri, Domenico Palli, Giovanna Masala, Carlotta Sacerdote, Paolo Vineis, Rosario Tumino, Maria Concetta Giurdanella, and Valeria Pala. 2013. "Italian Mediterranean Index and risk of colorectal cancer in the Italian section of the EPIC cohort." *Int J Cancer Res* 132 (6):1404–1411.

Åkesson, Agneta, Lene F. Andersen, Ása G. Kristjánsdóttir, Eva Roos, Ellen Trolle, Eeva Voutilainen, and Elisabet Wirfält. 2013. "Health effects associated with foods characteristic of the Nordic diet: A systematic literature review." *Food Nutr Res* 57 (1):22790. https://doi.org/10.3402/fnr.v57i0.22790.

Altonsy, M. O., T. N. Habib, and G. S. Mokhtar ElSA. 2015. "Diallyl disulphide protects against colon cancer in vitro, of HT-29 cells and in male rabbits of colon cancer model: An analysis of genetic and epigenetic variations." *J Mol Biol Hassanain* 4 (136):2.

Alvarez-Lesmes, J., J. R. Chapman, D. Cassidy, Y. Zhou, M. Garcia-Buitrago, E. A. Montgomery, I. S. Lossos, D. Sussman, and J. Poveda. 2021. "Gastrointestinal tract lymphomas." *Arch Pathol Lab Med* 145 (12):1585–1596. https://doi.org/10.5858/arpa.2020-0661-RA.

Armand, L., M. Fofana, K. Couturier-Becavin, M. Andriamihaja, and F. Blachier. 2022. "Dual effects of the tryptophan-derived bacterial metabolite indole on colonic epithelial cell metabolism and physiology: Comparison with its co-metabolite indoxyl sulfate." *Amino Acids* 54 (10):1371–1382. https://doi.org/10.1007/s00726-021-03122-4.

Arnold, Melina, Christian C. Abnet, Rachel E. Neale, Jerome Vignat, Edward L. Giovannucci, Katherine A. McGlynn, and Freddie Bray. 2020. "Global burden of 5 major types of gastrointestinal cancer." *Gastroenterology* 159 (1):335–349.e15. https://doi.org/10.1053/j.gastro.2020.02.068.

Åsli, Lene A., Anja Olsen, Tonje Braaten, Eiliv Lund, and Guri Skeie. 2017. "Potato consumption and risk of colorectal cancer in the norwegian women and cancer cohort." *Nutr Cancer* 69 (4):564–572. https://doi.org/10.1080/01635581.2017.1295086.

Augimeri, Giuseppina, and Daniela Bonofiglio. 2021. "The mediterranean diet as a source of natural compounds: Does Itit represent a protective choice against cancer?" *Pharmaceuticals* 14 (9):920.

Aune, Dagfinn, NaNa Keum, Edward Giovannucci, Lars T. Fadnes, Paolo Boffetta, Darren C. Greenwood, Serena Tonstad, Lars J. Vatten, Elio Riboli, and Teresa Norat. 2016. "Nut consumption and risk of cardiovascular disease, total cancer, all-cause and cause-specific mortality: A systematic review and dose-response meta-analysis of prospective studies." *J BMC Med* 14 (1):1–14.

Awedew, Atalel Fentahun, Zelalem Asefa, and Woldemariam Beka Belay. 2022. "Burden and trend of colorectal cancer in 54 countries of Africa 2010–2019: A systematic examination for Global Burden of Disease." *BMC Gastroenterol* 22 (1):204. https://doi.org/10.1186/s12876-022-02275-0.

Azeem, Salman, Syed Wasif Gillani, Ammar Siddiqui, Suresh Babu Jandrajupalli, Vinci Poh, and Syed Azhar Syed Sulaiman. 2015. "Diet and colorectal cancer risk in Asia-a systematic review." *Asian Pac J Cancer Prev* 16 (13):5389–5396.

Baidoun, F., K. Elshiwy, Y. Elkeraie, Z. Merjaneh, G. Khoudari, M. T. Sarmini, M. Gad, M. Al-Husseini, and A. Saad. 2021. "Colorectal cancer epidemiology: Recent trends and impact on outcomes." *Curr Drug Targets* 22 (9):998–1009. https://doi.org/10.2174/1389450121999201117115717.

Bamia, Christina, Pagona Lagiou, Genevieve Buckland, Sara Grioni, Claudia Agnoli, Aliki J. Taylor, Christina C. Dahm, Kim Overvad, Anja Olsen, and Anne Tjønneland. 2013. "Mediterranean diet and colorectal cancer risk: Results from a European cohort." *Eur J Epidemiol* 28:317–328.

Ban, Jung Ok, Dong Yeon Yuk, Koan Sik Woo, Tae Myoung Kim, Ung Soo Lee, Heon-Sang Jeong, Dae Joong Kim, Yeun Bok Chung, Bang Yeon Hwang, and Ki Wan Oh. 2007. "Inhibition of cell growth and induction of apoptosis via inactivation of NF-κB by a sulfurcompound isolated from garlic in human colon cancer cells." *J Pharmacol Sci* 104 (4):374–383.

Banks, Leah D, Priscilla Amoah, Mohammad S Niaz, Mary K Washington, Samuel E Adunyah, and Aramandla Ramesh. 2016. "Olive oil prevents benzo (a) pyrene [B (a) P]-induced colon carcinogenesis through altered B (a) P metabolism and decreased oxidative damage in ApcMin mouse model." *J Nutr Biochem* 28:37–50.

Barak, Yoram, and Dana Fridman. 2017. "Impact of mediterranean diet on cancer: Focused literature review." *Cancer Genomics & Proteomics* 14 (6):403–408.

Barsouk, Adam, Prashanth Rawla, Alexander Barsouk, and Krishna Chaitanya Thandra. 2019. "Epidemiology of cancers of the small intestine: Trends, risk factors, and prevention." *Medical Sciences* 7 (3):46.

Bielik, V., and M. Kolisek. 2021a. "Bioaccessibility and bioavailability of minerals in relation to a healthy gut microbiome." *Int J Mol Sci* 22 (13). https://doi.org/10.3390/ijms22136803.

Bielik, Viktor, and Martin Kolisek. 2021b. "Bioaccessibility and bioavailability of minerals in relation to a healthy gut microbiome." *International Journal of Moecular Sciences* 22 (13):6803.

Biesalski, H. K. 2016. "Nutrition meets the microbiome: Micronutrients and the microbiota." *Ann N Y Acad Sci* 1372 (1):53–64. https://doi.org/10.1111/nyas.13145.

Bigman, Galya, Linda Otieno, Sally N. Adebamowo, and Clement Adebamowo. 2022. "Dietary intake and cancer in Sub-Saharan Africa: A critical review of epidemiological studies." *Nutr Cancer* 74 (8):2803–2814. https://doi.org/10.1080/01635581.2022.2032217.

Bimonte, Sabrina, Antonio Barbieri, Giuseppe Palma, Domenica Rea, Antonio Luciano, Massimiliano D'Aiuto, Claudio Arra, and Francesco Izzo. 2015. "Dissecting the role of curcumin in tumour growth and angiogenesis in mouse model of human breast cancer." *BioMed Res Int* 2015:878134. https://doi.org/10.1155/2015/878134.

Bingham, Sheila A., Nicholas E Day, Robert Luben, Pietro Ferrari, Nadia Slimani, Teresa Norat, Françoise Clavel-Chapelon, Emmanuelle Kesse, Alexandra Nieters, and Heiner Boeing. 2003. "Dietary fibre in food and protection against colorectal cancer in the European prospective investigation into cancer and Nutrition (EPIC): An observational study." *The Lancet* 361 (9368):1496–1501.

Bingham, Sheila A., and Elio Riboli. 2004. "Diet and cancer—the European prospective investigation into cancer and nutrition." *Nat Rev Cancer* 4 (3):206–215.

Bleau, Christian, Antony D Karelis, David H St-Pierre, and Lucie Lamontagne. 2015. "Crosstalk between intestinal microbiota, adipose tissue and skeletal muscle as an early event in systemic low-grade inflammation and the development of obesity and diabetes." *Diabetes Metab Res Rev* 31 (6):545–561.

Bodalal, Z., and R. Bendardaf. 2014. "Colorectal carcinoma in a Southern Mediterranean country: The Libyan scenario." *World J Gastrointest Oncol* 6 (4):98–103. https://doi.org/10.4251/wjgo.v6.i4.98.

Bolla, Kusuma Neela, Santhi Sri K. V., Kazi Afnan, P. Krishna Veni, and M. Kusuma. 2015. "Effect of diet counseling on type 2 diabetes mellitus." *Int J Sci Technol Res* 4:112–118.

Bouvard, Véronique, Dana Loomis, Kathryn Z. Guyton, Yann Grosse, Fatiha El Ghissassi, Lamia Benbrahim-Tallaa, Neela Guha, Heidi Mattock, and Kurt Straif. 2015. "Carcinogenicity of consumption of red and processed meat." *Lancet Oncol* 16 (16):1599–1600.

Bruce, W. R., R. Furrer, N. Shangari, P. J. O'Brien, A. Medline, and Y. Wang. 2003. "Marginal dietary thiamin deficiency induces the formation of colonic aberrant crypt foci (ACF) in rats." *Cancer Lett* 202 (2):125–129. https://doi.org/10.1016/j.canlet.2003.08.005.

Burney, Ikram A., and Mansour S. Al-Moundhri. 2014. "Life-style modification and cancer prevention: Another weapon in war against cancer." *Int J Nutr Pharmacol Neurol Dis* 4 (1):1–3.

Butler, L. M., R. Wang, W. P. Koh, and M. C. Yu. 2008. "Prospective study of dietary patterns and colorectal cancer among Singapore Chinese." *Br J Cancer* 99 (9):1511–1516. https://doi.org/10.1038/sj.bjc.6604678.

Cai, J., L. Sun, and F. J. Gonzalez. 2022. "Gut microbiota-derived bile acids in intestinal immunity, inflammation, and tumorigenesis." *Cell Host Microbe* 30 (3):289–300. https://doi.org/10.1016/j.chom.2022.02.004.

Calder, P. C. 2012. "Long-chain fatty acids and inflammation." *Proc Nutr Soc* 71 (2):284–289. https://doi.org/10.1017/s0029665112000067.

Cambie, Richard C., and Lynnette R. Ferguson. 2003. "Potential functional foods in the traditional Maori diet." *Mutat Res Fundam Mol Mech Mutagen* 523–524:109–117. https://doi.org/10.1016/S0027-5107(02)00344-5.

Cani, Patrice D., Jacques Amar, Miguel Angel Iglesias, Marjorie Poggi, Claude Knauf, Delphine Bastelica, Audrey M. Neyrinck, Francesca Fava, Kieran M. Tuohy, and Chantal Chabo. 2007. "Metabolic endotoxemia initiates obesity and insulin resistance." *Diabetes* 56 (7):1761–1772.

Casaburi, Ivan, Francesco Puoci, Adele Chimento, Rosa Sirianni, Carmen Ruggiero, Paola Avena, and Vincenzo Pezzi. 2013. "Potential of olive oil phenols as chemopreventive and therapeutic agents against cancer: A review of in vitro studies." *Mol Nutr Food Res* 57 (1):71–83.

Cevikbas, Adile, Ersin Yemni, Fikret W. Ezzedenn, Turay Yardimici, Ugur Cevikbas, and S. J. Stohs. 1994. "Antitumoural antibacterial and antifungal activities of kefir and kefir grain." 8 (2):78–82. https://doi.org/10.1002/ptr.2650080205.

Chen, A, J. Xu, and A. C. Johnson. 2006. "Curcumin inhibits human colon cancer cell growth by suppressing gene expression of epidermal growth factor receptor through reducing the activity of the transcription factor Egr-1." *Oncogene* 25 (2):278–287.

Cho, I., and Blaser, M. J. 2012. "The human microbiome: at the interface of health and disease." *Nat Rev Genet.* 13 (4):260–270. doi: 10.1038/nrg3182.

Citronberg, Jessica, Roberd Bostick, Thomas Ahearn, D. Kim Turgeon, Mack T. Ruffin, Zora Djuric, Ananda Sen, Dean E. Brenner, and Suzanna M. Zick. 2013. "Effects of ginger supplementation on cell-cycle biomarkers in the normal-appearing colonic mucosa of patients at increased risk for colorectal cancer: Results from a pilot, randomized, and controlled trial." *Cancer Prev Res* 6 (4):271–281.

Clinton, S. K., E. L. Giovannucci, and S. D. Hursting. 2020. "The world cancer research fund/american institute for cancer research third expert report on diet, nutrition, physical activity, and cancer: Impact and future directions." *J Nutr* 150 (4):663–671. https://doi.org/10.1093/jn/nxz268.

Coker, O. O., C. Liu, W. K. K. Wu, S. H. Wong, W. Jia, J. J. Y. Sung, and J. Yu. 2022. "Altered gut metabolites and microbiota interactions are implicated in colorectal carcinogenesis and can be non-invasive diagnostic biomarkers." *Microbiome* 10 (1):35. https://doi.org/10.1186/s40168-021-01208-5.

Coletta, C., A. Papapetropoulos, K. Erdelyi, G. Olah, K. Módis, P. Panopoulos, A. Asimakopoulou, D. Gerö, I. Sharina, E. Martin, and C. Szabo. 2012. "Hydrogen sulfide and nitric oxide are mutually dependent in the regulation of angiogenesis and endothelium-dependent vasorelaxation." *Proc Natl Acad Sci U S A* 109 (23):9161–9166. https://doi.org/10.1073/pnas.1202916109.

Comín-Anduix, B., J. Boren, S. Martinez, C. Moro, J. J. Centelles, R. Trebukhina, N. Petushok, W. N. Lee, L. G. Boros, and M. Cascante. 2001. "The effect of thiamine supplementation on tumour proliferation. A metabolic control analysis study." *Eur J Biochem* 268 (15):4177–4182. https://doi.org/10.1046/j.1432-1327.2001.02329.x.

Corona, Giulia, Monica Deiana, Alessandra Incani, David Vauzour, M. Assunta Dessì, and Jeremy P. E. Spencer. 2007. "Inhibition of p38/CREB phosphorylation and COX-2 expression by olive oil polyphenols underlies their anti-proliferative effects." *Biochem Biophys Res Commun* 362 (3):606–611.

Costea, T., A. Hudiță, O. A. Ciolac, B. Gălățeanu, O. Ginghină, M. Costache, C. Ganea, and M. M. Mocanu. 2018. "Chemoprevention of colorectal cancer by dietary compounds." *Int J Mol Sci* 19 (12). https://doi.org/10.3390/ijms19123787.

Crawford-Williams, Fiona, Sonja March, Michael J. Ireland, Arlen Rowe, Belinda Goodwin, Melissa K. Hyde, Suzanne K. Chambers, Joanne F. Aitken, and Jeff Dunn. 2018. "Geographical variations in the clinical management of colorectal cancer in Australia: A systematic review." *Front Oncol* 8:116.

D'Amelio, P., and F. Sassi. 2018. "Gut microbiota, immune system, and bone." *Calcif Tissue Int* 102 (4):415–425. https://doi.org/10.1007/s00223-017-0331-y.

Dai, Z., O. O. Coker, G. Nakatsu, W. K. K. Wu, L. Zhao, Z. Chen, F. K. L. Chan, K. Kristiansen, J. J. Y. Sung, S. H. Wong, and J. Yu. 2018. "Multi-cohort analysis of colorectal cancer metagenome identified altered bacteria across populations and universal bacterial markers." *Microbiome* 6 (1):70. https://doi.org/10.1186/s40168-018-0451-2.

Dalal, N., R. Jalandra, N. Bayal, A. K. Yadav, Harshulika, M. Sharma, G. K. Makharia, P. Kumar, R. Singh, P. R. Solanki, and A. Kumar. 2021. "Gut microbiota-derived metabolites in CRC progression and causation." *J Cancer Res Clin Oncol* 147 (11):3141–3155. https://doi.org/10.1007/s00432-021-03729-w.

David, L. A., C. F. Maurice, R. N. Carmody, D. B. Gootenberg, J. E. Button, B. E. Wolfe, A. V. Ling, A. S. Devlin, Y. Varma, M. A. Fischbach, S. B. Biddinger, R. J. Dutton, and P. J. Turnbaugh. 2014. "Diet rapidly and reproducibly alters the human gut microbiome." *Nature* 505 (7484):559–563. https://doi.org/10.1038/nature12820.

De Filippo, Carlotta, Duccio Cavalieri, Monica Di Paola, Matteo Ramazzotti, Jean Baptiste Poullet, Sebastien Massart, Silvia Collini, Giuseppe Pieraccini, and Paolo Lionetti. 2010. "Impact of diet in shaping gut microbiota revealed by a comparative study in children from Europe and rural Africa." *Proc Natl Acad Sci* 107 (33):14691–14696.

De Stefani, Eduardo, Hugo Deneo-Pellegrini, Alvaro L. Ronco, Pelayo Correa, Paolo Boffetta, Dagfinn Aune, Gisele Acosta, María Mendilaharsu, María E. Luaces, and Gabriel Landó. 2011. "Dietary patterns and risk of colorectal cancer: A factor analysis in Uruguay." *Asian Pac J Cancer Prev* 12 (3):753–759.

de Vogel, S., V. Dindore, M. van Engeland, R. A. Goldbohm, P. A. van den Brandt, and M. P. Weijenberg. 2008. "Dietary folate, methionine, riboflavin, and vitamin B-6 and risk of sporadic colorectal cancer." *J Nutr* 138 (12):2372–2378. https://doi.org/10.3945/jn.108.091157.

den Besten, G., K. van Eunen, A. K. Groen, K. Venema, D. J. Reijngoud, and B. M. Bakker. 2013. "The role of short-chain fatty acids in the interplay between diet, gut microbiota, and host energy metabolism." *J Lipid Res* 54 (9):2325–2340. https://doi.org/10.1194/jlr.R036012.

Devarajan, A., and M. K. Mohanmarugaraja. 2017. "A comprehensive review on rasam: A south Indian traditional functional food." *Pharmacogn Rev* 11 (22):73–82. https://doi.org/10.4103/phrev.phrev_13_17.

Di Francesco, Andrea, Anastasia Falconi, Clara Di Germanio, Maria Vittoria Micioni Di Bonaventura, Antonio Costa, Stefano Caramuta, Michele Del Carlo, Dario Compagnone, Enrico Dainese, and Carlo Cifani. 2015. "Extravirgin olive oil up-regulates CB1 tumor suppressor gene in human colon cancer cells and in rat colon via epigenetic mechanisms." *J Nutr Biochem* 26 (3):250–258.

Di Mascio, P., S. Kaiser, and H. Sies. 1989. "Lycopene as the most efficient biological carotenoid singlet oxygen quencher." *Arch Biochem Biophys* 274 (2):532–538. https://doi.org/10.1016/0003-9861(89)90467-0.

Dixon, L. Beth, Amy F. Subar, Ulrike Peters, Joel L. Weissfeld, Robert S. Bresalier, Adam Risch, Arthur Schatzkin, and Richard B. Hayes. 2007. "Adherence to the USDA food guide, DASH eating plan, and mediterranean dietary pattern reduces risk of colorectal adenoma." *J Nutr* 137 (11):2443–2450.

Donaldson, M. S. 2004. "Nutrition and cancer: A review of the evidence for an anti-cancer diet." *Nutr J* 3:19. https://doi.org/10.1186/1475-2891-3-19.

Donovan, Micah G., Ornella I. Selmin, Tom C. Doetschman, and Donato F. Romagnolo. 2017. "Mediterranean diet: Prevention of colorectal cancer." 4. https://doi.org/10.3389/fnut.2017.00059.

Duizer, C., and M. R. de Zoete. 2023. "The role of microbiota-derived metabolites in colorectal cancer." *Int J Mol Sci* 24 (9). https://doi.org/10.3390/ijms24098024.

El Kinany, K., M. Mint Sidi Deoula, Z. Hatime, H. A. Boudouaya, I. Huybrechts, A. El Asri, A. Benider, M. Ahallat, S. Afqir, N. Mellas, M. Khouchani, and K. El Rhazi. 2020. "Consumption of modern and traditional Moroccan dairy products and colorectal cancer risk: A large case control study." *Eur J Nutr* 59 (3):953–963. https://doi.org/10.1007/s00394-019-01954-1.

Emmy De Caluwé, Kateřina Halamová Patrick Van Damme. 2010. "Tamarindus indica L. – A review of traditional uses, phytochemistry and pharmacology." *Afrika Focus* 23 (1):53–83.

Fang, Y., C. Yan, Q. Zhao, J. Xu, Z. Liu, J. Gao, H. Zhu, Z. Dai, D. Wang, and D. Tang. 2021. "The roles of microbial products in the development of colorectal cancer: A review." *Bioengineered* 12 (1):720–735. https://doi.org/10.1080/21655979.2021.1889109.

Fauser, J. K., L. D. Prisciandaro, A. G. Cummins, and G. S. Howarth. 2011. "Fatty acids as potential adjunctive colorectal chemotherapeutic agents." *Cancer Biol Ther* 11 (8):724–731. https://doi.org/10.4161/cbt.11.8.15281.

Ferguson, Lynnette R., Roy Lay Yee, Robert Scragg, Patricia A. Metcalf, and Philip J. Harris. 1995. "Differences in intake of specific food plants by Polynesians may explain their lower incidence of colorectal cancer compared with Europeans in New Zealand." *Nutr Cancer* 23 (1):33–42. https://doi.org/10.1080/01635589509514359.

Ferrari, L., S. A. Panaite, A. Bertazzo, and F. Visioli. 2022. "Animal- and plant-based protein sources: A scoping review of human health outcomes and environmental impact." *Nutrients* 14 (23). https://doi.org/10.3390/nu14235115.

Filippis, Francesca De, Nicoletta Pellegrini, Lucia Vannini, Ian B Jeffery, Antonietta La Storia, Luca Laghi, Diana I. Serrazanetti, Raffaella Di Cagno, Ilario Ferrocino, Camilla Lazzi, Silvia Turroni, Luca Cocolin, Patrizia Brigidi, Erasmo Neviani, Marco Gobbetti, Paul W. O'Toole, and Danilo Ercolini. 2016. "High-level adherence to a Mediterranean diet beneficially impacts the gut microbiota and associated metabolome." *Gut* 65 (11):1812–1821. https://doi.org/10.1136/gutjnl-2015-309957.

Flynn, K. J., M. T. Ruffin, D. K. Turgeon, and P. D. Schloss. 2018. "Spatial variation of the native colon microbiota in healthy adults." *Cancer Prev Res (Phila)* 11 (7):393–402. https://doi.org/10.1158/1940-6207.Capr-17-0370.

Fook, J. M., L. L. Macedo, G. E. Moura, F. M. Teixeira, A. S. Oliveira, A. F. Queiroz, and M. P. Sales. 2005. "A serine proteinase inhibitor isolated from Tamarindus indica seeds and its effects on the release of human neutrophil elastase." *Life Sci* 76 (25):2881–2891. https://doi.org/10.1016/j.lfs.2004.10.053.

Fung, Teresa T., Frank B. Hu, Kana Wu, Stephanie E. Chiuve, Charles S. Fuchs, and Edward Giovannucci. 2010. "The Mediterranean and Dietary Approaches to Stop Hypertension (DASH) diets and colorectal cancer." *AM J Clin Nutr* 92 (6):1429–1435.

Gagnière, J., Raisch, J., Veziant, J., Barnich, N., Bonnet, R., Buc, E., Bringer, M. A., Pezet, D., and Bonnet, M. 2016. "Gut microbiota imbalance and colorectal cancer." *World J Gastroenterol.* 22 (2):501–518. doi:10.3748/wjg.v22.i2.501.

Garcia-Mantrana, Izaskun, Marta Selma-Royo, Cristina Alcantara, and María C. Collado. 2018. "Shifts on gut microbiota associated to mediterranean diet adherence and specific dietary intakes on general adult population." 9. https://doi.org/10.3389/fmicb.2018.00890.

Genua, F., Raghunathan, V., Jenab, M., Gallagher, W.M., and Hughes, D.J. 2021. "The role of gut barrier dysfunction and microbiome dysbiosis in colorectal cancer development." *Frontiers in Oncology* 11:626349. doi:10.3389/fonc.2021.626349.

Gianfredi, Vincenza, Tania Salvatori, Milena Villarini, Massimo Moretti, Daniele Nucci, and Stefano Realdon. 2018. "Is dietary fibre truly protective against colon cancer? A systematic review and meta-analysis." *Int J Food Sci Nutr* 69 (8):904–915.

Giovannucci, E., M. J. Stampfer, G. A. Colditz, E. B. Rimm, D. Trichopoulos, B. A. Rosner, F. E. Speizer, and W. C. Willett. 1993. "Folate, methionine, and alcohol intake and risk of colorectal adenoma." *J Natl Cancer Inst* 85 (11):875–884. https://doi.org/10.1093/jnci/85.11.875.

Glanz, K., and L. N. Kolonel. 1998. "Culture, religion, diet and health: Challenges and opportunities." *Nutrition* 14 (2):238–240. https://doi.org/10.1016/s0899-9007(97)00424-3.

Gonzalez, Carlos A., and Elio Riboli. 2010. "Diet and cancer prevention: Contributions from the European prospective investigation into cancer and nutrition (EPIC) study." *Eur J Cancer* 46 (14):2555–2562.

Grajek, W., A. Olejnik, and A. Sip. 2005. "Probiotics, prebiotics and antioxidants as functional foods." *Acta Biochim Pol* 52 (3):665–671.

Grishina, Anna, Irina Kulikova, Ludmila Alieva, Andrew Dodson, Ian Rowland, Jing Jin. 2011. "Antigenotoxic effect of kefir and ayran supernatants on fecal water-induced DNA damage in human colon cells." *Nutr Cancer* 63 (1):73–79.

Grosso, Giuseppe, Francesca Bella, Justyna Godos, Salvatore Sciacca, Daniele Del Rio, Sumantra Ray, Fabio Galvano, and Edward L. Giovannucci. 2017. "Possible role of diet in cancer: Systematic review and multiple meta-analyses of dietary patterns, lifestyle factors, and cancer risk." *Nutr Rev* 75 (6):405–419.

Grosso, Giuseppe, Silvio Buscemi, Fabio Galvano, Antonio Mistretta, Stefano Marventano, Vanessa La Vela, Filippo Drago, Santi Gangi, Francesco Basile, and Antonio Biondi. 2013. "Mediterranean diet and cancer: Epidemiological evidence and mechanism of selected aspects." *BMC Surgery* 13 (2):S14. https://doi.org/10.1186/1471-2482-13-S2-S14.

Güllü, N., D. Kobelt, H. Brim, S. Rahman, L. Timm, J. Smith, A. Soleimani, S. Di Marco, S. Bisti, H. Ashktorab, and U. Stein. 2020. "Saffron crudes and compounds restrict MACC1-dependent cell proliferation and migration of colorectal cancer cells." *Cells* 9 (8). https://doi.org/10.3390/cells9081829.

Guo, Shuhong, Meghali Nighot, Rana Al-Sadi, Tarik Alhmoud, Prashant Nighot, and Thomas Y. Ma. 2015. "Lipopolysaccharide regulation of intestinal tight junction permeability is mediated by TLR4 signal transduction pathway activation of FAK and MyD88." *J Immunol* 195 (10):4999–5010.

Guo, Xiulan, Jinchao Li, Renyong Tang, Guodong Zhang, Huawei Zeng, Richard J. Wood, and Zhenhua Liu. 2017. "High fat diet alters gut microbiota and the expression of paneth cell-antimicrobial peptides preceding changes of circulating inflammatory cytokines." *Mediators Inflamm* 2017:9474896. https://doi.org/10.1155/2017/9474896.

Han, S., J. Gao, Q. Zhou, S. Liu, C. Wen, and X. Yang. 2018. "Role of intestinal flora in colorectal cancer from the metabolite perspective: A systematic review." *Cancer Manag Res* 10:199–206. https://doi.org/10.2147/cmar.S153482.

Blomhoff, R., Andersen, R., Arnesen, E.K., Christensen, J.J., Eneroth, H., Erkkola, M., Gudanaviciene, I., Halldorsson, T.I., Høyer-Lund, A., Lemming, E.W., Meltzer, H.M., Pitsi, T., Schwab, U., Siksna, I., Thorsdottir, I. and Trolle, E. 2023. *Nordic Nutrition Recommendations 2023*. Copenhagen: Nordic Council of Ministers

Hecker, M., N. Sommer, H. Voigtmann, O. Pak, A. Mohr, M. Wolf, I. Vadász, S. Herold, N. Weissmann, R. E. Morty, W. Seeger, and K. Mayer. 2014. "Impact of short- and medium-chain fatty acids on mitochondrial function in severe inflammation." *JPEN J Parenter Enteral Nutr* 38 (5):587–594. https://doi.org/10.1177/0148607113489833.

Hong, W. S., Y. P. Chen, and M. J. Chen. 2010. "The antiallergic effect of kefir Lactobacilli." *J Food Sci* 75 (8):H244–H253. https://doi.org/10.1111/j.1750-3841.2010.01787.x.

Huang, C. B., Y. Alimova, T. M. Myers, and J. L. Ebersole. 2011. "Short- and medium-chain fatty acids exhibit antimicrobial activity for oral microorganisms." *Arch Oral Biol* 56 (7):650–654. https://doi.org/10.1016/j.archoralbio.2011.01.011.

Imamura, Fumiaki, Renata Micha, Shahab Khatibzadeh, Saman Fahimi, Peilin Shi, John Powles, and Dariush Mozaffarian. 2015. "Dietary quality among men and women in 187 countries in 1990 and 2010: A systematic assessment." *The Lancet Global Health* 3 (3):e132–e142. https://doi.org/10.1016/S2214-109X(14)70381-X.

Irabor, David Omoareghan. 2011. "Colorectal Carcinoma: Why is there a lower incidence in nigerians when compared to caucasians?" *J Cancer Epidemiol* 2011:675154. https://doi.org/10.1155/2011/675154.

Islam, K. B. M. Saiful, Satoru Fukiya, Masahito Hagio, Nobuyuki Fujii, Satoshi Ishizuka, Tadasuke Ooka, Yoshitoshi Ogura, Tetsuya Hayashi, and Atsushi Yokota. 2011. "Bile acid is a host factor that regulates the composition of the cecal microbiota in rats." *Gastroenterology* 141 (5):1773–1781.

Islami, F., A. Goding Sauer, K. D. Miller, R. L. Siegel, S. A. Fedewa, E. J. Jacobs, M. L. McCullough, A. V. Patel, J. Ma, I. Soerjomataram, W. D. Flanders, O. W. Brawley, S. M. Gapstur, and A. Jemal. 2018. "Proportion and number of cancer cases and deaths attributable to potentially modifiable risk factors in the United States." *CA Cancer J Clin* 68 (1):31–54. https://doi.org/10.3322/caac.21440.

Itokawa, H., Q. Shi, T. Akiyama, S. L. Morris-Natschke, and K. H. Lee. 2008. "Recent advances in the investigation of curcuminoids." *Chin Med* 3:11. https://doi.org/10.1186/1749-8546-3-11.

Jaiswal, Sandeep, Tarun Pant, Mangesh Suryavanshi, and Usha Antony. 2023. "Microbiological diversity of fermented food Bhaati Jaanr and its antioxidant and anti-inflammatory properties: Effect against colon cancer." *Food Biosci* 55:102822. https://doi.org/10.1016/j.fbio.2023.102822.

Jenab, Mazda, H. Bas Bueno-de-Mesquita, Pietro Ferrari, Franzel J. B. van Duijnhoven, Teresa Norat, Tobias Pischon, Eugène H. J. M Jansen, Nadia Slimani, Graham Byrnes, and Sabina Rinaldi. 2010. "Association between pre-diagnostic circulating vitamin D concentration and risk of colorectal cancer in European populations: A nested case-control study." *BMJ* 340.

Jeyakumar, Arunan, Lakal Dissabandara, and Vinod Gopalan. 2017. "A critical overview on the biological and molecular features of red and processed meat in colorectal carcinogenesis." *J Gastroenterol* 52:407–418.

Jia, W., G. Xie, and W. Jia. 2018. "Bile acid-microbiota crosstalk in gastrointestinal inflammation and carcinogenesis." *Nat Rev Gastroenterol Hepatol* 15 (2):111–128. https://doi.org/10.1038/nrgastro.2017.119.

Jiang, Ai-Jun, Guan Jiang, Lian-Tao Li, and Jun-Nian Zheng. 2015. "Curcumin induces apoptosis through mitochondrial pathway and caspases activation in human melanoma cells." *Mol Biol Rep* 42:267–275.

Jin, Junzhe, Guofu Lin, Hong Huang, Dong Xu, Hao Yu, Xu Ma, Lisi Zhu, Dongyan Ma, and Honglei Jiang. 2014. "Capsaicin mediates cell cycle arrest and apoptosis in human colon cancer cells via stabilizing and activating p53." *Int J Biol Sci* 10 (3):285.

Jonoush, M., S. Fathi, N. Hassanpour Ardekanizadeh, G. Khalatbari Mohseni, N. Majidi, S. A. Keshavarz, S. Shekari, S. Nemat Gorgani, S. A. Torki, M. Sotoudeh, F. Habibi, M. Gholamalizadeh, A. Alizadeh, and S. Doaei. 2022. "The association between different types of dietary carbohydrates and colorectal cancer: A case-control study." *Front Nutr* 9:898337. https://doi.org/10.3389/fnut.2022.898337.

Joung, Eun-Joo, Mei-Hua Li, Hee Geum Lee, Nuntiya Somparn, Young Suk Jung, Hye-Kyung Na, Sung-Hoon Kim, Young-Nam Cha, Young-Joon Surh. 2007. "Capsaicin induces heme oxygenase-1 expression in HepG2 cells via activation of PI3K-Nrf2 signaling: NAD (P) H: Quinone oxidoreductase as a potential target." *Antioxid Redox Signal* 9 (12):2087–2098.

Kanehara, R., R. Katagiri, A. Goto, T. Yamaji, N. Sawada, M. Iwasaki, M. Inoue, and S. Tsugane. 2023. "Sugar intake and colorectal cancer risk: A prospective Japanese cohort study." *Cancer Sci* 114 (6):2584–2595. https://doi.org/10.1111/cas.15766.

Kasprzak, Aldona. 2021. "Insulin-like growth factor 1 (IGF-1) signaling in glucose metabolism in colorectal cancer." *International Journal of Molecular Sciences* 22 (12):6434.

Katona, Bryson W., and Jennifer M. Weiss. 2020. "Chemoprevention of colorectal cancer." *Gastroenterology* 158 (2):368–388. https://doi.org/10.1053/j.gastro.2019.06.047.

Katsidzira, Leolin, Innocent Gangaidzo, Sandie Thomson, Simbarashe Rusakaniko, Jonathan Matenga, and Raj Ramesar. 2017. "The shifting epidemiology of colorectal cancer in sub-Saharan Africa." *Lancet Gastroenterol Hepatol* 2 (5):377–383.

Katsidzira, Leolin, Ria Laubscher, Innocent T. Gangaidzo, Rina Swart, Rudo Makunike-Mutasa, Tadios Manyanga, Sandie Thomson, Raj Ramesar, Jonathan A. Matenga, and Simbarashe Rusakaniko. 2018. "Dietary patterns and colorectal cancer risk in Zimbabwe: A population based case-control study." *Cancer Epidemiol* 57:33–38. https://doi.org/10.1016/j.canep.2018.09.005.

Kim, Chu-Sook, Teruo Kawada, Byung-Sam Kim, In-Seob Han, Suck-Young Choe, Tadao Kurata, and Rina Yu. 2003. "Capsaicin exhibits anti-inflammatory property by inhibiting IkB-a degradation in LPS-stimulated peritoneal macrophages." *Cell Signal* 15 (3):299–306.

Kim, E. K., J. H. Cho, E. Kim, and Y. J. Kim. 2017. "Ursodeoxycholic acid inhibits the proliferation of colon cancer cells by regulating oxidative stress and cancer stem-like cell growth." *PLoS One* 12 (7):e0181183. https://doi.org/10.1371/journal.pone.0181183.

Kim, Mi Kyung, Satoshi Sasaki, Tetsuya Otani, Shoichiro Tsugane, and for the Japan Public Health Center-based Prospective Study Group. 2005. "Dietary patterns and subsequent colorectal cancer risk by subsite: A prospective cohort study." 115 (5):790–798. https://doi.org/10.1002/ijc.20943.

Kim, Sue Ok, Joydeb Kumar Kundu, Young Kee Shin, Jin-Hong Park, Myung-Haing Cho, Tae-Yoon Kim, and Young-Joon Surh. 2005. "[6]-Gingerol inhibits COX-2 expression by blocking the activation of p38 MAP kinase and NF-κB in phorbol ester-stimulated mouse skin." *Oncogene* 24 (15):2558–2567.

Kim, Y. I. 2007. "Folate and colorectal cancer: An evidence-based critical review." *Mol Nutr Food Res* 51 (3):267–292. https://doi.org/10.1002/mnfr.200600191.

Kirkland, James B., and Meyer-Ficca, Mirella L. 2018.

Kobayasi, Renata, Eliana H. Akamine, Ana P. Davel, Maria A. M. Rodrigues, Carla R. O. Carvalho, and Luciana V. Rossoni. 2010. "Oxidative stress and inflammatory mediators contribute to endothelial dysfunction in high-fat diet-induced obesity in mice." *J Hypertens* 28 (10):2111–2119.

Kohlenberg-Mueller, K., and L. Raschka. 2003. "Calcium balance in young adults on a vegan and lactovegetarian diet." *J Bone Miner Metab* 21 (1):28–33. https://doi.org/10.1007/s007740300005.

Kune, Gabriel A. 2010. "The melbourne colorectal cancer study: Reflections on a 30-year experience." 193 (11–12):648–652. https://doi.org/10.5694/j.1326-5377.2010.tb04093.x.

Kyrø, Cecilie, Guri Skeie, Steffen Loft, Kim Overvad, Jane Christensen, Anne Tjønneland, and Anja Olsen. 2013. "Adherence to a healthy Nordic food index is associated with a lower incidence of colorectal cancer in women: The Diet, Cancer and Health cohort study." *Br J Nutr* 109 (5):920–927. https://doi.org/10.1017/S0007114512002085.

La Vecchia, C., E. Negri, C. Pelucchi, and S. Franceschi. 2002. "Dietary folate and colorectal cancer." *Int J Cancer* 102 (5):545–547. https://doi.org/10.1002/ijc.10720.

Lam, Yan Y., Connie W.Y. Ha, Jenny M.A. Hoffmann, Jan Oscarsson, Anuwat Dinudom, Thomas J. Mather, David I. Cook, Nicholas H. Hunt, Ian D. Caterson, Andrew J. Holmes, and Len H. Storlien. 2015. "Effects of dietary fat profile on gut permeability and microbiota and their relationships with metabolic changes in mice." 23 (7):1429–1439. https://doi.org/10.1002/oby.21122.

Levy, M., Kolodziejczyk, A., Thaiss, C., et al. 2017. "Dysbiosis and the immune system." *Nat Rev Immunol* 17:219–232. https://doi.org/10.1038/nri.2017.7.

Lin, R., W. Liu, M. Piao, and H. Zhu. 2017. "A review of the relationship between the gut microbiota and amino acid metabolism." *Amino Acids* 49 (12):2083–2090. https://doi.org/10.1007/s00726-017-2493-3.

Liu, J. R., S. Y. Wang, Y. Y. Lin, and C. W. Lin. 2002. "Antitumor activity of milk kefir and soy milk kefir in tumor-bearing mice." *Nutr Cancer* 44 (2):183–187. https://doi.org/10.1207/s15327914nc4402_10.

Llor, X., E. Pons, A. Roca, M. Alvarez, J. Mane, F. Fernandez-Banares, and M. A. Gassull. 2003. "The effects of fish oil, olive oil, oleic acid and linoleic acid on colorectal neoplastic processes." *Clin Nutr* 22 (1):71–79.

Lu, Demin, Chi Pan, Chenyang Ye, Huijie Duan, Fei Xu, Li Yin, Wei Tian, and Suzhan Zhang. 2017. "Meta-analysis of soy consumption and gastrointestinal cancer risk." *Sci Rep* 7 (1):4048. https://doi.org/10.1038/s41598-017-03692-y.

LU'O'NG, K. V. Q., and Lan Thi Hoàng Nguyễn. 2013. "The role of thiamine in cancer: Possible genetic and cellular signaling mechanisms." *Cancer Genomics Proteomics* 10 (4):169–185.

Ma, Y., Y. Huangfu, L. Deng, P. Wang, L. Shen, and Y. Zhou. 2023. "High serum riboflavin is associated with the risk of sporadic colorectal cancer." *Cancer Epidemiol* 83:102342. https://doi.org/10.1016/j.canep.2023.102342.

Mahady, G. B., S. L. Pendland, G. Yun, and Z. Z. Lu. 2002. "Turmeric (Curcuma longa) and curcumin inhibit the growth of Helicobacter pylori, a group 1 carcinogen." *Anticancer Res* 22 (6c):4179–4181.

Mahfouz, Eman Mohamed, Refaat Raouf Sadek, Wafaa Mohamed Abdel-Latief, FA Mosallem, and Ebtesam Esmail Hassan. 2014. "The role of dietary and lifestyle factors in the development of colorectal cancer: Case control study in Minia, Egypt." *Cent Eur J Public Health* 22 (4):215–222.

Malesza, Ida Judyta, Michał Malesza, Jarosław Walkowiak, Nadiar Mussin, Dariusz Walkowiak, Raisa Aringazina, Joanna Bartkowiak-Wieczorek, and Edyta Mądry. 2021. "High-Fat, Western-style diet, systemic inflammation, and gut microbiota: A narrative review." *Cells* 10 (11):3164.

Mao, Jun J., Geetha Gopalakrishna Pillai, Carlos Jose Andrade, Jennifer A. Ligibel, Partha Basu, Lorenzo Cohen, Ikhlas A. Khan, Karen M. Mustian, Rammanohar Puthiyedath, Kartar Singh Dhiman, Lixing Lao, Ricardo Ghelman, Paulo Cáceres Guido, Gabriel Lopez, Daniel F. Gallego-Perez, and Luis Alejandro Salicrup. 2022. "Integrative oncology: Addressing the global challenges of cancer prevention and treatment." 72 (2):144–164. https://doi.org/10.3322/caac.21706.

Marandola, P., K. Bova, M. A. Labib, A. L. Bonghi, D. Tiscione, K. Christodoulakis, M. Gerardini, F. P. Tinozzi, and S. Watia. 2004. "Cancer and nutrition in Africa in the postgenomic and proteomics era." *Afr J Urol* 10 (3):178–197.

Martinez-Medina, Margarita, Jérémy Denizot, Nicolas Dreux, Frédéric Robin, Elisabeth Billard, Richard Bonnet, Arlette Darfeuille-Michaud, and Nicolas Barnich. 2014. "Western diet induces dysbiosis with increased E coli in CEABAC10 mice, alters host barrier function favouring AIEC colonisation." *Gut* 63 (1):116–124.

Mehta, R. S., M. Song, R. Nishihara, D. A. Drew, K. Wu, Z. R. Qian, T. T. Fung, T. Hamada, Y. Masugi, A. da Silva, Y. Shi, W. Li, M. Gu, W. C. Willett, C. S. Fuchs, E. L. Giovannucci, S. Ogino, and A. T. Chan. 2017. "Dietary patterns and risk of colorectal cancer: Analysis by tumor location and molecular subtypes." *Gastroenterology* 152 (8):1944–1953.e1. https://doi.org/10.1053/j.gastro.2017.02.015.

Miller, Dennis D., and Ross M. Welch. 2013. "Food system strategies for preventing micronutrient malnutrition." *Food Policy* 42:115–128. https://doi.org/10.1016/j.foodpol.2013.06.008.

Mitsou, E. K., A. Kakali, S. Antonopoulou, K. C. Mountzouris, M. Yannakoulia, D. B. Panagiotakos, and A. Kyriacou. 2017. "Adherence to the Mediterranean diet is associated with the gut microbiota pattern and gastrointestinal characteristics in an adult population." *Br J Nutr* 117 (12):1645–1655. https://doi.org/10.1017/s0007114517001593.

Mohanty, B. P., S. Ganguly, A. Mahanty, T. V. Sankar, R. Anandan, K. Chakraborty, B. N. Paul, D. Sarma, J. Syama Dayal, G. Venkateshwarlu, S. Mathew, K. K. Asha, D. Karunakaran, T. Mitra, S. Chanda, N. Shahi, P. Das, P. Das, M. S. Akhtar, P. Vijayagopal, and N. Sridhar. 2016. "DHA and EPA content and fatty acid profile of 39 food fishes from India." *Biomed Res Int* 2016:4027437. https://doi.org/10.1155/2016/4027437.

Mohseni, Amir Hossein, Sedigheh Taghinezhad-S, and Xiangsheng Fu. 2020. "Gut microbiota-derived metabolites and colorectal cancer: New insights and updates." *Microbial Pathogenesis* 149:104569. https://doi.org/10.1016/j.micpath.2020.104569.

Mosieniak, Grazyna, Marek Adamowicz, Olga Alster, Hubert Jaskowiak, Andrzej A Szczepankiewicz, Grzegorz M. Wilczynski, Iwona A. Ciechomska, and Ewa Sikora. 2012. "Curcumin induces permanent growth arrest of human colon cancer cells: Link between senescence and autophagy." *Mech Ageing Dev* 133 (6):444–455.

Muccioli, Giulio G., Damien Naslain, Fredrik Bäckhed, Christopher S. Reigstad, Didier M. Lambert, Nathalie M. Delzenne, and Patrice D. Cani. 2010. "The endocannabinoid system links gut microbiota to adipogenesis." *Mol Syst Biol* 6 (1):392.

Murakami, Yuki, Soichi Tanabe, and Takuya Suzuki. 2016. "High-fat diet-induced intestinal hyperpermeability is associated with increased bile acids in the large intestine of mice." *J Food Sci* 81 (1):H216–H222.

Murtaugh, Maureen A., Carol Sweeney, Anna R. Giuliano, Jennifer S. Herrick, Lisa Hines, Tim Byers, Kathy B. Baumgartner, and Martha L. Slattery. 2008. "Diet patterns and breast cancer risk in Hispanic and non-Hispanic white women: The Four-Corners Breast Cancer Study." *Am J Clin Nutr* 87 (4):978–984.

Nascimento, Roberto de Paula do, Lívia Mateus Reguengo, Ana Paula da Fonseca Machado, and Mario Roberto Marostica Junior. 2022. "The preventive and therapeutic potential of native Brazilian fruits on colorectal cancer." *Food Biosci* 46:101539. https://doi.org/10.1016/j.fbio.2021.101539.

Naudin, Sabine, Marta Solans Margalef, Fatemeh Saberi Hosnijeh, Alexandra Nieters, Cecilie Kyrø, Anne Tjønneland, Christina C. Dahm, Kim Overvad, Yahya Mahamat-Saleh, Caroline Besson, Marie-Christine Boutron-Ruault, Tilman Kühn, Federico Canzian, Matthias B. Schulze, Eleni Peppa, Anna Karakatsani, Antonia Trichopoulou, Sabina Sieri, Giovana Masala, Salvatore Panico, Rosario Tumino, Fulvio Ricceri, Sairah L. F. Chen, Leila L. Barroso, José M. Huerta, Maria-Jose Sánchez, Eva Ardanaz, Virginia Menéndez, Pilar Amiano Exezarreta, Florentin Spaeth, Mats Jerkeman, Karin Jirstom, Julie A. Schmidt, Dagfinn Aune, Elisabete Weiderpass, Elio Riboli, Roel Vermeulen, Delphine Casabonne, Marc Gunter, Paul Brennan, and Pietro Ferrari. 2020. "Healthy lifestyle and the risk of lymphoma in the European prospective investigation into cancer and Nutrition study." 147 (6):1649–1656. https://doi.org/10.1002/ijc.32977.

Nawaz, Javaria, Azhar Rasul, Muhammad Ajmal Shah, Ghulam Hussain, Ammara Riaz, Iqra Sarfraz, Saba Zafar, Muhammad Adnan, Abdul Haleem Khan, and Zeliha Selamoglu. 2020. "Cardamonin: A new player to fight cancer via multiple cancer signaling pathways." *Life Sciences* 250:117591. https://doi.org/10.1016/j.lfs.2020.117591.

Newsome, Rachel, Ye Yang, and Christian Jobin. 2023. "Western diet influences on microbiome and carcinogenesis." *Semin Immunol* 67:101756. https://doi.org/10.1016/j.smim.2023.101756.

Ngo, Suong NT, Desmond B. Williams, Lynne Cobiac, and Richard J. Head. 2007. "Does garlic reduce risk of colorectal cancer? A systematic review, 3." *J Nutr* 137 (10):2264–2269.

Niccolai, E., S. Baldi, F. Ricci, E. Russo, G. Nannini, M. Menicatti, G. Poli, A. Taddei, G. Bartolucci, A. S. Calabrò, F. C. Stingo, and A. Amedei. 2019. "Evaluation and comparison of short chain fatty acids composition in gut diseases." *World J Gastroenterol* 25 (36):5543–5558. https://doi.org/10.3748/wjg.v25.i36.5543.

Niccolai, E., F. Boem, G. Emmi, and A. Amedei. 2020. "The link 'Cancer and autoimmune diseases' in the light of microbiota: Evidence of a potential culprit." *Immunol Lett* 222:12–28. https://doi.org/10.1016/j.imlet.2020.03.001.

Norat, Teresa, Sheila Bingham, Pietro Ferrari, Nadia Slimani, Mazda Jenab, Mathieu Mazuir, Kim Overvad, Anja Olsen, Anne Tjønneland, and Françoise Clavel. 2005. "Meat, fish, and colorectal cancer risk: The European Prospective Investigation into cancer and nutrition." *J Natl Cancer* 97 (12):906–916.

O'Keefe, Stephen J. 2019. "The association between dietary fibre deficiency and high-income lifestyle-associated diseases: Burkitt's hypothesis revisited." *Lancet Gastroenterol Hepatol* 4 (12):984–996.

Ocvirk, S., and S. J. D. O'Keefe. 2021. "Dietary fat, bile acid metabolism and colorectal cancer." *Semin Cancer Biol* 73:347–355. https://doi.org/10.1016/j.semcancer.2020.10.003.

Ocvirk, Soeren, Annette S. Wilson, Corynn N. Appolonia, Timothy K. Thomas, and Stephen J. D. O'Keefe. 2019. "Fiber, fat, and colorectal cancer: New insight into modifiable dietary risk factors." *Curr Gastroenterol Rep* 21 (11):62. https://doi.org/10.1007/s11894-019-0725-2.

Ohashi, Koji, Rei Shibata, Toyoaki Murohara, Noriyuki Ouchi. 2014. "Role of anti-inflammatory adipokines in obesity-related diseases." *Trends Endocrinol Metab* 25 (7):348–355.

Onyoh, Elias F., Wen-Feng Hsu, Li-Chun Chang, Yi-Chia Lee, Ming-Shiang Wu, and Han-Mo Chiu. 2019. "The rise of colorectal cancer in Asia: Epidemiology, screening, and management." *Curr Gastroenterol Rep* 21 (8):36. https://doi.org/10.1007/s11894-019-0703-8.

Pan, S. Y., and H. Morrison. 2011. "Epidemiology of cancer of the small intestine." *World J Gastrointest Oncol* 3 (3):33–42. https://doi.org/10.4251/wjgo.v3.i3.33.

Pandey, Himani, Daryl W. T. Tang, Sunny H. Wong, and Devi Lal. 2023. "Gut microbiota in colorectal cancer: Biological role and therapeutic opportunities." *Cancers* 15 (3):866.

Park, Y., J. Lee, J. H. Oh, A. Shin, and J. Kim. 2016. "Dietary patterns and colorectal cancer risk in a Korean population: A case-control study." *Medicine (Baltimore)* 95 (25):e3759. https://doi.org/10.1097/md.0000000000003759.

Perez-Jimenez, F., J. Ruano, P. Perez-Martinez, F. Lopez-Segura, and J. Lopez-Miranda. 2007. "The influence of olive oil on human health: Not a question of fat alone." *Mol Nutr Food Res* 51: 1199–1208.

Peters, Ulrike, Rashmi Sinha, Nilanjan Chatterjee, Amy F. Subar, Regina G. Ziegler, Martin Kulldorff, Robert Bresalier, Joel L. Weissfeld, Andrew Flood, and Arthur Schatzkin. 2003. "Dietary fibre and colorectal adenoma in a colorectal cancer early detection programme." *Lancet* 361 (9368):1491–1495.

Peterson, C. T., D. A. Rodionov, A. L. Osterman, and S. N. Peterson. 2020. "B vitamins and their role in immune regulation and cancer." *Nutrients* 12 (11). https://doi.org/10.3390/nu12113380.

Psaltopoulou, Theodora, Rena I. Kosti, Dimitrios Haidopoulos, Meletios Dimopoulos, Demosthenes B. Panagiotakos. 2011. "Olive oil intake is inversely related to cancer prevalence: A systematic review and a meta-analysis of 13800 patients and 23340 controls in 19 observational studies." *Lipids in Health and Disease* 10:1–16.

Qiao, L., and Y. Feng. 2013. "Intakes of heme iron and zinc and colorectal cancer incidence: A meta-analysis of prospective studies." *Cancer Causes Control* 24 (6):1175–1183. https://doi.org/10.1007/s10552-013-0197-x.

Randi, Giorgia, Valeria Edefonti, Monica Ferraroni, Carlo La Vecchia, and Adriano Decarli. 2010. "Dietary patterns and the risk of colorectal cancer and adenomas." *Nutr Rev* 68 (7):389–408.

Reynolds, Andrew, Jim Mann, John Cummings, Nicola Winter, Evelyn Mete, and Lisa Te Morenga. 2019. "Carbohydrate quality and human health: A series of systematic reviews and meta-analyses." *Lancet* 393 (10170):434–445.

Riboli, Elio, K. J. Hunt, Nadia Slimani, Pietro Ferrari, Teresa Norat, M. Fahey, U. R. Charrondiere, B. Hemon, C. Casagrande, and J. Vignat. 2002. "European prospective investigation into cancer and nutrition (EPIC): Study populations and data collection." *Public Health Nutr* 5 (6b):1113–1124.

Rinninella, E., M. Cintoni, P. Raoul, L. R. Lopetuso, F. Scaldaferri, G. Pulcini, G. A. D. Miggiano, A. Gasbarrini, and M. C. Mele. 2019. "Food components and dietary habits: Keys for a healthy gut microbiota composition." *Nutrients* 11 (10). https://doi.org/10.3390/nu11102393.

Rizk, Sandra, Nathalie El Khoury, Stephany El-Hayek, Omayr Tarras, Mirvat El-sibai, and Marwan El-Sabban. 2014. "Kefir exhibits anti-proliferative and pro-apoptotic effects on colon adenocarcinoma cells with no significant effects on cell migration and invasion (647.24)." *FASEB J* 28:647.24.

Rock, Cheryl L., Cynthia Thomson, Ted Gansler, Susan M. Gapstur, Marjorie L. McCullough, Alpa V. Patel, Kimberly S. Andrews, Elisa V. Bandera, Colleen K. Spees, Kimberly Robien, Sheri Hartman, Kristen Sullivan, Barbara L. Grant, Kathryn K. Hamilton, Lawrence H. Kushi, Bette J. Caan, Debra Kibbe, Jessica Donze Black, Tracy L. Wiedt, Catherine McMahon, Kirsten Sloan, and Colleen Doyle. 2020. "American cancer society guideline for diet and physical activity for cancer prevention." 70 (4):245–271. https://doi.org/10.3322/caac.21591.

Roediger, W. E. 1980. "Role of anaerobic bacteria in the metabolic welfare of the colonic mucosa in man." *Gut* 21 (9):793–798.

Roncal, Carmen, Esther Martínez-Aguilar, Josune Orbe, Susana Ravassa, Alejandro Fernandez-Montero, Goren Saenz-Pipaon, Ana Ugarte, Ander Estella-Hermoso de Mendoza, Jose A. Rodriguez, Sebastián Fernández-Alonso, Leopoldo Fernández-Alonso, Julen Oyarzabal, and Jose A. Paramo. 2019. "Trimethylamine-N-Oxide (TMAO) predicts cardiovascular mortality in peripheral artery disease." *Sci Rep* 9 (1):15580. https://doi.org/10.1038/s41598-019-52082-z.

Rosato, Valentina, Valentina Guercio, Cristina Bosetti, Eva Negri, Diego Serraino, Attilio Giacosa, Maurizio Montella, Carlo La Vecchia, and Alessandra Tavani. 2016. "Mediterranean diet and colorectal cancer risk: A pooled analysis of three Italian case–control studies." *Br J Cancer* 115 (7):862–865.

Rosés, Carles, Amanda Cuevas-Sierra, Salvador Quintana, José I. Riezu-Boj, J. Alfredo Martínez, Fermín I. Milagro, and Anna Barceló. 2021. "Gut microbiota bacterial species associated with mediterranean diet-related food groups in a Northern Spanish population." *Nutrients* 13 (2):636.

Rubio-López, Nuria, Agustín Llopis-González, Yolanda Picó, and María Morales-Suárez-Varela. 2017. "Dietary calcium intake and adherence to the mediterranean diet in Spanish children: The ANIVA study." *International Journal of Environmental Research and Public Health* 14 (6):637.

S. Deoula, Meimouna, Khaoula El Kinany, Inge Huybrechts, Marc J. Gunter, Zineb Hatime, Hanae A. Boudouaya, Abdelilah Benslimane, Chakib Nejjari, Mohammed El Abkari, Wafaa Badre, Abdellah E. El Feydi, Saîd Afkir, Naima Abda, and Karima El Rhazi. 2020. "Consumption of meat, traditional and modern processed meat and colorectal cancer risk among the Moroccan population: A large-scale case–control study." 146 (5):1333–1345. https://doi.org/10.1002/ijc.32689.

Sánchez-Alcoholado, L., B. Ramos-Molina, A. Otero, A. Laborda-Illanes, R. Ordóñez, J. A. Medina, J. Gómez-Millán, and M. I. Queipo-Ortuño. 2020. "The role of the gut microbiome in colorectal cancer development and therapy response." *Cancers (Basel)* 12 (6). https://doi.org/10.3390/cancers12061406.

Sano, Renata, and John C. Reed. 2013. "ER stress-induced cell death mechanisms." *Biochim Biophys Acta Mol Cell Res* 1833 (12):3460–3470.

Saura-Calixto, Fulgencio, and Isabel Goñi. 2009. "Definition of the mediterranean diet based on bioactive compounds." *Crit Rev Food Sci Nutr* 49 (2):145–152. https://doi.org/10.1080/10408390701764732.

Schatzkin, Arthur, Yikyung Park, Michael F. Leitzmann, Albert R. Hollenbeck, and Amanda J. Cross. 2008. "Prospective study of dietary fiber, whole grain foods, and small intestinal cancer." *Gastroenterology* 135 (4):1163–1167. https://doi.org/10.1053/j.gastro.2008.07.015.

Schönfeld, P., and L. Wojtczak. 2016. "Short- and medium-chain fatty acids in energy metabolism: The cellular perspective." *J Lipid Res* 57 (6):943–954. https://doi.org/10.1194/jlr.R067629.

Schwingshackl, Lukas, and Georg Hoffmann. 2014. "Adherence to Mediterranean diet and risk of cancer: A systematic review and meta-analysis of observational studies." 135 (8):1884–1897. https://doi.org/10.1002/ijc.28824.

Schwingshackl, Lukas, and Georg Hoffmann. 2015. "Adherence to Mediterranean diet and risk of cancer: An updated systematic review and meta-analysis of observational studies." 4 (12):1933–1947. https://doi.org/10.1002/cam4.539.

Seow, Adeline, Stella R. Quah, Denis Nyam, Paulin T. Straughan, Terrence Chua, and Tar-Choon Aw. 2002. "Food groups and the risk of colorectal carcinoma in an Asian population." *Int J Am Cancer Soc* 95 (11):2390–2396.

Serra-Majem, Lluís, Maira Bes-Rastrollo, Blanca Román-Vinas, Karina Pfrimer, Almudena Sánchez-Villegas, and Miguel A. Martínez-González. 2009. "Dietary patterns and nutritional adequacy in a Mediterranean country." *Br J Nutr* 101 (S2):S21–S28.

Serra-Majem, Lluís, Reina García-Closas, Lourdes Ribas, Carmen Pérez-Rodrigo, and Javier Aranceta. 2001. "Food patterns of Spanish schoolchildren and adolescents: The enKid Study." *Public Health Nutrition* 4 (6a):1433–1438.

Sharifi, Mohammadreza, Abbas Moridnia, Deniz Mortazavi, Mahsa Salehi, Marzieh Bagheri, and Abdolkarim Sheikhi. 2017. "Kefir: A powerful probiotics with anticancer properties." *Medical Oncol* 34 (11):183. https://doi.org/10.1007/s12032-017-1044-9.

Shi, Hang, Maia V. Kokoeva, Karen Inouye, Iphigenia Tzameli, Huali Yin, and Jeffrey S. Flier. 2006. "TLR4 links innate immunity and fatty acid–induced insulin resistance." *J Clin Investig* 116 (11):3015–3025.

Shuwen, H., D. Miao, Q. Quan, W. Wei, Z. Zhongshan, Z. Chun, and Y. Xi. 2019. "Protective effect of the 'food-microorganism-SCFAs' axis on colorectal cancer: From basic research to practical application." *J Cancer Res Clin Oncol* 145 (9):2169–2197. https://doi.org/10.1007/s00432-019-02997-x.

Siegel, R. L., C. D. Jakubowski, S. A. Fedewa, A. Davis, and N. S. Azad. 2020. "Colorectal cancer in the young: Epidemiology, prevention, management." *Am Soc Clin Oncol Educ Book* 40:1–14. https://doi.org/10.1200/edbk_279901.

Sierra, Monica S., and David Forman. 2016. "Burden of colorectal cancer in Central and South America." *Cancer Epidemiol* 44:S74–S81. https://doi.org/10.1016/j.canep.2016.03.010.

Earticle, C. 2012. "Singapore cancer trends in the last decade." *Singapore Med J* 53 (1):3–10

Singh, Neetu, and Surender Singh Yadav. 2022. "Ethnomedicinal uses of Indian spices used for cancer treatment: A treatise on structure-activity relationship and signaling pathways." *Current Research in Food Science* 5:1845–1872. https://doi.org/10.1016/j.crfs.2022.10.005.

Slimani, N., M. Fahey, A. Welch, E. Wirfält, C. Stripp, E. Bergström, J. Linseisen, M. B. Schulze, C. Bamia, Y. Chloptsios, F. Veglia, S. Panico, H. B. Bueno-de-Mesquita, M. C. Ocké, M. Brustad, E. Lund, C. A. González, A. Barcos, G. Berglund, A. Winkvist, A. Mulligan, P. Appleby, K. Overvad, A. Tjønneland, F. Clavel-Chapelon, E. Kesse, P. Ferrari, W. A. Van Staveren, and E. Riboli. 2002. "Diversity of dietary patterns observed in the European prospective investigation into cancer and nutrition (EPIC) project." *Public Health Nutrition* 5 (6b):1311–1328. https://doi.org/10.1079/PHN2002407.

Sofi, Francesco, Rosanna Abbate, Gian Franco Gensini, and Alessandro Casini. 2010. "Accruing evidence on benefits of adherence to the Mediterranean diet on health: An updated systematic review and meta-analysis." *Am J Clin Nutr* 92 (5):1189–1196.

Sofi, Francesco, Francesca Cesari, Rosanna Abbate, Gian Franco Gensini, and Alessandro Casini. 2008. "Adherence to Mediterranean diet and health status: Meta-analysis." *BMJ* 337:a1344.

Solans, Marta, Yolanda Benavente, Marc Saez, Antonio Agudo, Sabine Naudin, Fatemeh Saberi Hosnijeh, Hwayoung Noh, Heinz Freisling, Pietro Ferrari, Caroline Besson, Yahya Mahamat-Saleh, Marie-Christine Boutron-Ruault, Tilman Kühn, Rudolf Kaaks, Heiner Boeing, Cristina Lasheras, Miguel Rodríguez-Barranco, Pilar Amiano, Jose Maria Huerta, Aurelio Barricarte, Julie A. Schmidt, Paolo Vineis, Elio Riboli, Antonia Trichopoulou, Christina Bamia, Eleni Peppa, Giovanna Masala, Claudia Agnoli, Rosario Tumino, Carlotta Sacerdote, Salvatore Panico, Guri Skeie, Elisabete Weiderpass, Mats Jerkeman, Ulrika Ericson, Florentin Späth, Lena Maria Nilsson, Christina C Dahm, Kim Overvad, Anne Katrine Bolvig, Anne Tjønneland, Silvia de Sanjose, Genevieve Buckland, Roel Vermeulen, Alexandra Nieters, and Delphine Casabonne. 2019. "Adherence to the mediterranean diet and lymphoma risk in the european prospective investigation into cancer and nutrition." 145 (1):122–131. https://doi.org/10.1002/ijc.32091.

Song, M., and A. T. Chan. 2019. "Environmental factors, gut microbiota, and colorectal cancer prevention." *Clin Gastroenterol Hepatol* 17 (2):275–289. https://doi.org/10.1016/j.cgh.2018.07.012.

Song, M., A. T. Chan, and J. Sun. 2020. "Influence of the gut microbiome, diet, and environment on risk of colorectal cancer." *Gastroenterology* 158 (2):322–340. https://doi.org/10.1053/j.gastro.2019.06.048.

Soong, Yean-Yean, and Philip J. Barlow. 2004. "Antioxidant activity and phenolic content of selected fruit seeds." *Food Chem* 88 (3):411–417. https://doi.org/10.1016/j.foodchem.2004.02.003.

Suez, J., T. Korem, D. Zeevi, G. Zilberman-Schapira, C. A. Thaiss, O. Maza, D. Israeli, N. Zmora, S. Gilad, A. Weinberger, Y. Kuperman, A. Harmelin, I. Kolodkin-Gal, H. Shapiro, Z. Halpern, E. Segal, and E. Elinav. 2014. "Artificial sweeteners induce glucose intolerance by altering the gut microbiota." *Nature* 514 (7521):181–186. https://doi.org/10.1038/nature13793.

Sun, Lijuan, Cheng Luo, and Jiankang Liu. 2014. "Hydroxytyrosol induces apoptosis in human colon cancer cells through ROS generation." *Food and Funct* 5 (8):1909–1914.

Sung, H., J. Ferlay, R. L. Siegel, M. Laversanne, I. Soerjomataram, A. Jemal, and F. Bray. 2021. "Global cancer statistics 2020: GLOBOCAN estimates of incidence and mortality worldwide for 36 cancers in 185 countries." *CA Cancer J Clin* 71 (3):209–249. https://doi.org/10.3322/caac.21660.

Swapna, B. . 2015. "An ethnobotanical survey of plants used by Yanadi tribe of Kavali, Nellore district, Andhra Pradesh, India." *J Sci Innov Res* 4:22–26. https://doi.org/10.31254/jsir.2015.4106.

Tapias, Victor, Xiaoping Hu, Kelvin C. Luk, Laurie H. Sanders, Virginia M. Lee, and J. Timothy Greenamyre. 2017. "Synthetic alpha-synuclein fibrils cause mitochondrial impairment and selective dopamine neurodegeneration in part via iNOS-mediated nitric oxide production." *Cell Mol Life Sci* 74:2851–2874.

Theodoratou, E., S. M. Farrington, A. Tenesa, G. McNeill, R. Cetnarskyj, R. A. Barnetson, M. E. Porteous, M. G. Dunlop, and H. Campbell. 2008. "Dietary vitamin B6 intake and the risk of colorectal cancer." *Cancer Epidemiol Biomarkers Prev* 17 (1):171–182. https://doi.org/10.1158/1055-9965.Epi-07-0621.

Thomson, B., and I. Shaw. 2002. "A comparison of risk and protective factors for colorectal cancer in the diet of New Zealand Maori and non-Maori." *Asian Pac J Cancer Prev* 3 (4):319–324.

Trichopoulou, Antonia, Pagona Lagiou, Hannah Kuper, Dimitrios %J Cancer Epidemiology Biomarkers Trichopoulos, and Prevention. 2000. "Cancer and Mediterranean dietary traditions." *Cancer Epidemiology, Biomarkers & Prevention: A Publication of the American Association for Cancer Research, Cosponsored by the American Society of Preventive Oncology* 9 (9):869–873.

Trimble, Edward L, and Preetha Rajaraman. 2017. "Integrating traditional and allopathic medicine: An opportunity to improve global health in cancer." *J Natl Cancer Inst Monogr* 2017 (52):lgx011.

Turner, Nancy D., and Shannon K. Lloyd. 2017. "Association between red meat consumption and colon cancer: A systematic review of experimental results." *Exp Biol Med* 242 (8):813–839.

Valerino-Perea, Selene, Laura Lara-Castor, Miranda Elaine Glynis Armstrong, and Angeliki Papadaki. 2019. "Definition of the traditional mexican diet and its role in health: A systematic review." *Nutrients* 11 (11):2803.

Veettil, Sajesh K., Tse Yee Wong, Yee Shen Loo, Mary C. Playdon, Nai Ming Lai, Edward L. Giovannucci, and Nathorn Chaiyakunapruk. 2021. "Role of diet in colorectal cancer incidence: Umbrella review of meta-analyses of prospective observational studies." *JAMA Network Open* 4 (2):e2037341–e2037341.

Velasquez, Manuel T. 2018. "Altered gut microbiota: A link between diet and the metabolic syndrome." *Metab Syndr Relat Disord* 16 (7):321–328.

Vernia, F., S. Longo, G. Stefanelli, A. Viscido, and G. Latella. 2021. "Dietary factors modulating colorectal carcinogenesis." *Nutrients* 13 (1). https://doi.org/10.3390/nu13010143.

Vieira, A. R., L. Abar, D. S. M. Chan, S. Vingeliene, E. Polemiti, C. Stevens, D. Greenwood, and T. Norat. 2017. "Foods and beverages and colorectal cancer risk: A systematic review and meta-analysis of cohort studies, an update of the evidence of the WCRF-AICR continuous update project." *Ann Oncol* 28 (8):1788–1802.

Vijayalakshmi, Subramani, Mark Faraday, A. Wilfred, and Manoharan Murali. 2016. "Exploration of nutrient content of traditional Tamil nadu recipe with Solanum trilobatum." *Int J Pharma Bio Sci* 7. https://doi.org/10.22376/ijpbs.2016.7.4.b364-368.

Wahrendorf, Jürgen. 1987. "An estimate of the proportion of colo-rectal and stomach cancers which might be prevented by certain changes in dietary habits." *Int J Cancer Res* 40 (5):625–628.

Wang, Caihua, Peiwei Li, Junmei Xuan, Chunpeng Zhu, Jingjing Liu, Lizhen Shan, Qin Du, Yuezhong Ren, and Jun Ye. 2017. "Cholesterol enhances colorectal cancer progression via ROS elevation and MAPK signaling pathway activation." *Cell Physiol Biochem* 42 (2):729–742. https://doi.org/10.1159/000477890.

Wang, Y., W. Wang, K. Z. Sanidad, P. A. Shih, X. Zhao, and G. Zhang. 2018. "Eicosanoid signaling in carcinogenesis of colorectal cancer." *Cancer Metastasis Rev* 37 (2–3):257–267. https://doi.org/10.1007/s10555-018-9739-8.

Wang, Z., Z. Lu, S. Lin, J. Xia, Z. Zhong, Z. Xie, Y. Xing, J. Qie, M. Jiao, Y. Li, H. Wen, P. Zhao, D. Zhang, P. Zhou, J. Qian, F. Luo, L. Wang, H. Yu, J. Liu, J. Gu, R. Liu, and Y. Chu. 2022. "Leucine-tRNA-synthase-2-expressing B cells contribute to colorectal cancer immunoevasion." *Immunity* 55 (6):1067–1081.e8. https://doi.org/10.1016/j.immuni.2022.04.017.

Weng, Zuyi, Arti B. Patel, Smaro Panagiotidou, and Theoharis C. Theoharides. 2015. "The novel flavone tetramethoxyluteolin is a potent inhibitor of human mast cells." *J Allergy Clin Immunol* 135 (4):1044–1052.e5. https://doi.org/10.1016/j.jaci.2014.10.032.

Willett, Walter C, Frank Sacks, Antonia Trichopoulou, Greg Drescher, Anna Ferro-Luzzi, Elisabet Helsing, and Dimitrios Trichopoulos. 1995. "Mediterranean diet pyramid: A cultural model for healthy eating." *Am J Clin Nutr* 61 (6):1402S–1406S.

Williams, Mark T., and Norman G. Hord. 2005. "The role of dietary factors in cancer prevention: Beyond fruits and vegetables." *Nutr Clin Pract* 20 (4):451–459.

Wu, Lang, Zhen Wang, Jingjing Zhu, Angela L. Murad, Larry J. Prokop, and Mohammad H. Murad. 2015. "Nut consumption and risk of cancer and type 2 diabetes: A systematic review and meta-analysis." *Nutr Rev* 73 (7):409–425.

Xu, F., H. L. Jiang, W. W. Feng, C. Fu, and J. C. Zhou. 2023. "Characteristics of amino acid metabolism in colorectal cancer." *World J Clin Cases* 11 (27):6318–6326. https://doi.org/10.12998/wjcc.v11.i27.6318.

Yatsunenko, Tanya, Federico E. Rey, Mark J. Manary, Indi Trehan, Maria Gloria Dominguez-Bello, Monica Contreras, Magda Magris, Glida Hidalgo, Robert N. Baldassano, and Andrey P. Anokhin. 2012. "Human gut microbiome viewed across age and geography." *Nature* 486 (7402):222–227.

Yee, Yuk Kei, Victoria P. Y. Tan, Pierre Chan, Ivan F. N. Hung, Roberta Pang, and Benjamin C. Y. Wong. 2009. "Epidemiology of colorectal cancer in Asia." 24 (12):1810–1816. https://doi.org/10.1111/j.1440-1746.2009.06138.x.

Yuan, Chen, and Edward L. Giovannucci. 2020. "Epidemiological evidence for dietary sugars and colorectal cancer." *Curr Colorectal Cancer Rep* 16 (3):55–63. https://doi.org/10.1007/s11888-020-00453-4.

Zhang, Kezhong, Xiaohua Shen, Jun Wu, Kenjiro Sakaki, Thomas Saunders, D Thomas Rutkowski, Sung Hoon Back, and Randal J. Kaufman. 2006. "Endoplasmic reticulum stress activates cleavage of CREBH to induce a systemic inflammatory response." *Cell* 124 (3):587–599.

Zheng, Y., L. Meng, H. Liu, L. Sun, Y. Nie, Q. Wu, D. Fan, and M. Li. 2022. "Let food be thy medicine: The role of diet in colorectal cancer: A narrative review." *J Gastrointest Oncol* 13 (4):2020–2032. https://doi.org/10.21037/jgo-22-32.

8 Liver-Associated Disorders and Gut-Friendly Food

Homa Fatma[1], *Mohd Jameel*[1], *and Hifzur R. Siddique*[2]

[1]Molecular Cancer Genetics and Translational Research Lab, Department of Zoology, Aligarh-202002, India
[2]Molecular Cancer Genetics and Translational Research Lab, Department of Zoology, Aligarh, Uttar Pradesh, India

8.1 INTRODUCTION

Every organism lives in close association with its surrounding flora and fauna. Human beings are no exception and also depend upon their association with their surrounding environment. Interestingly, the microbial community inhabited by the human body coexists with the host and is referred to as normal flora, microflora, or microbiota (Sekirov et al., 2010). The microbiota, encompassing bacteria, archaea, viruses, and eukaryotes within and external to the human body, substantially impacts physiological equilibrium and homeostasis. Variations in microbial abundance may modulate metabolic and immune functions, augmenting or impairing normal physiology and homeostasis (Ogunrinola et al., 2020; Singh et al., 2021). In recent decades, heightened focus has been directed towards the microbiome due to its profound implications for human health. Variations in the human microbiota composition have been associated with a spectrum of health conditions, including metabolic disorders, neurodegenerative diseases, and cancer. Analysis of metagenomic data derived from healthy individuals and those afflicted by diseases reveals compelling evidence supporting the protective role of microbial communities in mitigating the onset of diverse disorders and pathological conditions (Cani, 2018).

Ancient sayings like "death sits in the bowels" and "bad digestion is the root of all evils" suggest the important association between gut health and human health. Gut health not only depends upon the absence of gastrointestinal pathogens but is also associated with the microbes present in the gut and healthy eating habits (Sekirov et al., 2010). Furthermore, the bidirectional gut-brain axis is important for establishing cellular homeostasis, and gut microbiota are considered one of the axis's important regulators. The gut microbiota serves diverse functions within the human body, encompassing, but not restricted to, food fermentation, defense against pathogens,

DOI: 10.1201/9781003514244-11

and stimulation of immune responses. Furthermore, the gut microbiota significantly impacts fundamental biological processes, including nutrient extraction, and influences metabolic genes and biochemical pathways (Hou et al., 2022). Regarding the symbiotic relationship between gut microbiota and human health, it has been observed that the gut microbial population can affect various organs in addition to the brain. The lung-gut, liver-gut, pancreatic-gut, bone-gut, and gut-kidney axes are a few examples of how gut microbiota affect our body (Afzaal et al., 2022; Figure 8.1).

Accumulating research suggests a decline in beneficial gut bacteria and an increase in potentially harmful bacteria like *Staphylococcaceae*, *Enterobacteriaceae*, and *Enterococcaceae* in individuals with liver disease. This shift in the gut microbiome has been linked to the severity of liver disease, as well as the development of endotoxemia (a toxic condition caused by bacteria) and dysfunction in multiple organs (Woodhouse et al., 2018). Caussy et al. (2019) discovered a distinct pattern within the stool microbiome that is characteristic of nonalcoholic fatty liver disease (NAFLD)-induced cirrhosis. This microbiome signature can be used to accurately diagnose NAFLD cirrhosis, offering a non-invasive alternative to traditional methods for determining liver disease progression. This easily accessible biomarker could potentially complement existing invasive procedures and enhance the diagnosis and management of liver disease.

The constituents of the gut microbiota are subject to modulation based on dietary choices and lifestyle practices. In this context, the traditional dietary habits of a

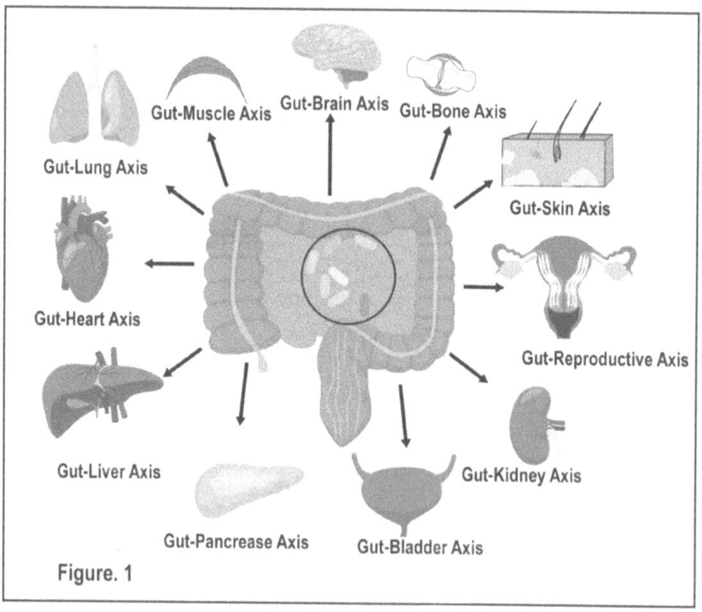

FIGURE 8.1 The microbiota relation with different organs in the human body.

society serve as a reflection of its cultural heritage, embodying historical patterns of sustenance. The Food and Agriculture Organization (FAO) characterizes traditional food as the synthesis of culinary culture and lifestyle, intricately interwoven with local climatic conditions and agricultural methodologies, which collectively govern production and processing procedures (Bhaskarachary et al., 2016). Altering the gut microbiome through dietary changes, probiotics, or fecal microbiota transplantation to encourage the growth of beneficial bacteria may improve the imbalance and potentially change the course of liver-associated disease (Woodhouse et al., 2018). The configuration and consumption patterns of traditional foods exert discernible effects on the health attributes of the populace (Trichopoulou et al., 2007).

8.2 GUT HEALTH AND DISEASES

The gut microbiota comprises a diverse collection of microorganisms living within the gastrointestinal tract. Interestingly, the microbiota composition is unique to each host and always changes throughout an individual's lifespan. Internal and external factors and modifications influence the composition of the microbial community. The current resurgence of interest in understanding the structure and functionality of this "organ" has highlighted its crucial role in health and disease (Sekirov et al., 2010). Microbial density within the gut varies along a gradient, with fewer bacteria in the upper small intestine and a concentration reaching up to 10^{12} bacteria/gram of feces in the distal colon. The microbiota is vastly diversified, typically consisting of 500–1,000 species, with a shared presence of at least 160 species among individuals. Dominant phyla include Bacteroidetes and Firmicutes, while less abundant phyla comprise Proteobacteria, Cyanobacteria, Actinobacteria, Fusobacteria, and Verrucomicrobia. Our co-evolution with the microbiota has resulted in a finely tuned symbiotic relationship featuring mutual benefits and mechanisms to prevent adverse effects on the host (Jakobsson et al., 2015).

Further, microbial colonization occurs in tandem with the immune system and plays a key role in shaping intestinal physiology and regulation. Emerging evidence indicates that the initial colonization of the human intestinal microbiota commences prenatally, with the maternal microbiota serving as the initial microbial inoculum. Postnatally, microbial diversity expands, converging towards an adult-like microbiota within the first 3–5 years of life. Perinatal factors, including delivery mode, dietary patterns, genetic factors, and intestinal mucin glycosylation, collectively influence microbial colonization dynamics. Upon establishment, the configuration of the gut microbiota exhibits relative stability throughout adulthood; however, it can be subject to alterations induced by factors such as bacterial infections, antibiotic treatments, lifestyle choices, surgical interventions, and sustained modifications in dietary patterns (Rodríguez et al., 2015). Long- and short-term dietary changes have discernible impacts on gut microbiota. Wu et al. observed a strong correlation between microbiota composition and long-term dietary habits.

High-fat diets positively associate with Bacteroides and Actinobacteria but negatively with fiber, while Firmicutes and Proteobacteria exhibit opposing trends. The Bacteroidetes-predominant enterotype aligns with Western diets rich in meat protein and saturated fats, whereas the Prevotella-prevailing enterotype links to carbohydrate-heavy diets typical of agrarian societies and vegetarians. Short-term dietary alterations induce shifts in microbial communities. An animal-based diet increases bile-tolerant microorganisms (*Alistipes*, *Bilophila*, and *Bacteroides*) and reduces Firmicutes involved in plant polysaccharide metabolism (*Roseburia, Eubacterium rectale*, and *Ruminococcus bromii*). These findings highlight fundamental differences between gut microbiota in vegetarians and carnivorous individuals, crucial for carbohydrate and protein fermentation (Bibbò et al., 2016). Fluctuations in nutrient loads and ingredient profiles notably impact the constituent makeup, composition, and metabolic outputs of gut microorganisms. This phenomenon is particularly evident in conditions such as obesity, colon cancer, and atherosclerosis, where alterations in nutritional intake and shifts in the microbiota populace contribute to an elevated risk of disease incidence. Specifically, a diet characterized by high fat or sugar content and low fiber intake can engender an aberrant microbial population, leading to intestinal inflammation and heightened permeability. The intricate interplay between dietary factors and microbiota underscores their substantial role in influencing the pathogenesis of various diseases (Aguirre et al., 2016).

The components of gut microbiota play a pivotal role in controlling the severity of malaria, as evidenced by research conducted by Villarino et al. (2016). Their study involved mice bred in distinct environments, revealing significant variations in pathology when these mice were infected with the malarial parasite, *Plasmodium*. Mice originating from different environments exhibited diverse gut microbiota compositions, leading to varying resistance levels against the malarial parasites. Notably, mice enriched with *Lactobacillus* and *Bifidobacterium* demonstrated heightened resistance to the parasite. Conversely, mice lacking these bacterial colonies exhibited increased Plasmodium burden. Remarkably, introducing *Lactobacillus* and *Bifidobacterium* to deficient mice resulted in a notable reduction in *Plasmodium* burden. This underscores the intricate relationship between the gut microbiota population, particularly the presence or absence of specific microbial species, and the host's susceptibility or resistance to malaria (Villarino et al., 2016). Accumulating evidence underscores the interconnectedness of gut microbiota patterns with depression, mediated through the intricate channel of the brain-gut microbiome axis. This association extends to the pivotal involvement of microbiota in the realms of anxiety and depression. Disparities in bacterial taxa manifest in these conditions, with discernible indications of higher abundance in proinflammatory species such as *Enterobacteriaceae* and *Desulfovibrio*, juxtaposed with a diminished presence of bacteria fostering the generation of short-chain fatty acids, exemplified by *Faecalibacterium*. The intricate taxonomy and mechanistic underpinnings of these microbial communities may be pivotal in elucidating the pathophysiology of anxiety and depression. The orchestration of these taxa, along with their intricate modes of action, suggests a potential avenue through which peripheral inflammation communicates with the brain, contributing to the nuanced interplay in the etiology of anxiety and depression (Simpson et al., 2021).

Fluctuations in gut microbiota composition are implicated in the development of several diseases, including diabetes. The metabolic pathways, gut microbes, and bacterial genes in individuals with type 2 diabetes (T2D) showed a substantial correlation among them, as revealed by a human metagenome-wide association study. T2D patients exhibited elevated *Lactobacillus* spp. levels in comparison to non-diabetic counterparts. Notably, *Lactobacillus* spp. positively correlated with glycated hemoglobin (HbA1c) levels and fasting glucose, while *Clostridium* spp. exhibited a negative correlation with fasting glucose levels (Blandino et al., 2016). Microbial composting is also implicated in cardiovascular disorders. Various studies have suggested the involvement of altered gut microbiota populations in developing hypertension, atherosclerosis, or myocardial infarctions (Zhou et al., 2020). The role of butyrate-producing bacteria and associated levels of butyrate have been demonstrated to be notably diminished in individuals with hypertension (HTN). This suggests a pertinent connection between imbalanced host-microbiome communication and systolic blood pressure. Additionally, a metagenome-wide association study involving 218 subjects with atherosclerosis and 187 healthy controls confirmed the gut microbiome's altered composition in these conditions. In the preclinical setup of acute myocardial infarction (AMI), distinct enrichments were observed in the Synergistetes phylum, *Lachnospiraceae* family, *Spirochaetes* phylum, *Syntrophomonadaceae* family, as well as the *Tissierella* and *Soehngenia* genera, in comparison to the sham group (Wu et al., 2017).

Regarding the pivotal role of gut microbiota in inflammation and immunity, it becomes imperative to delve into its influence on the initiation and progression of cancer. The altered composition of gut microbiota can either impede or catalyze cancer development. Notably, gut bacteria-produced butyrate and propionate exhibit the capacity to inhibit histone deacetylases in host tumor cells, thereby exerting a broad-spectrum anti-cancer effect. This mechanistic action underlies the observed anti-tumoral effects of butyrate in both *in vitro* and *in vivo* settings, particularly in colorectal cancer (CRC) and lymphoma (Wei et al., 2016). However, microbiota dysbiosis in the gut and the subsequent emergence of pathogenic communities may contribute to an array of pathologies. These repercussions extend beyond local manifestations, encompassing conditions ranging from bowel inflammation to neurodegenerative diseases such as Parkinson's disease and noncommunicable diseases such as cancer. In the context of cancer, the dysbiotic gut environment creates a milieu where specific bacterial pathogens can adversely impact host metabolism and compromise the functionalities of the host's gut and immune systems. Collectively, these perturbations contribute to the initiation and progression of tumor growth (Vivarelli et al., 2019).

The food we ingest nourishes our bodies and the extensive and varied bacterial population residing in our gastrointestinal tract. Theoretically, the gut microbiota could be seen as parasitic because it consumes our food. Still, the actual association is symbiotic, meaning both parties benefit. The gut microbiota plays a crucial role in maintaining the health and strength of the gut, which in turn enhances the host's immune system and helps the host respond better to medication. In return, the host provides the necessary nutrition and shelter for the gut microbiota. Nonetheless, research has shown that a number of health problems, including CRC, are primarily caused by a disturbed symbiotic association between the gut microbiota and the host (Vipperla & O'Keefe, 2016). The symbiotic gut microbiota flourishes on the remnants of undigested dietary substances and, in return, produces diverse metabolites that play a vital role in regulating the immune system in the gut and producing necessary nutrients like vitamins, including biotin and folic acid, which strengthen the protective barrier of the mucosa (O'Keefe et al., 2009; Wardwell et al., 2011).

In another study, it was found that individuals following a reduced carbohydrate diet showed higher levels of Firmicutes, Bacteroidetes, and Proteobacteria in their stool samples at the phylum level (Fragiadakis et al., 2020). Furthermore, the study revealed that individuals following a low-fat diet experienced a reduction in Firmicutes and Actinobacteria. These findings indicate that adopting a low-carbohydrate diet over a lengthy period can result in a higher prevalence of bacterial species linked to the development and advancement of CRC. The preference for low-carb diets among a significant number of people in the Western world may partially explain why these countries are experiencing a higher risk of CRC. This increased risk could be attributed to their consumption of less fiber and carbohydrates while consuming high levels of red and processed meat. This is also the reason why certain countries, like Burkina Faso, have a low incidence rate of CRC. Their customary diet consists of minimal or no animal fats, resulting in a scarcity of opportunistic bacterial phyla, such as Firmicutes. Research has also shown that switching from a traditional diet to a Western diet raises the likelihood of developing colorectal cancer (Vipperla & Keefe, 2016). For instance, the incidence of CRC among the first generation of Japanese migrants to Hawaii has risen to levels comparable to those of local Hawaiians, despite the fact that Japanese individuals typically have a low prevalence of CRC (Cologne et al., 2018). Hence, this research demonstrates that diet has the potential to induce dysbiosis in the gut microbiota. Furthermore, a prolonged or regular deviation from a typical diet can completely alter the gut microbiota composition. Research suggests that consistent and prolonged changes in diet, specifically in young adults adopting a Western diet, might result in a significant increase in opportunistic gut microorganisms, some of which may be associated with the development of colon cancer.

Notably, this microbial shift occurred concurrently with compromised gut barrier function indications. These findings underscore the intricate interplay between microbiota composition and the individual's health. The implication of gut microbiota in various cardiovascular conditions, such as atherosclerosis and the aftermath of acute myocardial infarction, and metabolic conditions, such as diabetes, mental health, and cancer, signifies the importance of maintaining a healthy gut condition and gut microflora composition.

8.3 LIVER-ASSOCIATED DISEASES

Liver is an important organ located in the right upper side of the body and is primarily involved in producing bile and detoxification of metabolites, as well as RBC regulation and glucose storage (Lala & Zubair, 2023). Liver disease (acute and chronic) is one of the major causes of death around the world, with liver cirrhosis being the most prevalent non-neoplastic and gastrointestinal disease in Europe and the US (Buechter & Gerken, 2022). Liver diseases entail continuous decline in the liver functions, resulting in reduced synthesis of clotting factors or other proteins, inability to detoxify harmful metabolic by-products, and diminished bile excretion. Chronic liver disease is a repeated process of distention, damage, and degeneration of liver parenchymal cells, which further results in liver fibrosis and cirrhosis. The underlying causes of liver disease include alcohol consumption, infection, autoimmune diseases, genetic and metabolic disorders, and idiopathic causes (Sharma & Nagalli, 2023).

Alcohol-associated liver disease (AALD) is a significant contributor to liver disease globally, with alcohol use disorders (AUDs) affecting 4.1% of the world's population and alcohol dependence affecting 3.0%. The progression to cirrhosis may be more rapid in individuals with AALD compared to those with NAFLD. NAFLD comprises two distinct conditions: nonalcoholic fatty liver (NAFL), characterized by the accumulation of fat in the liver with or without mild inflammation, and nonalcoholic steatohepatitis (NASH), which encompasses varying stages of fibrosis, cirrhosis, and hepatocellular carcinoma (HCC). Similar to the relationship between alcohol consumption and AALD, NAFLD often coexists with risk factors like obesity and diabetes (Asrani et al., 2019).

Drug-induced liver injury (DILI) is an adverse reaction to drugs or other xenobiotics that occurs either as a predictable event when an individual is exposed to toxic doses of some compounds or as an unpredictable event with many drugs in common use. Drugs can be harmful to the liver in susceptible individuals owing to genetic and environmental risk factors. These risk factors modify hepatic metabolism and excretion of the DILI-causative agent, leading to cellular stress, cell death, activation of an adaptive immune response, and a failure to adapt, with progression to overt liver injury. Idiosyncratic DILI is a relatively rare hepatic disorder but can be severe and, in some cases, fatal, presenting with a variety of phenotypes that mimic other hepatic diseases (Andrade et al., 2019; Singh et al., 2024). DILI is the primary cause of acute liver failure (ALF), with the exception of cases involving paracetamol overdose. The vast majority of DILI seen in clinical practice is idiosyncratic, meaning it occurs unpredictably and is not dose-dependent.

8.3.1 Liver Diseases and Gut-Microbiome

Multiple liver conditions, including alcoholic liver disease, nonalcoholic liver disease, and primary sclerosing cholangitis, have been linked to changes in the gut microbiome. This imbalance, known as dysbiosis, may impact the severity of fat accumulation, inflammation, and scarring in the liver. This occurs through interactions between the altered microbiome, the body's immune system, and other cell types (Tilg et al., 2016). Both animal and human studies reveal a connection between alcohol consumption and bacterial overgrowth in the intestines. Introducing alcohol directly into the stomachs of mice led to a rise in bacteria in the upper small intestine, and prolonged exposure resulted in excess bacteria in the large intestine, similar to rodents with advanced liver disease. In humans, both moderate alcohol use and alcohol-related cirrhosis are associated with bacterial overgrowth in the small intestine, with the severity of overgrowth correlating with the severity of liver disease (Hartmann et al., 2015).

Studies in mice show that ethanol consumption decreases beneficial bacteria like Firmicutes and *Lactobacillus* while increasing harmful bacteria such as *Enterococcus* in the gut. To assess this imbalance in humans, Bajaj et al. (2012) introduced the concept of the "cirrhosis dysbiosis ratio" (CDR). This ratio compares the levels of beneficial bacteria (*Lachnospiraceae*, *Ruminococcaceae*, and *Clostridiales* XIV Incertae Sedis) to potentially harmful ones (*Enterobacteriaceae* and *Bacteroidaceae*) in the intestines of cirrhosis patients. A lower CDR suggests a more severe disease state (Hartmann et al., 2015).

Similarly, research suggests that the gut microbiome can influence the development and progression of NAFLD through multiple pathways. Evidence indicates that increased intestinal permeability, bacterial overgrowth, and elevated lipopolysaccharides (LPS) levels in the bloodstream are associated with NAFLD in patients (Mokhtari et al., 2017). Evidence suggests a connection between gut bacteria and liver disease, particularly in the context of NAFLD. Studies in mice and humans with NAFLD have revealed elevated alcohol levels, potentially due to gut bacteria producing alcohol. Notably, *Klebsiella pneumoniae*, an alcohol-producing bacterium, has been found in the gut microbiome of individuals with NAFLD. Importantly, transferring these bacteria to healthy mice induced NAFLD, supporting the idea that gut bacteria may play a causal role in this liver condition (Jennison & Byrne, 2021).

Gut bacteria play a role in the development of fatty liver, partly by suppressing a protein that normally limits fat absorption. This can lead to small intestinal bacterial overgrowth (SIBO), which has been observed in patients with NAFLD. SIBO is associated with increased intestinal permeability and more severe fat accumulation in the liver. Additionally, SIBO in NASH patients triggers inflammatory pathways, further contributing to fat buildup by impairing insulin signaling. These findings highlight the complex relationship between gut bacteria, inflammation, and the development of fatty liver disease (Boursier et al., 2016). The liver, a key player in immune function, is impacted by gut bacteria in various liver diseases. In NAFLD, excessive activation of liver immune cells, triggered by gut bacteria components, is a major contributor. Early gut colonization from obese mothers may also increase NAFLD susceptibility due to impaired immune responses. In autoimmune liver diseases like AIH, altered gut microbiota composition highlights the intricate relationship between the gut and liver in immune regulation. These findings underscore the importance of understanding the gut-liver axis and its impact on immune function in liver diseases (Wang et al., 2021).

8.4 TRADITIONAL FOOD

Shifts in lifestyle behavior and socio-demographic configurations have increased the recognition of the impact of food quality on public health. Additionally, the understanding that environmental quality is contingent upon sustainable consumption patterns, coupled with recent incidents of food safety breaches, has led to heightened enthusiasm and consumption of both traditional and organic foods (Nikolić et al.,

2014). These factors have similarly piqued the interest of the biological research community. Different geographical regions have different traditional foods. An exemplar of a dietary regimen associated with favorable health outcomes is the Mediterranean diet, distinguished by six constituent elements: elevated consumption of vegetables, legumes, fruits, olive oil, nuts, and cereals; restricted intake of saturated lipids; limited to moderate consumption of dairy products, predominantly in the form of cheese and yogurt; diminished consumption of meat and poultry; moderately elevated consumption of fish; and moderate ingestion of ethanol, primarily in the form of wine (Panagou et al., 2013). European culture is a concoction of various cultural expressions reflected in the cuisines and practices of dietary intake. Over centuries, Greek traditional foods have undergone evolutionary development within the country, shaped by local culture and artisanal practices predominantly conducted in small-scale, family-owned establishments. These foods form a significant component of the renowned Mediterranean diet (Panagou et al., 2013).

In India, traditional food includes grains, fruits, dairy, meat, and their products. As India is a diverse country with vast geographical and cultural variations, the traditional food also shows a large variety (Bhaskarachary et a., 2016). The microbiotic profiles of children adhering to distinct traditional diets were systematically examined in an innovative investigation. Specifically, the microbial ecosystems of children from Burkina Faso, whose dietary practices center around carbohydrates, were juxtaposed with those of their Italian counterparts, whose alimentary habits are characterized by a Westernized regimen rich in animal fats and proteins. This comparative analysis revealed a higher prevalence of Bacteroidetes, particularly Prevotella, in the microbiomes of the Burkinabé children, contrasting with a predominance of Firmicutes in the Italian cohort (Dubois et al., 2017).

Extending this inquiry, the gut microbiome of Italian children was further juxtaposed with that of hunter-gatherers from the Hadza community in Tanzania. This comparative assessment aimed to delineate the impact of diverse traditional diets on microbial composition. Additionally, an exploration into the microbiome of the Inuit population, characterized by a traditional diet and lifestyle, elucidated the ramifications of such practices on microbial composition, diversity, and stability. Intriguingly, despite expectations of heightened seasonality in the Inuit diet, the observed influence on the gut microbiome appeared somewhat mitigated, potentially attributable to an evolving shift towards a more Westernized dietary pattern (Dubois et al., 2017; Schnorr et al., 2014).

Examining the dietary patterns within a population is crucial for a comprehensive understanding of essential micro- and macro-nutrient availability. Moreover, the deficiency of specific nutrients can have overarching implications for the overall health of the populace. For instance, an investigation into the traditional diets of three distinct Indigenous populations in the Canadian Arctic unveiled insufficiencies in vitamins A, D, and E (Kuhnlein et al., 2006). The traditional dietary practices in Egypt encompass the consumption of seafood. Given the escalating levels of environmental pollution, researchers have undertaken a systematic examination to assess the distribution of major and trace elements within the tissues of crustacean species. The findings suggest that incorporating marine crustaceans into the Egyptian diet may mitigate adverse sodium-related effects, such as hypertension, preeclampsia, and cardiovascular diseases (El-Said et al., 2021). Notably, given Egypt's location within the Mediterranean region, its culinary traditions align with the principles of the Mediterranean diet. Despite the richness of Egyptian foods in fiber, legumes, green vegetables, fruits, lean meat, and animal protein, the dietary profile is incomplete without incorporating white bread and polished rice, both exhibiting relatively high glycemic indices. This elevated glycemic index in the Egyptian diet correlates with a discernible rise in the prevalence of type 2 diabetes within the population (Hegazi et al., 2015).

Roasted duck is a cherished traditional dish in China, honed and perfected over centuries. The optimal roasting temperature hovers around 230–240°C, crucial for imparting the captivating aroma and flavor. The roasting process relies on the Maillard reaction, triggering the formation of lipid peroxides and other compounds. Within the Maillard reaction, the interaction between amino and carbonyl groups of a compound with oxidized lipid products yields N-substituted glycosylamine. This intermediary transforms into the Amadori product, subsequently decomposing into hydroxymethylfurfural and furfural. While these intricate chemical processes contribute to the culinary appeal, it is essential to acknowledge potential health concerns. The decomposition products, alongside N-(carboxymethyl)lysine and acrylamide, generated during the thermal processing of the duck have been linked to significant toxicity in humans. Notably, acrylamide, a by-product of duck roasting, has been identified for its potential carcinogenic properties.

Awareness of these chemical reactions and their consequences underscores the importance of balancing culinary tradition with considerations for human health (Zhou et al., 2016).

Ginger plays a pivotal role in South Asian culinary traditions, where its various components contribute to a wide array of dishes. The unique flavor of ginger oil enhances the taste of soft drinks, while ginger powder elevates the flavors of baked goods, sweets, pickles, tea, sauces, and other culinary delights. In addition to its culinary uses, traditional Chinese practitioners have thoroughly explored the potential health benefits of ginger. Abundant in phenols and flavonoids, ginger possesses antioxidant and anti-inflammatory properties, adding value to the kitchen and holistic health practices (Shahrajabian et al., 2019). The implications of traditional food for human health is an important field that should be thoroughly researched for the betterment of society and to understand the prevalence of different diseases and disorders. Indian traditional foods are full of cereal grains, and the staple foods include various types of lentils, rice, millet, etc. Indians also consume milk, honey, fresh fruits, and cooked vegetables as their staple foods. Grain-based traditional foods such as Idli, Dosa, Ambali, and Ragi hurihittu are nutrient-rich fermented foods that are rich in aroma and taste (Sarkar et al., 2015).

8.4.1 Dietary Fiber Forms the Fundamental Structure of Bacterial Food Webs

Introducing fiber into the diet modifies the specific conditions in the gut, creating an environment that supports the growth of microorganisms. This enables the proliferation of microbial species that can utilize these specific substances as a source of nutrition (Deehan et al., 2017). The gut microbiome has 22 polysaccharide lyase, 16 carbohydrate esterase, and 130 glycoside hydrolase families. These families allow the microbiome to adapt and utilize different types of fibers as energy sources (Flint et al., 2012).

The primary beneficiaries of dietary fiber are species classified under the phyla Firmicutes and Actinobacteria (Deehan et al., 2017). However, these species possess a very low number of enzymes responsible for metabolizing fiber per organism. Their roles typically include more specialized tasks, such as initiating the destruction of complex substrates. For instance, the administration of resistant starch (RS) has been demonstrated to enhance the presence of *Eubacterium rectale*, *Ruminococcus bromii*, *Bifidobacterium adolescentis*, and *Parabacteroides distasonis* in a specific group of persons (Walker et al., 2011). On the other hand, the intake of galacto-oligosaccharides primarily stimulates the growth of Bifidobacterium species that have the necessary enzymes to metabolize this substance effectively (Davis et al., 2011). The ability of a microbe to benefit from dietary fiber depends on its ability to adhere to a substrate, tolerate changes in environmental conditions caused by fiber fermentation (such as increased acidity), and its enzymatic capacity to degrade fiber and carbohydrate metabolites through cross-feeding (Deehan et al., 2017).

Primary fiber degraders trigger the breakdown of complex fibers through a group of species known as a "guild" (Zhao et al., 2018). *R. bromii* is regarded as a keystone species in the breakdown of RS and plays a crucial role in butyrate formation in the colon, although it does not directly produce butyrate itself. There are probably more keystone species that have a similar role in different forms of dietary fiber, although they have not been discovered yet.

The influence of dietary fiber on microbiota makeup exhibits some consistent features. The alterations caused by non-digestible carbohydrates in humans, whether they are recognized as prebiotics or not, are only seen in a small number of organisms (Davis et al., 2011; Walker et al., 2011). Furthermore, the extent of the induced alterations might be significant, with certain species accounting for over 30% of the overall sequences acquired using amplicon sequencing of the fecal microbiota. However, these alterations are sustained only as long as the substrate is utilized. Furthermore, the impact of dietary fiber on microbial activity varies significantly among individuals. The cause of this distinctiveness remains unknown. Some individuals may not have keystone species or may not have strains that possess the enzymatic ability to use a particular substrate (Zhao et al., 2018).

8.4.2 Dietary Fiber Regulates the Composition of the Gut Microbiota

The microbiome's diversity, composition, and abundance are directly influenced by dietary fiber. It offers a wide range of substances that can be fermented by specialized microorganisms with the required enzymes to break down these complex polysaccharides. Multiple species of microbiota in the large intestine are classified as fiber fermenters. Although there are various forms of dietary fiber, all of them are broken down by these microbiotas. Dietary fiber intake changes the nutritional environment within the colon, enabling the bacteria to increase their populations (Walter, 2015). People who consume diets that are not rich in fiber tend to have a decreased diversity in the microbial community and are likely to host microorganisms that flourish on fats and amino acids, as the decreased intake of dietary fiber is likely to be substituted in the diet with animal protein and fat (Makki et al., 2018). Multiple studies elucidate how ingesting various diets in diverse geographic regions and socio-economic contexts can modify microbial communities in the gastrointestinal tract (Clemente et al., 2015; De Filippo et al., 2017). One shared characteristic throughout this research is that individuals from less developed and rural nations consume a considerably greater amount of fiber in their diet than individuals from industrialized nations (Sonnenburg & Sonnenburg, 2014). Notably, persons from unindustrialized cultures have a significantly lower occurrence of metabolic and inflammatory illnesses, such as obesity and inflammatory bowel disease (IBD).

A study conducted in 2015 examined the microbiota of individuals from Papua New Guinea and compared it to that of Americans. The study discovered that the ratio of *Prevotella* to *Bacteroides* in the fecal microbiota differed between the two groups (Wu et al., 2017). The study found that Papua New Guineans who followed a high-fiber diet had a significant presence of *Prevotella* in their microbiota but a reduced presence of *Ruminococcus*, *Faecalibacterium*, *Bifidobacterium*, *Blautia*, *Bilophila*, *Bacteroides*, and *Alistipes* (Wu et al., 2017). Yatsunenko et al. (2012) observed comparable results when examining the microbiome communities of Malawi, Venezuela, and individuals in the U.S.A. In a study by Clemente et al. (2015), the researchers examined the microbiome of the Yanomami tribe in rural Venezuela. An important discovery from this study was that the individuals in question had a greater prevalence of Prevotella and a lower prevalence of Bacteroides than individuals from the U.S.A. (Clemente et al., 2015). Furthermore, this population exhibited a reduced prevalence of additional constituents belonging to the Bacteroidetes family, such as Bacteroidales, Mollicutes, and Verrucomicrobia. Finally, it was revealed that these individuals had a higher prevalence of the Phascolarctobacterium genus, which is recognized for producing short-chain fatty acids (SCFAs) (Wu et al., 2017). A separate study also compared the microbiota of the Tanzanian Hadza tribe and Italians. Both populations were discovered to possess a plentiful amount of widely recognized bacteria that may break down fibers, including the Firmicutes family such as *Lachnospiraceae*, *Ruminococcaceae*, *Veillonellaceae*, Clostridiales Incertae Sedis XIV, and Clostridiaceae. Consistent with similar research, the Hadza microbiome exhibited a higher presence of Prevotella, whereas a lower prevalence of Bacteroides was noted compared to their Italian counterparts (Schnorr et al., 2014). A separate study conducted in the United States highlighted the significant impact of a diet rich in plant fiber compared to a diet rich in animal protein. The study found that people with greater levels of Prevotella, a type of bacteria, were associated with the breakdown of fiber in the body (David et al., 2014). In addition, there were observed associations between the plant-based diet and the microorganisms *Eubacterium rectale*, *Roseburia*, and *Faecalibacterium prausnitzii* (David et al., 2014), along with higher overall levels of SCFAs.

Hence, it is evident that the dietary fiber component can impact the microbiota. There is a noticeable pattern in the individual microbiota from industrialized nations, indicating that these people have a greater amount of Bacteroides, Bifidobacterium, Ruminococcus, Faecalibacterium, Alistipes, Bilophila, and Blautia (Martínez et al., 2015; De Filippo et al., 2017). Prior analyses of the microbiota in industrialized countries indicate minimal disparities, suggesting that cultural and ethnic variations likely have minimal impact on the host microbiota (Arumugam et al., 2011). Curiously, all of these countries adhere to a modified version of the Western diet, characterized by a high percentage of saturated fat and a low dietary fiber intake. Another significant discovery from these studies is that the prevalence of *Prevotella* spp. is greater in non-industrialized persons who ingest substantial amounts of dietary fiber. Notably, numerous species discovered in the microbiome of Westerners have been associated with the extensive intake of fats and amino acids, which are the main components of the Western diet (David et al., 2014; Wu et al., 2017). The microbiota of non-industrialized cultures exhibits similarities to the microbiome of vegans and vegetarians in Western society, as both groups ingest substantial amounts of dietary fiber (David et al., 2014). Consuming a significant amount of dietary fiber and the presence of particular types of microorganisms capable of breaking it down may result in many positive effects on the host's health, such as the synthesis of a large amount of SCFAs.

8.5 TRADITIONAL FOODS CONTAINING HIGH FIBER AND GUT HEALTH

A diet encompasses macronutrients (proteins, carbohydrates, and fats), micronutrients (minerals and vitamins), and phytochemicals (bioactive molecules that do not provide nutrients). The different components, especially phytochemicals, play a crucial role against different diseases (Siddique et al., 2012; Fatma et al., 2022; Jameel et al., 2023). Recent findings indicate that the aforementioned dietary components can directly influence the makeup and functioning of the gut microbiota in mammals, thus influencing both physical and mental well-being. There is an increasing consensus that the increase in the burden of chronic diseases in Western countries may be attributed to the gradual decline of advantageous bacteria and microbial diversity. The traditional dietary method involves ingesting a diverse range of plant-based meals of various colors, which supply fiber and phytonutrients. This beneficial bacteria aids in the restoration of microbial diversity in the gut and can potentially prevent or reverse chronic illnesses, such as type 2 diabetes, cardiovascular disease, or colon cancer.

It is essential to include the intestinal flora while discussing food. Billions of bacteria are found in the gastrointestinal tract and play an important role in digestion and maintaining intestinal balance (Ulluwishewa et al., 2011). Different variables have a significant influence on the composition of every community. For instance, a low pH hinders the colonization of the stomach and upper small intestine by numerous harmful microorganisms. The large intestine provides an optimal environment for several anaerobic organisms. The gut microbiota is a dynamic assemblage of living organisms that can change in response to food, disease, and environmental influences (David et al., 2014).

The association between fluctuations in the intestinal microbiota and sickness was initially observed in 1681 by Anton van Leeuwenhoek, who noted that the microbial composition of his diarrhea was distinct from that of normal fecal samples (de Vos & De Vos, 2012). Subsequently, much research has been conducted on the intestinal microbiome to demonstrate its involvement in various illnesses, including obesity and colon cancer. A substantial amount of research on microbiota has been conducted in the past decade. Several of these reported alterations lead to a general decrease in bacterial diversity in the microbiota, suggesting that the variety of species is linked to good health (Willing et al., 2010). However, the inverse may be true for cause-effect relationships, but insufficient study has been elucidated on the matter.

The use of high-throughput technology has facilitated the discovery of a vast number of microorganisms and their corresponding metabolic activities in the colon. This discovery revealed that the gut plays an equally important part in metabolism as the liver. The proximity of these microorganisms to the lining of the intestines and the immune tissue in the gut elucidates their crucial function in maintaining good health and contributing to the development of diseases. Undoubtedly, dysbiosis is a crucial factor in the pathogenesis of inflammatory bowel illness, obesity, and colon cancer. Recent findings indicate that nutrition directly impacts the content and makeup of gut bacteria. Therefore, understanding the complex interrelationship among nutrition, gut microbiota, and the host is pivotal for preventing and treating the chronic diseases prevalent in our society (Turnbaugh et al., 2009). Research conducted on mice has demonstrated swift alterations in the gut microbiota when transitioning from a regular diet to a high-calorie diet and then reverting to a regular one (Turnbauh et al., 2009). Various clinical studies have demonstrated a positive association between high-fiber diets and increased microbial diversity, as well as a decrease in Enterobacteriaceae populations, including species such as *Escherichia* and *Shigella* (De Filippo et al., 2017). Noticeable disparities are also observed when comparing the intake of diets based on animals against those based on plants (David et al., 2014).

The presence of some compounds in food and the nutrients in the diet can also impact gut microorganisms. For instance, most anthocyanins are not assimilated into circulation in the small intestine, resulting in their retention in the gastrointestinal tract until they reach the colon (Ferrario et al., 2017). There, they can influence the composition of the intestinal microbiota through various mechanisms. To begin with, anthocyanins possess antioxidant properties that can mitigate the oxidative stress caused by inflammation of the gut (Li et al., 2012). Furthermore, anthocyanins can serve as a carbon source, which can be metabolized by bacteria, leading to the enhanced proliferation of specific microorganisms (Lacombe et al., 2012). Finally, bacterial metabolism breaks down anthocyanins and creates several by-products, some of which possess antibacterial properties against enteric pathogens such as Escherichia coli (Ferrario et al., 2017).

8.6 DIETARY BEANS AND PULSES, AS WELL AS OVERALL GUT HEALTH

The recent progress in understanding the connection between dietary fibers and their ability to improve various illnesses has sparked greater interest in dishes made from pulses. Pulses include a variety of fibers, such as galacto-oligosaccharides, resistant starch, and long-chain soluble and insoluble polysaccharides. Furthermore, these components can serve as prebiotic precursors, undergoing digestion by advantageous microbes in the gastrointestinal tract. Consuming pulses in the diet has been associated with lowering serum cholesterol, reducing post-prandial blood glucose levels, and increasing satiety. This helps mitigate the risk of many metabolic disorders, such as cardiovascular diseases, obesity, and diabetes (Ha et al., 2014). Multiple meta-analyses have determined that consuming around 2/3 cups of pulses daily will reduce total and LDL cholesterol levels (Padhi et al., 2017). The low glycemic response of pulses is attributed to the intact cell wall of whole pulses, which acts as a physical barrier between the starch and digestive enzymes, preventing rapid digestion after cooking. Moreover, the consumption of pulses is strongly linked to reducing blood pressure and providing a defense against reactive oxygen species (ROS) due to the abundance of polyphenols (Jayalath et al., 2014). Recently, there has been an increased interest in studying pulses as a potential sustainable alternative to animal protein. This research addresses the challenges of feeding a growing population and tackling food insecurity. Moreover, the high content of plant-based protein and dietary fibers in whole pulses supports the theory that they benefit the gut flora. Recently, there has been an increased focus on studying pulses as a potential sustainable alternative to animal protein for feeding the expanding population and addressing food insecurity issues (Ferreira et al., 2021). Moreover, the high concentration of plant-based protein and dietary fibers in whole pulses supports the theory that they benefit gut flora. A study conducted on pulse flour showed greater proliferation of *Faecalibacterium*, *Bifidobacterium*, *Eubacterium*, *Clostridium*, and *Roseburia* spp., coupled with increased synthesis of butyrate and acetate (Gullón et al., 2015). Multiple studies have documented that including pulses in the diet leads to an increase in the prevalence of Prevotella, Dorea, and *Ruminococcus flavefaciens*, while decreasing the prevalence of *Ruminococcus gnavus* in mice models (Graf et al., 2019). Prevotella is a genus with a wide range of glycoside hydrolases and is recognized for its capacity to generate short-chain fatty acids through the fermentation of carbohydrates (Monk et al., 2017). The presence of *Ruminococcus flavefaciens* has been observed to decrease in individuals who are obese (BMI: > 30.0) and overweight (BMI: 25.0–29.9) (Schwartz et al., 2010). An elevated presence of *Ruminococcus gnavus*, a species known for its ability to break down mucus, has been associated with a higher occurrence of gut-barrier diseases in individuals with obesity and IBD (Monk et al., 2017). Another beneficial outcome of consuming pulses is the heightened occurrence of *Akkermansia muciniphila* in the gastrointestinal tract, which is frequently classified as an advanced probiotic of the future. This bacterium can also break down mucus, although it shows an opposite relationship with *R. gnavus* (Hou et al., 2020). The majority of the research presented here showed a significant increase in Bacteroidetes and a decrease in the Firmicutes population. The decrease in the proportion of two primary groups of organisms has been linked to the improvement of obesity, potentially because of changes in how the colon processes carbohydrates for energy (Hou et al., 2020). Within the Bacteroidales group, the pulse-based diets are represented by Rikenellaceae, Muribaculaceae (S24-7), and *B. acidifaciens* (Lutsiv et al., 2021). A study discovered a correlation between lentil eating and a higher occurrence of *Roseburia* in mouse feces (Graf et al., 2019). *Roseburia* has a role in butyrate formation and is negatively associated with various disorders, including colitis and Crohn's disease (Tamanai-Schcoori et al., 2017). These studies have shown that pulses have positive effects on the gut microbiome. However, the impact on specific gut microbiota is complicated and can be influenced by different factors such as the type and amount of pulses consumed, the age and health status of the participants, the duration of the study, and the sequencing method used.

8.7 VEGETARIAN OR VEGAN DIET AND GUT HEALTH

A vegetarian diet is characterized by the complete or partial avoidance of products derived from animals. This dietary practice can be categorized into seven distinct types based on the inclusion or exclusion of specific animal products: veganism, ovo-vegetarianism, lacto-ovo vegetarianism, lacto-vegetarianism, raw veganism, semi-vegetarianism, and pescetarianism. These classifications delineate variations in the consumption of seafood, eggs, and dairy within the broader spectrum of vegetarian dietary choices (Xiao et al., 2022). As the popularity of vegetarian diets surges, veganism is increasingly acknowledged as healthful and disease-preventive. Since the 1970s, researchers have scrutinized the impact of diet on gut microflora, often juxtaposing meat-rich diets (termed "mixed Western diet") with loosely defined non-meat diets. Comparative analyses have revealed that omnivores exhibit an anaerobic microflora enriched in Bacteroides, *Lactobacillus*, *Bifidobacterium*, and *Peptococcus* in contrast to their non-meat-consuming counterparts. Conversely, vegans manifest a higher proportion of *Faecalibacterium prausnitzii*, recognized as an anti-inflammatory bacterium and prolific producer of butyrate within the Clostridia class (phylum Firmicutes). This class is purported to play a protective role for colonocytes. Additionally, vegetarian people exhibit an increased ratio of *C. clostridioforme* within the *C. coccoides* grouping (Glick-Bauer & Yeh, 2014).

F. prausnitzii, a prominent bacterium within the Firmicutes phylum and the most prevalent species in the intestines of healthy adults, was observed by Matijašić et al. to be notably more abundant in vegans compared to vegetarians. This bacterium emerges as a key player in conferring protection against metabolic diseases. Diminished levels of *F. prausnitzii* are associated with conditions such as intestinal disorders, inflammation, obesity, and type 2 diabetes, highlighting its significant role in maintaining intestinal health and metabolic balance (Glick-Bauer & Yeh, 2014; Matijašić et al., 2014). Temporal fluctuations in the microbiome of the Hutterite Anabaptist community in the United States, characterized by a traditional diet contingent upon seasonal food availability, have been discerned. Notably, an emphasis on substantial quantities of fruits and vegetables exclusively accessible during the summer months has prompted noteworthy alterations in species composition and a concomitant reduction in microbial diversity during this season as compared to winter. The augmented ingestion of fibrous fruits and vegetables in the summer months is posited as a potential causative factor for these observed variations, potentially fostering the proliferation of bacterial taxa specializing in the degradation of dietary fiber (Dubois et al., 2017).

The α-diversity index of the gut microbiome was found to be significantly higher in individuals adhering to a vegetarian diet compared to omnivores ($P < 0.05$); however, no significant differences were observed in the β-diversity index. Consistent with these findings, numerous studies have reported elevated α-diversity of the gut microbiota in vegetarians when compared to omnivorous individuals. The heightened α-diversity in vegetarians is thought to be attributed to the characteristics of a vegetarian diet that facilitate the intact delivery of nutrients to the colon, where colonic microorganisms effectively utilize them. Specifically, a typical vegetarian diet provides the colonic microbiota with approximately 5–15 g of protein, 60 g of fermentable carbohydrates, and 5–10 g of lipids. Furthermore, investigations have demonstrated that vegetarians exhibit lower levels of intestinal inflammation markers, such as intestinal lipocalin-2 and IgE, compared to omnivores. This suggests a potential association between dietary patterns and reduced intestinal inflammation among individuals adhering to a vegetarian diet (Xiao et al., 2022). Kim et al. (2013) have reported that a strict vegetarian diet assigned to obese subjects with hypertension and type 2 diabetes can improve the overall risk factors associated with it. Furthermore, a strict vegetarian diet can help reduce body weight, triglycerides, cholesterol, hemoglobin A1C, and the Firmicutes to Bacteroidetes ratio in the gut. An increase in *Bacteroides fragilis* inside the intestine showed a significant decrease in intestinal lipocalin-2 and short-chain fatty acids (Kim et al., 2013). Thus, it can be said that the vegetarian diet and its derivatives help maintain good gut bacteria and promote gut health.

8.8 FERMENTED FOOD AND GUT HEALTH

Fermented foods have been pivotal in human dietary practices since ancient times, as evidenced by deliberate fermentation in Chinese pottery vessels dating back to 7000 BC with substances such as rice, honey, and fruit. Initially employed to enhance food safety and extend shelf-life, contemporary perspectives emphasize the manifold health benefits of fermented foods. Recognition of the positive correlation between fermented foods and human health dates back to the early 1900s. Proposed mechanisms elucidating these health advantages encompass: (i) direct nutritional value – fermented foods offer a direct nutritional contribution, featuring bioactive compounds produced through the fermentation process; (ii) provision of nutrients – fermented foods provide nutrients that foster the growth of indigenous gut microbes; and (iii) microbial survival – microbes inherent in fermented foods exhibit the ability to endure gastric transit, potentially integrating into the gut microbiome or exerting inhibitory effects on existing members of the gut microbial community. These intricate mechanisms underscore the diverse health-promoting attributes associated with consuming fermented foods (Leeuwendaal et al., 2022). The matrix of certain fermented foods facilitates the delivery of abundant microorganisms to the gastrointestinal tract, exerting a beneficial influence on the human gut microbiota. Existing studies have consistently demonstrated that daily consumption of fermented foods primarily impacts the immune system and metabolic function. Notably, these foods have been associated with reducing the incidence and duration of respiratory infections, and enhancing bone density, liver function, body mass, and blood pressure indices. Additionally, fermented foods contribute to the prevention of both diarrhea and constipation while also exhibiting positive effects on skin health. Emerging findings further establish a connection between fermented foods and beverages and their positive influences on the gut microbiota, suggesting a potential long-term impact on gut-brain communication (Mota de Carvalho et al., 2018).

The bacterial diversity and microbial evolution of idli were examined at various stages of fermentation using 16S rRNA amplicon sequencing. The outcomes underscored that the predominant microbiota primarily comprised members of the Firmicutes (70%) and Proteobacteria (22%) phyla. Notably, during a crucial fermentation phase, there was an enhancement in the major metabolic activities of these microbial fermenters. This phase was characterized by increased activity of hydrolases responsible for degrading starch and phytins and the production of B vitamins. Identifying microorganisms possessing beneficial properties suggests that idli may be considered a potentially advantageous functional food (Mandhania et al., 2019). *Lactobacillus fermentum* CQPC06 is naturally present in fermented pickles and has demonstrated efficacy in stabilizing high-fat diet/fructose-induced nonalcoholic fatty liver disease (NAFLD) in mice. At the molecular biology level, intervention with *L. fermentum* notably affected the mRNA and protein expression of key factors in NAFLD mice. Specifically, CQPC06 significantly increased the expression of PPAR-α, CYP7A1, CPT1, and LPL while decreasing the expression of PPAR-γ and C/EBP-α. Moreover, at the molecular level in the small intestine of NAFLD mice, CQPC06 increased and facilitated the higher expression of tight junction proteins, including Occludin, ZO-1, and claudin-1, contributing to improved intestinal barrier function. Simultaneously, the expression of CD36, a marker associated with lipid uptake, was reduced. This study suggests the potential therapeutic role of *L. fermentum* CQPC06 in mitigating NAFLD, suggesting molecular mechanisms involving modulation of hepatic lipid metabolism and enhancement of intestinal barrier integrity (Mu et al., 2020).

The strain *Lactobacillus plantarum* ZJ316 (ZJ316) was introduced during the fermentation of *Stachys sieboldii* Miq. (SSM), and its impact on the quality and bacterial community dynamics of SSM throughout fermentation was studied. It was reported that the inoculation of ZJ316 played a pivotal role in preventing the occurrence of nitrite peaks, thereby maintaining the nitrite content of fermented SSM at a low level. Furthermore, the influence of ZJ316-fermented SSM on gut microbiota modulation was assessed using an in vitro fecal fermentation system. Results indicated that ZJ316 exerted a relatively subtle effect on intestinal communities, displaying a potentially positive influence on probiotics such as *Bifidobacterium* and *Lactobacillus* while negatively impacting *Enterobacteriaceae*. Additionally, SSM fermented with ZJ316 facilitated the production of short-chain fatty acids in the human intestine. These findings suggest a dual benefit of ZJ316 in mitigating nitrite levels during fermentation and contributing to a favorable modulation of gut microbiota, ultimately promoting the production of beneficial short-chain fatty acids in the human intestine (Hang et al., 2022). Further, consistent inclusion of pickled vegetables and/or fermented bean curd in one's diet has been associated with a notable reduction in the long-term risk of developing diabetes. Regular consumption of these fermented foods has demonstrated potential health benefits, suggesting a protective effect against the onset of diabetes over an extended period. Incorporating pickled vegetables and fermented bean curd into a balanced diet may contribute to a comprehensive strategy for mitigating the risk of diabetes in the long term (Cai et al., 2023).

In another study, *Lactobacillus plantarum* LD1, isolated from dosa batter, underwent comprehensive evaluation to assess its probiotic potential through various standard parameters. Strain LD1 exhibited resilience to bile salts and demonstrated bile salt hydrolase activity, indicative of its adaptability to gastrointestinal conditions. Moreover, the culture supernatant of LD1 showcased the ability to inhibit a spectrum of food-borne and clinically relevant pathogenic microorganisms, including *Staphylococcus aureus*, *Pseudomonas aeruginosa*, *Shigella flexneri*, *Salmonella typhimurium*, urogenic *Escherichia coli*, and *Vibrio* spp. These findings suggest that Lactobacillus plantarum LD1 holds promise as a probiotic candidate, given its proficiency in withstanding bile salt exposure and its inhibitory effects on a range of pathogenic microorganisms, both of which are indicative of its potential application in promoting digestive health (Gupta & Tiwari, 2014).

8.9 CONCLUSION AND PERSPECTIVES

In conclusion, the traditional Indian diet, characterized by its rich inclusion of high-fiber foods, aromatic spices, a diverse array of vegetables, and fermented foods, stands as a beacon of holistic nourishment for gut health. The synergistic impact of these dietary elements contributes significantly to fostering a balanced and thriving gut microbiome. The ample presence of dietary fiber supports digestive regularity and microbial diversity, while the vibrant array of spices enhances flavor and harbors potential anti-inflammatory and antimicrobial properties. A diverse range of vegetables provides essential nutrients and antioxidants crucial for gut well-being, promoting a dynamic and resilient microbial ecosystem. Crucially, the incorporation of fermented foods introduces live beneficial microorganisms, bolstering the population of probiotics in the gut. This symbiotic relationship between traditional Indian foods and gut health extends beyond mere sustenance, offering a comprehensive approach to fostering a resilient and harmonious gut environment. By embracing these traditional dietary practices, individuals may fortify their gut health, potentially mitigating the risk of gastrointestinal disorders and promoting overall well-being. The time-tested wisdom embedded in Indian culinary traditions underscores the profound connection between dietary choices and the flourishing ecosystem within the gut, reinforcing the notion that food can indeed be medicine for a thriving digestive system, which may then help prevent multiple liver-associated disorders.

REFERENCES

Afzaal, M., Saeed, F., Shah, Y. A., et al. (2022). Human gut microbiota in health and disease: Unveiling the relationship. *Frontiers in Microbiology, 26*(13), 999001.

Aguirre, M., Eck, A., Koenen, M. E., et al. (2016). Diet drives quick changes in the metabolic activity and composition of human gut microbiota in a validated in vitro gut model. *Research in Microbiology, 167*(2), 114–125.

Andrade, R. J., Chalasani, N., Bjornsson, E. S., et al. (2019). Drug-induced liver injury. *Nature Review Disease Primers, 58*, 2019.

Asrani, S. K., Devarbhavi, H., Eaton, J., et al. (2019). Burden of liver diseases in the world. *Journal of Hepatology, 70*(1), 151–171.

Arumugam, M., Raes, J., Pelletier, E., et al. (2011). Enterotypes of the human gut microbiome. *Nature, 473*(7346), 174–180.

Bajaj, J. S., Hylemon, P. B., Ridlon, J. M., et al. (2012). Colonic mucosal microbiome differs from stool microbiome in cirrhosis and hepatic encephalopathy and is linked to cognition and inflammation. *American Journal of Physiology-Gastrointestinal and Liver Physiology, 303*(6), G675–G685.

Bhaskarachary, K., Vemula, S. R., Gavaravarapu, S. R. M., et al. (2016). Traditional foods, functional foods and nutraceuticals. *Proceedings of the Indian National Science Academy, 82*(5), 1565–1577.

Bibbò, S., Ianiro, G., Giorgio, V., et al. (2016). The role of diet on gut microbiota composition. *European Review for Medical & Pharmacological Sciences, 20*(22), 4742–4749.

Blandino, G., Inturri, R., Lazzara, F., et al. (2016). Impact of gut microbiota on diabetes mellitus. *Diabetes & Metabolism, 42*(5), 303–315.

Boursier, J., & Diehl, A. M. (2016). Nonalcoholic fatty liver disease and the gut microbiome. *Clinics in Liver Disease, 20*(2), 263–275.

Buechter, M., & Gerken, G. (2022). Liver function–How to screen and to diagnose: Insights from personal experiences, controlled clinical studies and future perspectives. *Journal of Personalized Medicine, 12*(10), 1657.

Cai, Y., Yang, X., Chen, S., et al., (2023). Regular consumption of pickled vegetables and fermented bean curd reduces the risk of diabetes: A prospective cohort study. *Frontiers in Public Health, 11*, 1155989.

Cani, P. D. (2018). Human gut microbiome: Hopes, threats and promises. *Gut, 67*(9), 1716–1725.

Caussy, C., Tripathi, A., Humphrey, G., et al. (2019). A gut microbiome signature for cirrhosis due to nonalcoholic fatty liver disease. *Nature Communications, 10*(1), 1406.

Clemente, J. C., Pehrsson, E. C., Blaser, M. J., et al. (2015). The microbiome of uncontacted Amerindians. *Science Advances, 1*(3), e1500183.

Cologne, J., Loo, L., Shvetsov, Y. B., et al. (2018). Stepwise approach to SNP-set analysis illustrated with the Metabochip and colorectal cancer in Japanese Americans of the multiethnic cohort. *BMC Genomics, 19*, 1–10.

David, L. A., Maurice, C. F., Carmody, R. N., et al. (2014). Diet rapidly and reproducibly alters the human gut microbiome. *Nature, 505*(7484), 559–563.

Davis, L. M., Martínez, I., Walter, J., et al. (2011). Barcoded pyrosequencing reveals that consumption of galactooligosaccharides results in a highly specific bifidogenic response in humans. *PLoS One 6*(9), e25200.

De Filippo, C., Di Paola, M., Ramazzotti, M., Albanese, D., Pieraccini, G., Banci, E., Miglietta, F., Cavalieri, D., & Lionetti, P. (2017). Diet, environments, and gut microbiota. A preliminary investigation in children living in rural and urban Burkina Faso and Italy. *Frontiers in Microbiology*, *8*, 1979.

de Vos, W. M., & de Vos, E. A. (2012). Role of the intestinal microbiome in health and disease: From correlation to causation. *Nutrition Reviews*, *70*(suppl_1), S45–S56.

Deehan, E. C., Duar, R. M., Armet, A. M., et al. (2017). Modulation of the gastrointestinal microbiome with nondigestible fermentable carbohydrates to improve human health. *Microbiology Spectrum*, *5*(5), 10–1128.

Dubois, G., Girard, C., Lapointe, F.-J., et al. (2017). The Inuit gut microbiome is dynamic over time and shaped by traditional foods. *Microbiome*, *5*(1), 151.

El-Said, G. F., El-Sadaawy, M. M., Shobier, A. H., et al. (2021). Human health implication of major and trace elements present in commercial crustaceans of a traditional seafood marketing region, Egypt. *Biological Trace Element Research*, *199*(1), 315–328.

Fatma, H., & Siddique, H. R. (2022) Herbal medicine to cure male reproductive dysfunction. In M. Sarwat & H. R. Siddique (Eds.), *Herbal medicines* (pp. 409–435). Academic Press.

Ferrario, C., Statello, R., Carnevali, L., et al. (2017). How to feed the mammalian gut microbiota: Bacterial and metabolic modulation by dietary fibers. *Frontiers in Microbiology*, *8*, 1749.

Ferreira, H., Vasconcelos, M., Gil, A. M., et al. (2021). Benefits of pulse consumption on metabolism and health: A systematic review of randomized controlled trials. *Critical Reviews in* Food Science *and Nutrition*, *61*(1), 85–96.

Flint, H. J., Scott, K. P., Duncan, S. H., et al. (2012). Microbial degradation of complex carbohydrates in the gut. *Gut Microbes 3*, 289–306.

Fragiadakis, G. K., Wastyk, H. C., Robinson, J. L., et al. (2020). Long-term dietary intervention reveals resilience of the gut microbiota despite changes in diet and weight. *The American Journal of Clinical Nutrition*, *111*(6), 1127–1136.

Glick-Bauer, M., & Yeh, M.-C. (2014). The health advantage of a vegan diet: Exploring the gut microbiota connection. *Nutrients*, *6*(11), Article 11.

Graf, D., Monk, J. M., Lepp, D., et al. (2019). Cooked red lentils dose-dependently modulate the colonic microenvironment in healthy C57Bl/6 male mice. *Nutrients*, *11*(8), 1853.

Gullón, P., Gullón, B., Tavaria, F., et al. (2015). In vitro fermentation of lupin seeds (Lupinus albus) and broad beans (Vicia faba): Dynamic modulation of the intestinal microbiota and metabolomic output. *Food & Function*, *6*(10), 3316–3322.

Gupta, A., & Tiwari, S. K. (2014). Probiotic potential of lactobacillus plantarum LD1 Isolated from batter of dosa, a South Indian fermented food. *Probiotics and Antimicrobial Proteins*, *6*(2), 73–81.

Ha, V., Sievenpiper, J. L., De Souza, R. J., et al. (2014). Effect of dietary pulse intake on established therapeutic lipid targets for cardiovascular risk reduction: A systematic review and meta-analysis of randomized controlled trials. *CMAJ*, *186*(8), E252–E262.

Hartmann, P., Seebauer, C. T., & Schnabl, B., (2015). Alcoholic liver disease: The gut microbiome and liver cross talk. *Alcoholism, Clinical and Experimental Research*, 39(5), 763–775.

Hang, S., Zeng, L., Han, J., et al. (2022). Lactobacillus plantarum ZJ316 improves the quality of Stachys sieboldii Miq. Pickle by inhibiting harmful bacteria growth, degrading nitrite and promoting the gut microbiota health in vitro. *Food & Function*, *13*(3), 1551–1562.

Hegazi, R., El-Gamal, M., Abdel-Hady, N., et al. (2015). Epidemiology of and risk factors for type 2 diabetes in Egypt. *Annals of Global Health*, *81*(6), 814–820.

Hou, D., Zhao, Q., Yousaf, L., et al. (2020). Whole mung bean (Vigna radiata L.) supplementation prevents high-fat diet-induced obesity and disorders in a lipid profile and modulates gut microbiota in mice. *European Journal of Nutrition*, *59*, 3617–3634.

Hou, K., Wu, Z. X., Chen, X.-Y., et al. (2022). Microbiota in health and diseases. *Signal Transduction and Targeted Therapy*, *7*(1), 135.

Jakobsson, H. E., Rodríguez-Piñeiro, A. M., Schütte, A., et al. (2015). The composition of the gut microbiota shapes the colon mucus barrier. *EMBO Reports*, *16*(2), 164–177.

Jameel, M., Fatma, H., Nadtochii, L. A., & Siddique, H. R., (2023). Molecular insight into prostate cancer: Preventive role of selective bioactive molecules. *Life (Basel)*, *13*(10), 1976.

Jayalath, V. H., De Souza, R. J., Sievenpiper, J. L., et al. (2014). Effect of dietary pulses on blood pressure: A systematic review and meta-analysis of controlled feeding trials. *American Journal of Hypertension*, *27*(1), 56–64.

Jennison, E., & Byrne, C. D. (2021). The role of the gut microbiome and diet in the pathogenesis of nonalcoholic fatty liver disease. *Clinical and Molecular Hepatology*, *27*(1), 22.

Kim, M. S., Hwang, S. S., Park, E. J., et al. (2013). Strict vegetarian diet improves the risk factors associated with metabolic diseases by modulating gut microbiota and reducing intestinal inflammation. *Environmental Microbiology Reports*, *5*(5), 765–775.

Kuhnlein, H. V., Barthet, V., Farren, A., et al. (2006). Vitamins A, D, and E in Canadian Arctic traditional food and adult diets. *Journal of Food Composition and Analysis*, *19*(6), 495–506.

Lacombe, A., Wu, V. C., White, J., et al. (2012). The antimicrobial properties of the lowbush blueberry (Vaccinium angustifolium) fractional components against foodborne pathogens and the conservation of probiotic Lactobacillus rhamnosus. *Food Microbiology*, *30*(1), 124–131.

Lala, V., & Zubair, M. (2023). *Liver function tests* (D. A. Minter, Eds.). Retrieved May 19, 2024, from https://www.statpearls.com/point-of-care/20995/

Leeuwendaal, N. K., Stanton, C., O'Toole, P. W., et al. (2022). Fermented foods, health and the gut microbiome. *Nutrients*, *14*(7), 1527.

Li, H., Deng, Z., Zhu, H., et al. (2012). Highly pigmented vegetables: Anthocyanin compositions and their role in antioxidant activities. *Food Research International*, *46*(1), 250–259.

Lutsiv, T., Weir, T. L., McGinley, J. N., et al. (2021). Compositional changes of the high-fat diet-induced gut microbiota upon consumption of common pulses. *Nutrients*, *13*(11), 3992.

Makki, K., Deehan, E. C., Walter, J., et al. (2018). The impact of dietary fiber on gut microbiota in host health and disease. *Cell Host & Microbe*, *23*(6), 705–715.

Mandhania, M. H., Paul, D., Suryavanshi, M. V., et al. (2019). Diversity and succession of microbiota during fermentation of the traditional indian food idli. *Applied and Environmental Microbiology*, *85*(13), e00368.

Martínez, I., Stegen, J. C., Maldonado-Gómez, M. X., et al. (2015). The gut microbiota of rural papua new guineans: Composition, diversity patterns, and ecological processes. *Cell Reports*, *11*(4), 527–538.

Matijašić, B. B., Obermajer, T., Lipoglavšek, L., et al. (2014). Association of dietary type with fecal microbiota in vegetarians and omnivores in Slovenia. *European Journal of Nutrition*, *53*(4), 1051–1064.

Mokhtari, Z., Gibson, D. L., & Hekmatdoost, A. (2017). Nonalcoholic fatty liver disease, the gut microbiome, and diet. *Advances in Nutrition* (Bethesda, Md.), *8*(2), 240–252.

Monk, J. M., Lepp, D., Wu, W., et al. (2017). Navy and black bean supplementation primes the colonic mucosal microenvironment to improve gut health. *The Journal of Nutritional Biochemistry*, *49*, 89–100.

Mota de Carvalho, N., Costa, E. M., Silva, S., et al. (2018). Fermented foods and beverages in human diet and their influence on gut microbiota and health. *Fermentation*, *4*(4), 90.

Mu, J., Tan, F., Zhou, X., et al. (2020). Lactobacillus fermentum CQPC06 in naturally fermented pickles prevents nonalcoholic fatty liver disease by stabilizing the gut–liver axis in mice. *Food & Function*, *11*(10), 8707–8723.

Nikolić, A., Uzunović, M., & Spaho, N. (2014). Lifestyle pattern underlying organic and traditional food consumption. *British Food Journal*, *116*(11), 1748–1766.

O'Keefe, S. J., Ou, J., Aufreiter, S., et al. (2009). Products of the colonic microbiota mediate the effects of diet on colon cancer risk. *The Journal of Nutrition*, *139*(11), 2044–2048.

Ogunrinola, G. A., Oyewale, J. O., Oshamika, O. O., et al. (2020). The Human Microbiome and Its Impacts on Health. *International Journal of Microbiology*, *5*, 8045646.

Padhi, E. M., & Ramdath, D. D. (2017). A review of the relationship between pulse consumption and reduction of cardiovascular disease risk factors. *Journal of Functional Foods*, *38*, 635–643.

Panagou, E. Z., Nychas, G. J. E., Sofos, J. N. (2013). Types of traditional Greek foods and their safety. *Food Control*, *29*(1), 32–41.

Rodríguez, J. M., Murphy, K., Stanton, C., Ross, R. P., et al. (2015). The composition of the gut microbiota throughout life, with an emphasis on early life. *Microbial Ecology in Health and Disease*, *26*(1), 26050.

Sarkar, P., Lohith Kumar, D. H., Dhumal, C., et al. (2015). Traditional and ayurvedic foods of Indian origin. *Journal of Ethnic Foods*, *2*(3), 97–109.

Schnorr, S. L., Candela, M., Rampelli, S., et al. (2014). Gut microbiome of the Hadza hunter-gatherers. *Nature Communications*, *5*(1), 3654.

Schwiertz, A., Taras, D., Schäfer, K., Beijer et al. (2010). Microbiota and SCFA in lean and overweight healthy subjects. *Obesity*, *18*(1), 190–195.

Sekirov, I., Russell, S. L., Antunes, L. C. M., et al. (2010). Gut microbiota in health and disease. *Physiological Reviews*, *90*(3), 859–904.

Shahrajabian, M. H., Sun, W., & Cheng, Q. (2019). Clinical aspects and health benefits of ginger (Zingiber officinale) in both traditional Chinese medicine and modern industry. *Acta Agriculturae Scandinavica, Section B — Soil & Plant Science*, *69*(6), 546–556.

Sharma, A., & Nagalli S. (2023, May 17). *Chronic liver disease*. Statpearls (Internet). Chronic Liver Disease – StatPearls – NCBI Bookshelf (nih.gov).

Simpson, C. A., Diaz-Arteche, C., Eliby, D., et al. (2021). The gut microbiota in anxiety and depression – A systematic review. *Clinical Psychology Review*, *83*, 101943.

Singh, D., Khan, M. A., & Siddique, H. R. (2021). Therapeutic implications of probiotics in microbiota dysbiosis: A special reference to the liver and oral cancers. *Life Sciences*, *285*, 120008.

Singh, D., Khan, M. A., & Siddique, H. R. (2024). Unveiling the therapeutic promise of natural products in alleviating drug-induced liver injury: Present advancements and future prospects. *Phytotherapy Research*, *38*(1), 22–41.

Siddique, H. R., Nanda, S., Parray, A., & Saleem, M. (2012). Androgen receptor in human health: A potential therapeutic target. *Current Drug Targets*, *13*(14), 1907–1916.

Sonnenburg, E. D., & Sonnenburg, J. L. (2014). Starving our microbial self: The deleterious consequences of a diet deficient in microbiota-accessible carbohydrates. *Cell Metabolism*, *20*(5), 779–786.

Tamanai-Shacoori, Z., Smida, I., Bousarghin, L., et al. (2017). Roseburia spp: A marker of health?. *Future Microbiology*, *12*(2), 157–170.

Trichopoulou, A., Soukara, S., Vasilopoulou, E. (2007). Traditional foods: A science and society perspective. *Trends in Food Science & Technology*, *18*(8), 420–427.

Turnbaugh, P. J., Ridaura, V. K., Faith, J. J., et al. (2009). The effect of diet on the human gut microbiome: A metagenomic analysis in humanized gnotobiotic mice. *Science Translational Medicine*, *1*(6), 6ra14–6ra14.

Tilg, H., Cani, P. D., & Mayer, E. A., (2016). Gut microbiome and liver disease. *Gut*, *65*, 2035–2044.

Ulluwishewa, D., Anderson, R. C., McNabb, W. C., et al. (2011). Regulation of tight junction permeability by intestinal bacteria and dietary components. *The Journal of Nutrition*, *141*(5), 769–776.

Vipperla, K., & O'Keefe, S. J. (2016). Diet, microbiota, and dysbiosis: A 'recipe' for colorectal cancer. *Food & Function*, *7*(4), 1731–1740.

Villarino, N. F., LeCleir, G. R., Denny, J. E., et al. (2016). Composition of the gut microbiota modulates the severity of malaria. *Proceedings of the National Academy of Sciences*, *113*(8), 2235–2240.

Vivarelli, S., Salemi, R., Candido, S., et al. (2019). Gut microbiota and cancer: From pathogenesis to therapy. *Cancers*, *11*(1), 38.

Walker, A. W., Ince, J., Duncan, S. H., et al. (2011). Dominant and diet-responsive groups of bacteria within the human colonic microbiota. *The ISME Journal*, *5*(2), 220–230.

Walter, J. (2015). Murine gut microbiota—Diet trumps genes. *Cell Host & Microbe*, *17*(1), 3–5.

Wardwell, L. H., Huttenhower, C., & Garrett, W. S. (2011). Current concepts of the intestinal microbiota and the pathogenesis of infection. *Current Infectious Disease Reports, 13*, 28–34.

Wang, R., Tang, R., Li, B., et al. (2021). Gut microbiome, liver immunology, and liver diseases. *Cellular & Molecular Immunology*, *18*(1), 4–17

Wei, W., Sun, W., Yu, S., et al. (2016). Butyrate production from high-fiber diet protects against lymphoma tumor. *Leukemia & Lymphoma*, *57*(10), 2401–2408.

Willing, B. P., Dicksved, J., Halfvarson, J., et al. (2010). A pyrosequencing study in twins shows that gastrointestinal microbial profiles vary with inflammatory bowel disease phenotypes. *Gastroenterology*, *139*(6), 1844–1854.

Woodhouse, C. A., Patel, V. C., Singanayagam, A., et al. (2018). Review article: The gut microbiome as a therapeutic target in the pathogenesis and treatment of chronic liver disease. *Alimentary Pharmacology & Therapeutics*, *47*(2), 192–202.

Wu, F., Guo, X., Zhang, J., et al. (2017). Phascolarctobacterium faecium abundant colonization in human gastrointestinal tract. *Experimental and Therapeutic Medicine*, *14*(4), 3122–3126.

Wu, Z. X., Li, S. F., Chen, H., et al. (2017). The changes of gut microbiota after acute myocardial infarction in rats. *PLoS One*, *12*(7), e0180717.

Xiao, W., Zhang, Q., Yu, L., et al. (2022). Effects of vegetarian diet-associated nutrients on gut microbiota and intestinal physiology. *Food Science and Human Wellness*, *11*(2), 208–217.

Yatsunenko, T., Rey, F. E., Manary, M. J., et al. (2012). Human gut microbiome viewed across age and geography. *Nature*, *486*(7402), 222–227.

Zhao, L., Zhang, F., Ding, X., et al. (2018). Gut bacteria selectively promoted by dietary fibers alleviate type 2 diabetes. *Science*, *359*, 1151–1156.

Zhou, W., Cheng, Y., Zhu, P., et al. (2020). Implication of gut microbiota in cardiovascular diseases. *Oxidative Medicine and Cellular Longevity*, *2020*, 1–14.

Zhou, Y., Xie, F., Zhou, X., et al. (2016). Effects of maillard reaction on flavor and safety of Chinese traditional food: Roast duck. *Journal of the Science of Food and Agriculture*, *96*(6), 1915–1922.

9 Role of Biopolymers in Cancer Treatment

Jeetesh Sharma[1] *and Shiwangi Gupta*[2]

[1]Department of Clinical Science (k2), University of Bergen, Bergen, Norway
[2]Scientist-C, Department of Health Research MoHFW, Delhi, India

9.1 INTRODUCTION

Cancers are one of the major contributors to the disease burden. A systematic analysis conducted in 2021 reported 19.3 million new cancer cases and almost 10.0 million deaths globally in 2020 (Sung et al. 2021). Normally, cells divide in a highly controlled and regulated manner, but due to a series of successive mutations, the cell's function changes and leads to cancer. Currently, there are many therapeutic strategies for cancer, but due to the emergence of drug resistance and associated side effects, they lack efficacy. Therefore, many research groups are working at the forefront toward more efficient and less toxic alternatives. In the last decades, there has been significant development in the field of nanoparticles, which hold potential as an oncology treatment modality. One of the fundamental mechanisms inherent in a biological system is to respond to stimuli. A type of nanoparticle, the responsive polymer, can exploit this property of living organisms and use it to fight against cancer. There are different types of polymers and their nanocomposites that have significant potential in drug delivery, diagnosis, and as active anticancer agents.

Polymer in its simplest form can be designated as an unspecified number of monomer units. A few examples of polymers in living systems are protein, DNA, and cellulose, which plants and animals use, whereas nylon, Teflon, polyethylene, and polyester are synthetic polymers. The previous chapters have already discussed biopolymers, their types, molecular shape and structure, and their application in the food industry. In this chapter, we will deal with the potential of biopolymers in the field of biomedicine and the advancement of biopolymers as a treatment strategy for cancer.

For the past decade, nanoparticles (NPs) have been used in biomedicine. Through enhanced permeability and retention (EPR) effects in tumor tissues, therapeutic efficacy and selectivity could be improved via drug delivery through nanoparticles or nano-carriers (Zhao et al. 2018). Biopolymers are an interesting option in various applications (Figure 9.1) where sustainable and ecological solutions are needed. The

DOI: 10.1201/9781003514244-12

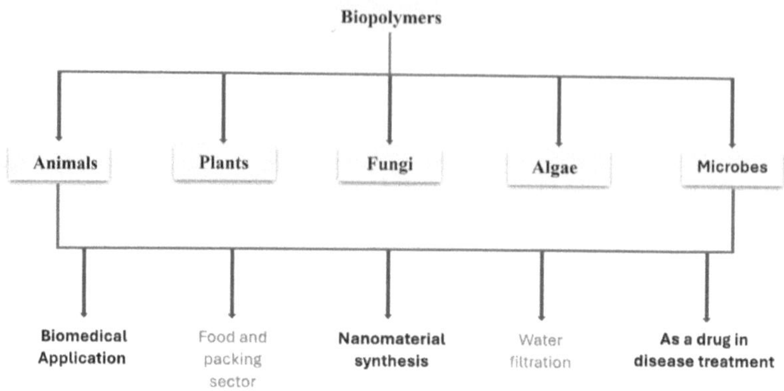

FIGURE 9.1 Different applications of biopolymers are extracted from various sources.

application of drug delivery methods to enhance the enactment of bioactive molecules remains a crucial technique for successful cancer therapy, and advancements in this area have been acute (Yao et al. 2020).

Several biopolymers are derived from plants, algae, bacteria, fungi, and yeast sources. Biopolymers are frequently derived from non-toxic monomers and are carbon-neutral. Nanomaterials from biopolymers are easy to make since they are deposited at the nanoscale. Starch, alginate, chitosan, dextran, and chitin are some of the macromolecules in use. Compounds like pectin, cellulose, lignin, and hyaluronic acid are used to form nanomaterial composites. In cancer treatment, nano biocomposites have shown promising results in recent years. Biodegradable substances and nanoparticles such as chitosan are employed to deliver anticancer drugs (Kanaoujiya et al. 2022). Biopolymers are made up of different components and have varied physiological properties. Biopolymeric nanomaterials can be made by conjugating metals to biopolymers. Starch, for example, can be used to create polymeric nanocomposites by combining metals or metal oxides. Chitosan is being used in various applications of nanotechnology because of its wide compatibility (Morin-Crini et al. 2019).

Given the scarcity of research on the health-promoting qualities of biopolymers, this chapter focuses on providing more significant information about biopolymers, with a particular focus on their applications.

9.2 TYPES OF BIOPOLYMERS

Ecologically friendly biopolymers are now being highlighted in many biomedical applications because they have outstanding biodegradability and biocompatibility (Wei et al. 2021). Examples of biopolymers are polysaccharides such as chitin/chitosan, starch, alginate, pectin, and konjac glucomannan; peptides such as collagen, gelatin, and fibrin gel; polyesters; and polyphenols.

Polysaccharides are natural and renewable polymers that provide a low-cost and eco-friendly source of raw materials. Polysaccharides are frequently used as initial products to produce high-performance macromolecules such as starch, chitosan, chitin, cellulose, gums, and konjac glucomannan and their by-products. For biomedical applications, polysaccharides can be considered flexible macromolecules that can be used as drug-delivery vehicles by increasing drug delivery and as templates for developing specific therapeutic substances (Ngwuluka 2018).

Cellulose is a plant-originated polysaccharide polymer that is still combined in raw form with certain foreign components such as lignin, fatty resins, and minerals (Kalász, Báthori, and Valkó 2020). Cellulose is made up of linear chains of glucose monomers connected by glycosidic bonds. Cellulose derivatives and polymer production have become an important step toward the use of biopolymers, which are utilized in biomedical applications.

After cellulose, ***chitin*** is the most common polysaccharide in nature and is derived from the cell walls of fungi and exoskeletons of arthropods such as crustaceans and insects, mollusks, and squid. The biopolymer is formed from N-acetylglucosamine and glucosamine. Chitosan may be produced by deacetylating chitin using sodium hydroxide heating in the microwave. Chitin is converted into chitosan after this heating, which is the last step of chitosan extraction (El Knidri et al. 2018). Chitosan is being used for medical purposes and has been extensively studied as a regenerative medicine (Jiang et al. 2021), including in the wound healing process (Mansouri et al. 2022) and drug delivery (Kumari, Narvi, and Dutta 2021). Chitosan is used as a vaccine vector with many benefits, such as load for the therapeutic drugs, reduced toxicity of drugs and adverse effects, and increased vaccination efficiency (Meng et al. 2021).

Two glucose polymers, amylopectin and amylose, together form starch. Amylopectin is a highly branched molecule consisting of several D-glucosyl units linked by 1,4- and 1,6-glycosidic bonds. Starch can be obtained from carbohydrate-rich plants such as corn, cassava, rice, potatoes, and wheat. According to review studies, srch produced with enzymes has been found to be widely applicable in daily diets due to its anti-glycemic activity (Himat et al. 2021). Starch provides a matrix for the release of phenolic chemicals that are regulated in the protection of functional food components (Fonseca et al. 2021).

Konjac glucomannan (KGM) is isolated from the tuber of Amorphophallus konjac and is formed by a linear carbohydrate polymer comprised of 1,4-linked d-mannosyl and d-glucosyl residues. Because of its outstanding water-binding and thickening capacity, KGM has long been researched as a possible biodegradable excipient in the food, pharmaceutical, and biomedical sectors (Abbasi and Bera 2021). KGM has been active in various studies as a potential drug delivery source in different disorders, either alone or in a biocomposite with other polymeric materials. KGM can lower oxidative stress levels via the nuclear erythroid 2-associated factor 2 pathway, as well as the nuclear factor-kappa B (NF-κB) route, which acts as a biomarker of anti-inflammatory and antioxidant activity in diabetic rats (Zhao, Jayachandran, and Xu 2020).

Pectin is a biological polymer that contains galacturonic acid units (Lipnizki 2017), and it is a form of structural fiber that is present in primary cell walls and intracellular layers of plant cells. Pectin is capable of inhibiting and treating intestinal infections, atherosclerosis, cancer, and obesity (Khotimchenko 2020; Zhao et al. 2022).

Many algae-based polysaccharides are attained from marine algae because they include a huge number of polysaccharides that are specific to the kind of seaweed, ranging from 4% to 75% of the total dry weight (Usman et al. 2017). Algae-based biopolymers can be exploited as environmental pollution remediation agents, antioxidants, and adsorbents (Kartik et al. 2021). The three components of algal polysaccharides are storage polysaccharides such as starch and laminarian, structural polysaccharides such as cellulose and alginate, and sulfate polysaccharides such as agar, carrageenan, furcellaran, porphyran, ulvan, and fucoidans (Mišurcová, Orsavová, and Ambrožová 2015). As a result of their biomedical applications, algal polysaccharides are receiving a lot of interest, particularly in the fields of health, functional food production, and pharmacy. Algae-based microparticles are being produced for drug delivery in the biomedical and pharmaceutical areas because of their stable sensitivity of pH to target medicine and ability for optimal absorption in the large intestinal area.

Fucoidan, which is a sulfur-based polysaccharide acquired from marine algae, has been confirmed for its efficacy against cervical cancer by blocking the action of histone deacetylase (HDAC) inhibitor receptors (Mustafa, Pawar, and Ghosh 2021).

Exopolysaccharide (EPS) and polyhydroxyalkanoates are microbial biopolymers produced by various microbiomes and are now being investigated as alternatives to contribute to more effective and environmentally friendly pollutant bioremediation (Gupta et al. 2020). Based on their antibacterial and antioxidant properties, they are also widely used in biomedical applications.

Peptides are biomolecules produced from proteins and contain between 2 and 20 amino acids, possessing bioactive qualities such as antihypertensive, antioxidant, antibacterial, and antiviral properties (Cruz-Casas et al. 2021). Peptides derived from vegetable proteins are used in the treatment of various diseases and risk factors. These plant-derived peptides affect energy and food balance via hypothalamic signaling molecules, which may be potential targets for promoting a healthy diet (Kaneko 2021). A report published in 2012 showed that a catalytic peptide linked with an antimicrobial peptide caused cell death in breast cancer cells (MCF-7 and MDA-MB-231) mostly by necrosis, while a fused branched peptide with DNA oligonucleotides induced apoptosis via caspase-3 activation (Sioud and Mobergslien 2012).

Biopolyesters are a major class of polymers made from biological monomers such as polylactic acid, polyhydroxy butyric acid, and polycaprolactone. Polyesters from microbes such as polyhydroxyalkanoate and polyhydroxy acids have attracted interest due to their potential as sustainable alternatives to nonrenewable fossil fuel-based plastics/polymers. Moreover, they have been recognized for their potential for development in the pharmacology, biomedical, and agricultural sectors (Scaffaro et al. 2018).

Polyphenols are found in vegetables, fruits, beverages, and cereals. They are also known as plant secondary metabolites, and some epidemiological studies and meta-analyses indicated that a longer intake of polyphenol-rich foods might provide protection against the growth of cancerous cells (Graf, Milbury, and Blumberg 2005).

Polyphenols can be classified as **flavonoids** and are commonly studied today. More than 4,000 variants of flavonoids have been found in various regions. Common flavonoids are quercetin, myricetin, and catechins, which are found in nature. Flavonoids are directly connected with the health sector because of their potential to treat different inflammatory disorders, such as nephritis, arthritis, hepatitis, ulcerative colitis, Alzheimer's disease, and atherosclerosis (Widhiantara and Jawi 2021).

Curcumin (diferuloylmethane) is a secondary metabolite found in *Curcuma longa* and Curcuma spp. Curcumin is commonly used as a natural food coloring in Indonesia and has also shown a number of medicinal properties (Lestari and Indrayanto 2014). Curcumin is long known for its anticancer potential by affecting several intracellular signaling pathways in cancer cells. These signaling pathways include PI3K/Akt, JAK/STAT, mitogen-activated protein kinase (MAPK), Wnt/β-catenin, p53, NF-κB, and apoptotic activity-related signaling pathways (Wang et al. 2021). The anticancer properties of curcumin are also associated with its molecular structure, in particular the presence of its diketone moiety in the keto-enol tautomer and tautomerism, which stimulates the interaction and binding of many enzymes.

9.3 BIOPOLYMERS IN DRUG DELIVERY

The use of drug delivery systems (DDSs) is an approach to overcome the disadvantages of chemotherapy for cancer treatment. In this section, the various types of natural biopolymers are reviewed and their potential use for cancer therapy is discussed.

9.3.1 Biopolymers and Their Mechanisms in Drug-Delivery

9.3.1.1 Chitosan

Chitosan is an ideal drug delivery carrier with abundant sources, easy modification, and non-toxicity. Chitosan has slowly become a wide area of research because it can be normally processed in the organism and removed from the body. Chitosan can be prepared into microparticles and hydrogels as drug carriers, which can load several anticancer drugs such as curcumin (CUR), doxorubicin (DOX), and resveratrol (Bratovcic 2020a). Chitosan anticancer DDSs are mainly divided into nanogels, nano-microspheres, nanoparticles, and nano-fibers (Tian, Liu, and Liu 2021; Bratovcic 2020b).

A mechanism of anticancer activity of chitosan is related to its capacity to increase the bio-distribution level. The drug accumulation in tumor cells holds high drug entrapment capacities. It brings together the functional groups of acetate subject to cross-linking, forming the projected matrix for competent encapsulation and sustained discharge of different drugs. Chitosan also has efficacy in the trans-mucosal delivery of peptides and drugs (Seferian and Martinez 2000). Chitosan nanoparticles can also recover the physicochemical properties of drugs and aid the advanced apoptosis of distinctive carcinomas due to their improved permeability and retention effect and low cytotoxicity. The treatment of human breast carcinoma cells (MDA-MB-231) with elevated chitosan concentration has been shown to prevent these cells' migration through a matrigel-coated membrane (Nam and Shon 2009).

One study showed that chitosan opens the tight junction between epithelial cells, thereby enhancing the permeability of carried drugs, causing tight link distraction at the molecular level (Liu, Jiang, and Hunziker 2016). Nandgude and Pagar in 2021 used breast cancer cell lines for efficient drug delivery through chitosan. Chitosan also demonstrated excellent competencies in the packaging of genes (Nandgude and Pagar 2021). Chitosan derivatives are favorable materials for targeted and non-viral gene delivery in the treatment of breast cancer. As carriers, drugs are normally delivered using chitin or chitosan. Machalowski et al. (2019) stated that chitin was developed to transfer the aceclofenac drug by the transdermal routes (Machałowski et al. 2019). Likewise, Panonnummal and colleagues used chitin in the form of nanogels to deliver 5-fluorouracil and clobetasol through the topical and transdermal routes (Panonnummal, Jayakumar, and Sabitha 2017). Chitosan can be used in many ways to deliver drugs.

The use of chitosan in biomedical applications as a carrier for cancer treatment is becoming very popular. Chitosan is a carrier for drugs with good biocompatibility and can be easily altered during synthesis. It has a significant role in cancer treatment. Current limitations will be solved when new technologies arise and an understanding of the mechanism of action of chitosan-based drug delivery carriers evolves.

9.3.1.2 Gelatin

Gelatin nanoparticles (GNPs) are known as protein nanostructures that have been widely used in the development of DDSs. GNPs are hydrophilic in nature and thus can improve the cargo permeation into the nanostructure, thereby raising the diffusion-based drug delivery and the subsequent release of the drug. Breast cancer is a globally spreading aggressive cancer in women, with a 75% survival rate (Yazdi Rouholamini et al. 2018).

Gelatin is a denatured protein derived from collagen hydrolysis (acidic or basic). Gelatin molecules are charged with positive charge, negative charge, and hydrophobic amino acids (50%). Gelatin also consists of glycine, hydroxyproline, and proline. Proline and hydroxyproline play an important role in gel formation, particularly in the strength of the gel (Liu et al. 2015). Gelatin contains a sequence such as arginine-lysine-glycine in its primary structure, which exists in many extracellular matrix proteins and plays an important role in the binding of cells and subsequent signaling (Davidenko et al. 2016). Gelatin has active molecules that are engaged in a diversity of chemical reactions, either directly or by using different linkers, especially when producing targeted drugs. Toshio and colleagues demonstrated that the selective targeting of mitomycin C to the spleen, lung, and liver is executed through nano or microspherical gelatin (Toshio et al. 1981). Gelatin nanoparticles with doxorubicin (DOX) stabilized by glutaraldehyde can be used as a potential DDS, as suggested by Leo and colleagues (Leo et al. 1997). In the case of GNP compatibility, modification on the surface is done using various surface coatings so that biomedical applications and stability can be achieved. Once nanoparticles have been introduced into the body, they can be used to transfer various compounds, such as drugs. Also, gelatin can work as a linker and provide support to both adsorb anticancer drugs and mediate their adsorption on the surface of NPs. For example, for targeted drug delivery, imaging, and reducing prostate cancer growth, gelatin-DOX conjugate-coated polyphenol functionalized gold nanoparticles and mercaptopurine (MER) loaded GFe3O4 NPs (Sirivat and Paradee 2019) have been manipulated. Studies have shown that the conjugated NPs significantly reduced cancer cell growth. Redox-responsive GNPs can be used as promising DDSs against a wide range of cancer cells.

9.3.1.3 Dextran

The physicochemical properties of dextran in drug delivery, release mechanisms, and therapeutic effects have been investigated through animal experiments in great detail to justify their wide biomedical applications, especially in the treatment of cancer. Although a lot of work has been done to produce dextran-based nanoparticles as delivery systems, some scientific challenges have also arisen. First of all, thrombocytopenia and hepatotoxicity are major side effects caused by dextran (Danhauser-Riedl et al. 1993) Also, most of the dextran derivatives are formed in the presence of organic solvents and toxic chemicals, which can produce potential safety issues. There is an urgency to explore eco-friendly fabrication methods. Secondly, the behavior of dextran-based nanoparticles inside living organisms is still unknown. There is an urgent need to conduct in vivo studies to clarify the changes in nanoparticle structure, physiological conditions, and metabolism, which will eventually help to understand dextran-based delivery systems. Lastly, for the therapeutic effectiveness of drug-loaded dextran-based nanoparticles, intravenous injection was a major route instead of oral delivery.

9.3.1.4 Starch

Starch is one of the best carriers due to its large quantity, as it is available from different plant sources (Mendes et al. 2016). Starch could be a better DDS due to its morphology, solubility, low toxicity, and easy digestion (Jacob et al. 2018). Starch easily targets the area of drug release in small doses over a long time – reducing the collateral effects of the drug compared to conventional systems (Onnainty et al. 2016). Drugs can be directly released by the sensitivity of the polymeric matrix, which degrades the base material and allows the drug to transfer to the surface (Fu and Kao 2010). Queiroz and colleagues in 2020 developed a long-term drug delivery system using chlorhexidine (CHX) with starch as a biodegradable polymer base (Queiroz et al. 2020). Several studies have reported that a starch-based DDS is best for cancer treatments. Starch nanoparticles help to overcome the problems associated with chemotherapy. Targeted starch DDSs are used to avoid high-concentration intake of drugs for chemotherapy, thus decreasing the possible side effects (Alexiou et al. 2000). Starch-based drug delivery can reduce tumor buildup of chemotherapeutics, avoiding immunogenic responses (Hu et al. 2016). Xu and colleagues (2020) developed nanoparticles in response to pH for anticancer drug delivery using starch oxidation modified by cholesterol and imidazole to obtain amphiphilic cholesterol/imidazole-modified oxidized starch (Cho-Imi-OS) nanoparticles loaded with curcumin. The results suggested that the starch nanoparticles slowly swell and break due to the imidazole group protonation at acidic pH, leading to the triggered release of encapsulated curcumin (Xu et al. 2020).

9.3.1.5 Cellulose

Cellulose is the most abundant polysaccharide; it is available globally, and it combines hydrophilicity with good mechanical properties. Both of these competitive characteristics are due to the several hydroxyl groups that interact favorably through hydrogen bonds with water or with hydroxyl groups of adjacent polymer chains (Klemm et al. 1998). Different variations of cellulose have found exciting uses in drug delivery. There are different types of cellulose-based systems reported to deliver drugs transdermally, including esters, cellulose from bacteria, nanocrystals, hydroxypropyl methylcellulose (HPMC), and ethyl cellulose (EC). They are either used alone or in combination with each other, or in combination with other biopolymers like chitosan for developing drug delivery devices.

Cellulose and chitosan nanoparticles are used for the oral release of repaglinide (Abo-Elseoud et al. 2018), and for targeting colon-diseased sites, calcium alginate beads with carboxymethylcellulose loaded with 5-fluoroalkyl have been used (Agarwal et al. 2015). Four cellulose derivatives – methylcellulose, sodium carboxymethylcellulose, hydroxypropyl methylcellulose, and cationic hydroxyethyl cellulose – have been used as DDSs in the nasal mucosa (Hansen et al. 2015). The impact of hydroxypropyl methylcellulose-based nanocomposites with cellulose nanofibrils for drug release in the form of thin films was explored (Orasugh et al. 2018). For in vitro anti-colon cancer treatment, magnetic-response drug carriers filled with cellulose-containing curcumin were utilized instead of Fe3O4. In vitro drug release of ciprofloxacin showed a 96.6% success rate over 5 hours using gold nanoparticles with cellulose-grafted polyacrylamide hydrogel (Low et al. 2019). Colon cancer treatment with 5-fluorouracil-encapsulated carboxymethyl chitosan is widely documented.

9.3.1.6 Pectin

Pectin is one of the few polysaccharides with biomedical activity and drug-delivery applications. Pectins are considered the most favorable for drug dose forms related to the colon, as they constantly change the media of the gastrointestinal tract and are easily degraded by pectinases produced by colonic microflora. A range of pectin delivery systems were established involving active drug molecules in higher concentrations, in particular with regard to sites inside the intestine and in blood, resulting in the reduced risk of severe side effects (Khotimchenko 2020). Consumption of pectin was shown to be a defense against cancer development. Nanoparticles usually have very small sizes with a diameter between 350 and 500 nm, allowing them to access the cell via membranes by pinocytosis. This mechanism provides very important benefits for antitumor therapy. Biopolymers with larger sizes can access cells only through pinocytosis, giving them a selective activity against malignant cells. They can perform this activity only because this absorption mechanism is mostly typical of tumor cells, whereas simple diffusion typical of normal cells occurs much less frequently. Zhu and colleagues in 2019 developed a simple, colon-aimed DDS by using pectin, porous starch (PS), and chitosan (Zhu et al. 2019). The anticancer drug doxorubicin was successfully loaded into PS granules, which were then coated with a pectin/chitosan complex solution. The use of pectin coating is an active method for colon-targeted drug delivery compared with the simple polysaccharide system. Pectin-based nanomaterials are one approach to delivering drugs to the colon, known as colon-targeted DDS. Pectin has swelling properties, and its capacity to tolerate gastrointestinal degradation has made it a popular carrier for colon-specific medication delivery. The biodegradability and gel-forming nature of this polysaccharide are the characteristics that drive its selection as a carrier for specific drug delivery.

9.4 BIOPOLYMERS AS ANTICANCER AGENTS

According to the characteristics and stage of the tumor, cancer is typically treated with a combination of therapies, including surgery, radiation therapy, chemotherapy, and immunotherapy. In all cases, the purpose is to destroy the tumor-forming cells and reduce the growth of the tumor. Natural products from marine and land organisms have drawn research attention in recent years, mainly marine organisms because of their large habitat, high diversity, and the conditions under which some organisms thrive.

Extensive research is ongoing in the study of biopolymers as a drug delivery system in many diseases including cancer (discussed later in this chapter), but there are not many biopolymers that have been explored in the context of directly targeting cancer. In this section, we will discuss three biopolymers in detail and their roles as anticancer agents.

9.4.1 Fucoidan

Fucoidan is an edible alga used in East Asian diets for many years. Recently, it has been discovered to have a variety of pharmacological properties, including antimicrobial, anti-inflammatory, antiviral, and anticancer properties (Etman, Elnaggar, and Abdallah 2020). Many reports highlight the importance of fucoidan as a DDS for many diseases (Etman, Elnaggar, and Abdallah 2020). Fucoidan is a sulfated polysaccharide and was first isolated by Kylin in 1913 from the extracellular matrix of brown seaweeds (Senthilkumar et al. 2013; Kylin 1913). Fucoidan has an extremely complex structure mainly composed of L-fucose sugar, but it may also contain other monosaccharides such as glucose, mannose, arabinose, and xylose (Bilan et al. 2002). Fucoidan has two main backbones, 1-3 linked-L-fucose residue or 1-3 and 1-4 linked-L-fucose residue (Cumashi et al. 2007). The most investigated type of fucoidan is from *Fucus vesiculosus*, used in the defense against various types of tumors, including liver (Roshan et al. 2014), bladder (Park et al. 2015), lymphoma (Yang et al. 2015), colon (Park et al. 2017), and gastric (Yoshimoto et al. 2015).

9.4.1.1 Anticancer Activity of Fucoidan and Possible Mode of Action

Many reports have confirmed the ability of fucoidan to kill cancer by arresting the cell cycle at the G1 phase (Kim, Kwon, and Nam 2017). Fucoidan induces arrest by reducing CDK2, CDK4, cyclin D1, and cyclin E, which leads to apoptosis.

Fucoidan promotes apoptosis in hepatocellular carcinoma and MCF-7 (Zhang et al. 2011) breast cancer cells in a dose-dependent manner. In contrast, fucoidan did not cause apoptosis in normal cells; on the contrary, it protected them from damage by preventing apoptosis (Li et al. 2016). Fucoidan also affects angiogenesis in tumor masses, although results regarding its effects conflict. Some studies have shown that fucoidan causes an anti-angiogenic effect, while other groups have shown its pro-angiogenic effects. It wasn't until 2014 that a review article investigating the effect of fucoidan on angiogenesis confirmed that high molecular weight fucoidan showed anti-angiogenic effects, while low molecular weight fucoidan showed pro-angiogenic effects. Another factor that determines fucoidan's angiogenic activity is the degree of sulfation: over-sulfated fucoidans were shown to have anti-angiogenic effects (Koyanagi et al. 2003). However, these conflicting results indicate the complexity of this biopolymer, and further research is required to find out more about the angiogenic effect of fucoidan. Matrix metalloproteinase (MMP) is one of the critical proteins involved during the metastasis of tumor cells, and inhibition of this protein might prevent cell migration and metastasis. There are reports that fucoidan inhibits MMP-2 (Lee, Kim, and Kim 2012) and MMP-9 (Cho, Kim, and Moon 2014) and can prevent metastasis in a dose-dependent manner in A549 (human lung cancer cell line). A research group investigated fucoidan's effect on C57BL/6 mice, transplanted with Lewis lung adenocarcinoma, and found that it produced an anti-metastatic effect at a dose of 10 mg/kg (Alekseyenko et al. 2007). Another study conducted on human bladder cancer cells (5637) showed decreased expression of MMP9 by down-regulating NF-kappa B and AP-1 binding activity after fucoidan treatment (Cho, Kim, and Moon 2014). In addition to inhibiting cell migration, fucoidan has been able to enhance the activity of macrophages, dendritic cells, T-cells, and natural killer cells (Zhang et al. 2015). A report published in 2019 showed that acetylated fucoidan induces macrophage activation via membrane receptors, including scavenger receptor class A, Toll-like receptor-4, and CD-14 on macrophages (Teruya et al. 2009). This immunomodulating activity in fucoidan among different species might be due to the presence of uronic acid. The content of uronic acid affects the immunomodulating activity of fucoidan as published in a report that shows that fucoidan from *Ascophyllum nodosum* with high uronic content had a stronger effect on the immune system than *Fucus vesiculosus* (Zhang et al. 2014; Zhang et al. 2015).

In terms of toxicity, LMW fucoidan from *Laminaria japonica* was considered to be safe at a dose of 2000 mg/kg (Hwang et al. 2016). Fucoidan from *Undaria pinnatifida* was tested on rats to evaluate oral toxicity (Kim et al. 2010). A dose of 2000 mg/kg bw/day for 4 weeks showed no side effects compared to the control group. Mathew et al. in 2017 investigated the interaction between different chemotherapeutic drugs and fucoidan, and they reported that fucoidan does not affect the metabolism of chemotherapeutic drugs negatively but in fact had synergistic effects with paclitaxel and tamoxifen in eight different types of orthotopic cancer models (Mathew et al. 2017).

9.4.1.2 Clinical Trials

LMW fucoidan was included in a clinical trial (double-blind controlled randomized) to test its efficacy as a food supplement in metastatic colorectal carcinoma patients. A total of 54 patients were enrolled in the study. Results showed a significant increase in disease control rate of 92.8% and 69.2% in the control groups, respectively (Tsai et al. 2017). Another clinical trial (single-blind randomized placebo-controlled) was conducted in Australia with 37 non-smoker volunteers. Fucoidan used in the study was extracted from *Undaria pinnatifida*. Volunteers were given three capsules (0.33 g) three times a day. Results showed that fucoidan results in better mobilization of hematopoietic progenitor cells by increasing the expression of CXCR4, which can be beneficial for the treatment of different types of tumors (Irhimeh, Fitton, and Lowenthal 2007).

9.4.1.3 Future Implications

The efficiency of fucoidan against cancer via different mechanisms indicates its potential as a therapeutic agent. It's also encouraging to mention that clinical trials conducted with fucoidan, along with other therapeutic agents, show positive effects. However, discrepancies in results between human trials and animal experiments need serious consideration, in addition to in-depth research on fucoidan structure, its absorption in the body, its metabolism, etc. Fucoidan's potential is not only limited to being an anticancer agent, but it can also be used as a drug delivery vehicle. Fucoidan/chitosan-based nanoparticles encapsulated with gemcitabine tested on a human breast cancer cell line resulted in 84% of drug release in the first 4 hours. Fucoidan has also been tested as an antiviral agent. A study conducted in 2017 reported that fucoidan was able to inhibit the influenza A virus infection by targeting the cellular EGFR pathway and viral neuraminidase (Wang et al. 2017). Another study on a potential anti-hepatitis B virus treatment conducted on mice shows its potential as a therapeutic agent not only for cancer but also infectious diseases. However, being a natural polymer, fucoidan needs a tremendous amount of treatment before it can be used as a therapeutic agent, and beyond that, a significant amount of research needs to be done on its structure, which greatly depends on its source.

9.4.2 Propolis

Also known as bee glue, propolis is a resin-like mixture produced by honeybees by mixing beeswax and saliva with exudate collected from various parts of plants, such as tree bark, leaves, and flower buds. Based on its source, propolis can be found in various colors such as yellow, brown, red, and green, with a specific herbaceous aromatic scent. Propolis has extensively been used and explored in traditional medicine throughout the centuries. In ancient Egypt, it was used to embalm cadavers to protect them from bacterial and fungal overgrowth. Hippocrates used it to cure wounds and internal and external ulcers. The use of propolis developed over time; in World War II, it was used as an anti-inflammatory and antibacterial agent (Forma and Bryś 2021). Nowadays, it is currently being supplemented in various health care products such as gels, cough syrups, lotions, creams, soap, toothpaste, tinctures, and mouthwash preparations. Propolis is mainly classified into seven major types based on the origin of the plant components. The physical properties and chemical composition of propolis depend on the solvent used for extraction. Ethanol is mainly used for extraction, but other solvents such as water, hexane, chloroform, glycolic solution, and seed oil are also used. In general, propolis mainly contains resin and balms (50–60%), fatty acids and waxes (30–40%), essential and aromatic oil (5–10%), pollens (5–10%), and other micro- and macronutrients (5%) (Forma and Bryś 2021).

Different extracts of propolis are enriched with many chemical compounds, such as phenols, flavonoids, aliphatic and aromatic acids, alkaloids, beta-steroids, fatty acids, alcohols, terpenes, and many more. These chemical compounds are responsible for the pharmacological activity of propolis. It has been shown that propolis-based active compounds affect cancer development and can be used by patients undergoing chemotherapy to limit side effects.

9.4.2.1 Anticancer Activity of Propolis and Possible Mode of Action

The controlled cell cycle is a crucial feature of cell division that is maintained by several key regulatory proteins. A cell that fails to regulate the cell cycle in a proper way results in uncontrolled cell division, which may lead to neoplastic foci. There are many reports available on the effect of propolis on modulating cell cycle regulators. In a study, Du145 and PC3 were arrested at the G0/G1 phase when treated with ethanol extract of Cameroonian propolis via downregulation of CDK1, cyclin A, and cyclin B in both cell lines, while only CDK2 was downregulated in PC3 (Zingue et al. 2020). Caffeic acid (CA) and caffeic acid phenylethyl ester (CAPE), the chemical group present in propolis, induce S-phase arrest in breast cancer cells (MDA-MB-231) in a dose and time-dependent manner. CAPE is also shown to have S6 beta-1 ribosomal protein kinase (p70S6K) inhibitory properties, which are necessary for protein synthesis and proliferation of cells (Kabała-Dzik et al. 2017). Another component, genistein, can inhibit the cell cycle at the G2/M phase in prostate cancer cells by reducing cyclin B expression and increasing p21 in a p53-independent manner (Zabaiou et al. 2017). During cell division in mitosis, chromosomes become separated with the help of microtubules, which are composed of alpha-tubulin and tubulin subunits. A Brazilian red propolis isoflavonoid called vetitol downregulated these tubulins and disrupted the normal cell division (Nani et al. 2018). In addition to disrupting the cell cycle, propolis and its components induce apoptosis, as well proven by many scientific studies. Artepillin C, a p-coumaric derivative from Brazilian green propolis, induces apoptosis in castration-resistant prostate cancer cells by increasing cleaved caspase-3 and poly ADP-ribose polymerase. Chrysin, one of the well-understood components of propolis, is a pro-apoptotic compound that initiates apoptosis via the mitochondrial pathway. It has been reported that chrysin treatment led to the loss of mitochondrial membrane potential, with an increase in cytoplasmic Ca^{2+} levels, production of reactive oxygen species (ROS), and lipid peroxidation (Naz et al. 2019). Another component, quercetin, present in propolis, induces apoptosis by reducing a key S-phase enzyme called thymidylate synthase in a dose and concentration-dependent manner (Forma and Bryś 2021). In addition to acting as a pro-apoptotic molecule, components from propolis, such as caffeic acid and CAPE, also affect angiogenesis by decreasing the phosphorylation of JNK-1 and inhibiting VEGF production (Espíndola et al. 2019; Chiu et al. 2020). FAK and Wnt2 are proteins that play important roles in cell migration and metastasis. A study conducted on mammary gland cancer-induced mice showed the anti-metastatic activity of propolis by suppressing the expression of Wnt2 and FAK (Tan and Hayati 2017).

9.4.2.2 Clinical Trials

Numerous studies have been conducted to evaluate the biological effects of natural products in cancer treatments. Chemotherapy and radiation therapy are the most widely used treatments for cancer patients and are associated with several side effects. Some studies include propolis and its components to either directly act on cancer cells or as a supplement to minimize side effects that are associated with the treatment. One such study conducted by Darvishi and colleagues, a randomized double-blind clinical trial on breast cancer patients, studied the antioxidant and anti-inflammatory effects of propolis. It was observed that the group taking propolis had no increase in TNF-alpha, IL-2 (pro-inflammatory), and protein carbonyl (oxidative stress biomarker). Oral mucositis is one of the major side effects of chemotherapy. Piredda and colleagues showed that rinsing the mouth with propolis (dry extract) is effective in the reduction of oral mucositis-related symptoms in patients with breast cancer during chemotherapy (Piredda et al. 2017). A chemical test conducted on a group of 135 patients with breast cancer undergoing radiotherapy was supplemented with propolis because of the ionizing radiation of leukocytes. The study showed a significant reduction in radiation-induced DNA damage, which suggests that propolis increases the efficiency of serum to neutralize free radicals (Ebeid et al. 2016).

9.4.2.3 Future Implications

Although propolis is tolerated by cancer patients in clinical trials, it can be allergenic and may cause gastric problems. Propolis has a highly variable chemical composition that depends on the origin and extraction methods used. Thus, it is necessary to develop a standardized method that is combined with a specific compound, its biological activity, and the recommended use of its type. Propolis has a wide variety of compounds with several different targets. Therefore, it is necessary to investigate these bioactive compounds and their targeted molecules to create novel therapeutics against cancer and overcome the problem of chemotherapy resistance. Multidrug resistance (MDR) is a huge problem in chemotherapy. The involvement of P-glycoprotein (P-gp), a membrane transporter that effluxes chemotherapeutic drugs, has a significant role in the MDR of cancer cells. A study conducted on doxorubicin-resistant lung cancer cells (A549) treated with propolis resulted in a decreased proliferation of A549 cells by inhibiting the P-gp efflux pump (Kebsa et al. 2018). This shows the potential of propolis and its bioactive compounds against MDR cancers, although in-depth study is required along with clinical trials. With sufficient research, propolis could be used to prepare novel, low-cost cancer therapies or be used as a combination therapy.

9.4.3 Pectin

Pectin is a very important polysaccharide consisting of a backbone of D-galacturonic acid residues linked at α-1,4. It is associated with cellulose and hemicellulose in plant cell walls. Around 70% of pectin molecules are made up of acetylated or methyl-esterified galacturonic acid monomers. Based on structural composition, pectin can be divided into two regions: the smooth and hairy regions. Homogalacturonan is a linear molecule responsible for the smooth region of pectin, whereas xylogalacturonan, rhamnogalacturonan-I, and rhamnogalacturonan-II are referred to as hairy regions. For several decades, pectin has been used in the food and pharmaceutical industries for gelling, thickening, health benefits, and emulsification. Pectin's application in biomedical fields and drug delivery is primarily due to the presence of major functional groups such as hydroxyl, carboxyl, methyl, and amide (Li et al. 2021).

9.4.3.1 Anticancer Activity of Pectin and Possible Mode of Action

Cancer occurs due to the alteration in many functional capabilities of cells. It was found that pectin affects multiple rate-limiting steps in cancer progression, including inhibition of anti-apoptotic function, interaction with cell adhesion molecules, and inhibition of cell proliferation. The anticancer potential of pectin has mostly been studied in colorectal cancer. However, pectin has also been assessed against cancer cell lines that do not belong to colorectal cancer. A study conducted on glioblastoma cells showed that purified pectin from *Campomanesia xanthomata*, a Brazilian Myrtaceae family species, increases cellular ROS while showing no cytotoxicity toward normal NIH 3T3 cells (Amaral et al. 2019). Pectin is also shown to have galectin-modulating activity. Galectin is a protein that binds to glycan and has a role in cancer development, progression, and metastasis. Many types of galectin have been studied, but galectin-3 has been most extensively studied. Galectin-3 is shown to cause upregulation of the MAPK pathway by inducing phosphorylation of ERK1/2 (Fukumori, Kanayama, and Raz 2007) which leads to the activation of several genes related to cell survival, proliferation, and metastasis. Therefore, pectin binding to the carbohydrate-binding domain of galectin-3 hinders its involvement in cancer progression (Picot-Allain and Neergheen 2023). The binding of galectin-3 with modified pectin-derived galactan side chains was shown to facilitate an inhibitory role as a prometastatic regulatory protein. Pectin is also found to induce caspase-dependent apoptosis. Caspase, a cysteine-dependent aspartate protease, plays a pivotal role in apoptosis, and its aberrant inactivation leads to the onset and progression of cancer. Caspase-dependent apoptosis is governed by the initiator (caspase-2, -8, -9, and -10) and effector caspases (caspase-3, -6, and -7). A study conducted on human breast cancer cells showed that treatment with pectin from apples causes the overexpression of caspase-3 (Delphi et al. 2015). In another report, colon cancer-induced male rats obtained from Charles River Laboratory were treated with citrus pectin. Results showed apoptosis in rat colonic tissue due to the activation and initiation of a protease cascade, such as caspase-3 that leads to cleavage of PARP, a DNA repair enzyme (Avivi-Green, Madar, and Schwartz 2000). Pectin is also shown to induce autophagy and cell cycle arrest in cancer cells. Modified citrus pectin can induce G1 cell cycle arrest in ovarian cancer cells (SKOV-3) (Hossein et al. 2013).

9.4.3.2 Clinical Trials

A recent phase II clinical study conducted on biochemically relapsed prostate cancer (BRPC) showed the potential benefit of modified citrus pectin in lowering disease progression with no significant toxicity (Keizman et al. 2021). Though there have not been many clinical trials conducted to study pectin as an anticancer agent, there are several reports on the extraction of pectin using non-conventional methods from different plant sources, suggesting its potential as an anticancer agent.

9.4.3.3 Future Implications

Several studies have been conducted to reveal pectin's benefit as a treatment enhancer not only for cancer but also for other diseases. Irinotecan in its active form SN-38 (glucuronidated) is used to treat colon cancer, but gut bacterial beta-glucuronidase removes glucuronic acid and produces SN-38, a toxic compound. Apple pectin is found to increase the cytostatic action of irinotecan and reduce the activity of beta-glucuronidase, highlighting its potential to be used as an adjunct in cancer treatment (Palko-Łabuz et al. 2021). In addition, as an anticancer agent, pectin has shown promising results for use as a drug delivery system. Since pectin is not hydrolyzed by the upper gastrointestinal tract but at the level of the colon by colon microflora, it could be used as a good strategy for treating colon cancer (Wong, Colombo, and Sonvico 2011). Many studies have suggested its importance in developing delayed response formulations, but in combination with other polymers, it has shown colon-specific successful therapeutic agent delivery (Das 2021).

9.5 CONCLUSIONS

In this chapter, we have highlighted the role of biopolymers in cancer treatment because they offer certain benefits over preexisting therapies. Biopolymers offer biodegradability, biocompatibility, and ecologically friendly qualities, and have been shown to hold promise against cancer treatment. Natural biopolymers are non-toxic to normal cells. However, extensive research on biopolymers is needed to fully assess their anticancer potential because the same biopolymer derived from different sources and different extraction methods can have different structures that may affect their activity significantly. The metabolic changes of the drug carrier in the organism must be taken into account when applied in clinical practice. Another aspect that needs attention is that, being natural products, they need to be purified to a level that can be used for clinical purposes, which is cost- and time-intensive. Therefore, it is necessary to fund more projects in this rapidly growing area of the biomedical sector. In the future of cancer research, cutting-edge technologies should be actively and flexibly introduced to prepare a drug carrier that can accurately load the drug with a therapeutic dose. With sufficient research, biopolymers could be used to prepare novel low-cost cancer therapies or could be used in combination with existing therapies.

REFERENCES

Abbasi, Yasir Faraz, and Hriday Bera. 2021. "Chapter 5 – Konjac glucomannan-based nanomaterials in drug delivery and biomedical applications." In *Biopolymer-Based Nanomaterials in Drug Delivery and Biomedical Applications*, edited by Hriday Bera, Chowdhury Mobaswar Hossain, and Sudipta Saha, 119–141. Academic Press.

Abo-Elseoud, Wafaa S., Mohammad L. Hassan, Magdy W. Sabaa, Mona Basha, Enas A. Hassan, and Shaimaa M. Fadel. 2018. "Chitosan nanoparticles/cellulose nanocrystals nanocomposites as a carrier system for the controlled release of repaglinide." *Int J Biol Macromol* 111:604–613.

Agarwal, Tarun, S. N. Gautham Hari Narayana, Kunal Pal, Krishna Pramanik, Supratim Giri, and Indranil Banerjee. 2015. "Calcium alginate-carboxymethyl cellulose beads for colon-targeted drug delivery." *Int J Biol Macromol* 75:409–417.

Alekseyenko, T. V., S. Y. Zhanayeva, A. A. Venediktova, T. N. Zvyagintseva, T. A. Kuznetsova, N. N. Besednova, and T. A. Korolenko. 2007. "Antitumor and antimetastatic activity of fucoidan, a sulfated polysaccharide isolated from the Okhotsk Sea Fucus evanescens brown alga." *Bull Exp Biol Med* 143 (6):730–732. doi: 10.1007/s10517-007-0226-4.

Alexiou, Christoph, Wolfgang Arnold, Roswitha J. Klein, Fritz G. Parak, Peter Hulin, Christian Bergemann, Wolfgang Erhardt, Stefan Wagenpfeil, and Andreas S. Lubbe. 2000. "Locoregional cancer treatment with magnetic drug targeting." *Cancer Res* 60 (23):6641–6648.

Amaral, S. D. C., S. F. Barbieri, A. C. Ruthes, J. M. Bark, S. M. Brochado Winnischofer, and J. L. M. Silveira. 2019. "Cytotoxic effect of crude and purified pectins from Campomanesia xanthocarpa Berg on human glioblastoma cells." *Carbohydr Polym* 224:115140. doi: 10.1016/j.carbpol.2019.115140.

Avivi-Green, C., Z. Madar, and B. Schwartz. 2000. "Pectin-enriched diet affects distribution and expression of apoptosis-cascade proteins in colonic crypts of dimethylhydrazine-treated rats." *Int J Mol Med* 6 (6):689–698. doi: 10.3892/ijmm.6.6.689.

Bilan, M. I., A. A. Grachev, N. E. Ustuzhanina, A. S. Shashkov, N. E. Nifantiev, and A. I. Usov. 2002. "Structure of a fucoidan from the brown seaweed Fucus evanescens C.Ag." *Carbohydr Res* 337 (8):719–730. doi: 10.1016/s0008-6215(02)00053-8.

Bratovcic, Amra. 2020a. "Antioxidant enzymes and their role in preventing cell damage." *Acta Sci Nutr Health* 4:01–07.

Bratovcic, Amra. 2020b. "Nanocomposite hydrogels reinforced by carbon nanotubes." *Int J Eng Res Appl* 10 (5):30–41.

Chiu, H. F., Y. C. Han, Y. C. Shen, O. Golovinskaia, K. Venkatakrishnan, and C. K. Wang. 2020. "Chemopreventive and chemotherapeutic effect of propolis and its constituents: A mini-review." *J Cancer Prev* 25 (2):70–78. doi: 10.15430/jcp.2020.25.2.70.

Cho, T. M., W. J. Kim, and S. K. Moon. 2014. "AKT signaling is involved in fucoidan-induced inhibition of growth and migration of human bladder cancer cells." *Food Chem Toxicol* 64:344–52. doi: 10.1016/j.fct.2013.12.009.

Cruz-Casas, D. E., C. N. Aguilar, J. A. Ascacio-Valdés, R. Rodríguez-Herrera, M. L. Chávez-González, and A. C. Flores-Gallegos. 2021. "Enzymatic hydrolysis and microbial fermentation: The most favorable biotechnological methods for the release of bioactive peptides." *Food Chem (Oxf)* 3:100047. doi: 10.1016/j.fochms.2021.100047.

Cumashi, A., N. A. Ushakova, M. E. Preobrazhenskaya, A. D'Incecco, A. Piccoli, L. Totani, N. Tinari, G. E. Morozevich, A. E. Berman, M. I. Bilan, A. I. Usov, N. E. Ustyuzhanina, A. A. Grachev, C. J. Sanderson, M. Kelly, G. A. Rabinovich, S. Iacobelli, and N. E. Nifantiev. 2007. "A comparative study of the anti-inflammatory, anticoagulant, antiangiogenic, and antiadhesive activities of nine different fucoidans from brown seaweeds." *Glycobiology* 17 (5):541–52. doi: 10.1093/glycob/cwm014.

Danhauser-Riedl, Susanne, Edith Hausmann, Hans-D Schick, Rita Bender, Hermann Dietzfelbinger, Johann Rastetter, and Axel-R Hanauske. 1993. "Phase I clinical and pharmacokinetic trial of dextran conjugated doxorubicin (AD–70, DOX-OXD)." *Invest New Drugs* 11:187–195.

Das, S. 2021. "Pectin based multi-particulate carriers for colon-specific delivery of therapeutic agents." *Int J Pharm* 605:120814. doi: 10.1016/j.ijpharm.2021.120814.

Davidenko, Natalia, Carlos F. Schuster, Daniel V. Bax, Richard W. Farndale, Samir Hamaia, Serena M. Best, and Ruth E. Cameron. 2016. "Evaluation of cell binding to collagen and gelatin: A study of the effect of 2D and 3D architecture and surface chemistry." *J Mater Sci Mater Med* 27:1–14.

Delphi, L., H. Sepehri, M. R. Khorramizadeh, and F. Mansoori. 2015. "Pectic-oligoshaccharides from apples induce apoptosis and cell cycle arrest in MDA-MB-231 cells, a model of human breast cancer." *Asian Pac J Cancer Prev* 16 (13):5265–5271. doi: 10.7314/apjcp.2015.16.13.5265.

Ebeid, Samia A., Nadia A. Abd El Moneim, Sanaa A. El-Benhawy, Nabila G. Hussain, and Mervat I. Hussain. 2016. "Assessment of the radioprotective effect of propolis in breast cancer patients undergoing radiotherapy. New perspective for an old honey bee product." *J Radiat Res Appl Sci* 9 (4):431–440.

El Knidri, H., R. Belaabed, A. Addaou, A. Laajeb, and A. Lahsini. 2018. "Extraction, chemical modification and characterization of chitin and chitosan." *Int J Biol Macromol* 120 (Pt A):1181–1189. doi: 10.1016/j.ijbiomac.2018.08.139.

Espíndola, K. M. M., R. G. Ferreira, L. E. M. Narvaez, A. C. R. Silva Rosario, A. H. M. da Silva, A. G. B. Silva, A. P. O. Vieira, and M. C. Monteiro. 2019. "Chemical and pharmacological aspects of caffeic acid and its activity in hepatocarcinoma." *Front Oncol* 9:541. doi: 10.3389/fonc.2019.00541.

Etman, S. M., Y. S. R. Elnaggar, and O. Y. Abdallah. 2020. ""Fucoidan, a natural biopolymer in cancer combating: From edible algae to nanocarrier tailoring"." *Int J Biol Macromol* 147:799–808. doi: 10.1016/j.ijbiomac.2019.11.191.

Fonseca, Laura Martins, Francine Tavares da Silva, Graziella Pinheiro Bruni, Caroline Dellinghausen Borges, Elessandra da Rosa Zavareze, and Alvaro Renato Guerra Dias. 2021. "Aerogels based on corn starch as carriers for pinhão coat extract (Araucaria angustifolia) rich in phenolic compounds for active packaging." *Int J Biol Macromol* 169:362–370. https://doi.org/10.1016/j.ijbiomac.2020.12.110.

Forma, Ewa, and Magdalena Bryś. 2021. "Anticancer activity of propolis and its compounds." *Nutrients* 13 (8):2594.

Fu, Yao, and Weiyuan John Kao. 2010. "Drug release kinetics and transport mechanisms of non-degradable and degradable polymeric delivery systems." *Expert Opin Drug Deliv* 7 (4):429–444.

Fukumori, T., H. O. Kanayama, and A. Raz. 2007. "The role of galectin-3 in cancer drug resistance." *Drug Resist Updat* 10 (3):101–8. doi: 10.1016/j.drup.2007.04.001.

Graf, B. A., P. E. Milbury, and J. B. Blumberg. 2005. "Flavonols, flavones, flavanones, and human health: Epidemiological evidence." *J Med Food* 8 (3):281–90. doi: 10.1089/jmf.2005.8.281.

Gupta, Juhi, Rashmi Rathour, Kristina Medhi, Bhawna Tyagi, and Indu Shekhar Thakur. 2020. "3 – Microbial-derived natural bioproducts for a sustainable environment: a bioprospective for waste to wealth." In *Refining Biomass Residues for Sustainable Energy and Bioproducts*, edited by R. Praveen Kumar, Edgard Gnansounou, Jegannathan Kenthorai Raman, and Gurunathan Baskar, 51–85. Academic Press.

Hansen, Kellisa, Gwangseong Kim, Kashappa-Goud H. Desai, Hiren Patel, Karl F. Olsen, Jaime Curtis-Fisk, Elizabeth Tocce, Susan Jordan, and Steven P Schwendeman. 2015. "Feasibility investigation of cellulose polymers for mucoadhesive nasal drug delivery applications." *Mol Pharmaceutics* 12 (8):2732–2741.

Himat, Abdelhamid S., Sandhya Gautam, Juan Pablo Chavez Garcia, Ana Xóchitl Vidrio-Sahagún, Zhengjie Liu, David Bressler, and Thava Vasanthan. 2021. "Starch-based novel ingredients for low glycemic food formulation." *Bioact Carbohydr Diet Fibre* 26:100275. https://doi.org/10.1016/j.bcdf.2021.100275.

Hossein, G., M. Keshavarz, S. Ahmadi, and N. Naderi. 2013. "Synergistic effects of PectaSol-C modified citrus pectin an inhibitor of Galectin-3 and paclitaxel on apoptosis of human SKOV-3 ovarian cancer cells." *Asian Pac J Cancer Prev* 14 (12):7561–7568. doi: 10.7314/apjcp.2013.14.12.7561.

Hu, Hang, Yihui Li, Qing Zhou, Yanxiao Ao, Chan Yu, Ying Wan, Huibi Xu, Zifu Li, and Xiangliang Yang. 2016. "Redox-sensitive hydroxyethyl starch–doxorubicin conjugate for tumor targeted drug delivery." *ACS Appl Mater Interfaces* 8 (45):30833–30844.

Hwang, P. A., M. D. Yan, H. T. Lin, K. L. Li, and Y. C. Lin. 2016. "Toxicological evaluation of low molecular weight fucoidan in vitro and in Vivo." *Mar Drugs* 14 (7). doi: 10.3390/md14070121.

Irhimeh, M. R., J. H. Fitton, and R. M. Lowenthal. 2007. "Fucoidan ingestion increases the expression of CXCR4 on human CD34+ cells." *Exp Hematol* 35 (6):989–994. doi: 10.1016/j.exphem.2007.02.009.

Jacob, Joby, Józef T. Haponiuk, Sabu Thomas, and Sreeraj Gopi. 2018. "Biopolymer based nanomaterials in drug delivery systems: A review." *Mater Today Chem* 9:43–55.

Jiang, Z., K. Zhang, L. Du, Z. Cheng, T. Zhang, J. Ding, W. Li, B. Xu, and M. Zhu. 2021. "Construction of chitosan scaffolds with controllable microchannel for tissue engineering and regenerative medicine." *Mater Sci Eng C Mater Biol Appl* 126:112178. doi: 10.1016/j.msec.2021.112178.

Kabała-Dzik, A., A. Rzepecka-Stojko, R. Kubina, Ż Jastrzębska-Stojko, R. Stojko, R. D. Wojtyczka, and J. Stojko. 2017. "Comparison of two components of propolis: Caffeic Acid (CA) and caffeic acid phenethyl ester (CAPE) induce apoptosis and cell cycle arrest of breast cancer cells MDA-MB-231." *Molecules* 22 (9). doi: 10.3390/molecules22091554.

Kalász, Huba, Mária Báthori, and Klára L. Valkó. 2020. "Chapter 10 – Basis and pharmaceutical applications of thin-layer chromatography." In *Handbook of analytical separations*, edited by Klára l Valkó, 523–585. Elsevier Science B.V.

Kanaoujiya, Rahul, Shruti Kumari Saroj, Shekhar Srivastava, and Manoj Kumar Chaudhary. 2022. "Renewable polysaccharide and biomedical application of nanomaterials." *Journal of Nanomaterials* 1:1050211. https://doi.org/10.1155/2022/1050211

Kaneko, K. 2021. "Appetite regulation by plant-derived bioactive peptides for promoting health." *Peptides* 144:170608. doi: 10.1016/j.peptides.2021.170608.

Kartik, Ashokkumar, Dilipkumar Akhil, Divya Lakshmi, Kannappan Panchamoorthy Gopinath, Jayaseelan Arun, Ramachandran Sivaramakrishnan, and Arivalagan Pugazhendhi. 2021. "A critical review on production of biopolymers from algae biomass and their applications." *Bioresour Technol* 329:124868. https://doi.org/10.1016/j.biortech.2021.124868.

Kebsa, W., M. Lahouel, H. Rouibah, M. Zihlif, M. Ahram, B. Abu-Irmaileh, E. Mustafa, H. J. Al-Ameer, and M. Al Shhab. 2018. "Reversing multidrug resistance in chemo-resistant human lung adenocarcinoma (A549/DOX) cells by algerian propolis through direct inhibiting the P-gp Efflux-pump, G0/G1 cell cycle arrest and apoptosis induction." *Anticancer Agents Med Chem* 18 (9):1330–1337. doi: 10.2174/1871520618666180808100800.

Keizman, D., M. Frenkel, A. Peer, I. Kushnir, E. Rosenbaum, D. Sarid, I. Leibovitch, R. Mano, O. Yossepowitch, D. Margel, I. Wolf, R. Geva, H. Dresler, K. Rouvinov, N. Rapoport, and I. Eliaz. 2021. "Modified citrus pectin treatment in non-metastatic biochemically relapsed prostate cancer: Results of a prospective phase II study." *Nutrients* 13 (12). doi: 10.3390/nu13124295.

Khotimchenko, M. 2020. "Pectin polymers for colon-targeted antitumor drug delivery." *Int J Biol Macromol* 158:1110–1124. https://doi.org/10.1016/j.ijbiomac.2020.05.002.

Kim, I. H., M. J. Kwon, and T. J. Nam. 2017. "Differences in cell death and cell cycle following fucoidan treatment in high-density HT-29 colon cancer cells." *Mol Med Rep* 15 (6):4116–4122. doi: 10.3892/mmr.2017.6520.

Kim, K. J., O. H. Lee, H. H. Lee, and B. Y. Lee. 2010. "A 4-week repeated oral dose toxicity study of fucoidan from the Sporophyll of Undaria pinnatifida in Sprague-Dawley rats." *Toxicology* 267 (1–3):154–158. doi: 10.1016/j.tox.2009.11.007.

Klemm, Dieter, B. Philpp, Thomas Heinze, Ute Heinze, and Wolfgang Wagenknecht. 1998. *Comprehensive cellulose chemistry. Volume 1: Fundamentals and analytical methods.* Wiley-VCH Verlag GmbH.

Koyanagi, S., N. Tanigawa, H. Nakagawa, S. Soeda, and H. Shimeno. 2003. "Oversulfation of fucoidan enhances its anti-angiogenic and antitumor activities." *Biochem Pharmacol* 65 (2):173–179. doi: 10.1016/s0006-2952(02)01478-8.

Kumari, Ruby, S. S. Narvi, and P. K. Dutta. 2021. "Thiol modified chitosan-silica nanohybrid for antibacterial, antioxidant and drug delivery application." *Journal of the Indian Chemical Society* 98 (8):100108. doi: https://doi.org/10.1016/j.jics.2021.100108.

Kylin, Harald. 1913. "Zur Biochemie der Meeresalgen." *Hoppe Seylers Z Physiol Chem* 83 (3):171–197.

Lee, H., J. S. Kim, and E. Kim. 2012. "Fucoidan from seaweed Fucus vesiculosus inhibits migration and invasion of human lung cancer cell via PI3K-Akt-mTOR pathways." *PLoS One* 7 (11):e50624. doi: 10.1371/journal.pone.0050624.

Leo, Eliana, Maria Angela Vandelli, Riccardo Cameroni, and Flavio Forni. 1997. "Doxorubicin-loaded gelatin nanoparticles stabilized by glutaraldehyde: Involvement of the drug in the cross-linking process." *International journal of Pharmaceutics* 155 (1):75–82.

Lestari, M. L., and G. Indrayanto. 2014. "Curcumin." *Profiles Drug Subst Excip Relat Methodol* 39:113–204. doi: 10.1016/b978-0-12-800173-8.00003-9.

Li, D. Q., J. Li, H. L. Dong, X. Li, J. Q. Zhang, S. Ramaswamy, and F. Xu. 2021. "Pectin in biomedical and drug delivery applications: A review." *Int J Biol Macromol* 185:49–65. doi: 10.1016/j.ijbiomac.2021.06.088.

Li, J., K. Chen, S. Li, T. Liu, F. Wang, Y. Xia, J. Lu, Y. Zhou, and C. Guo. 2016. "Pretreatment with fucoidan from fucus vesiculosus protected against ConA-induced acute liver injury by inhibiting both intrinsic and extrinsic apoptosis." *PLoS One* 11 (4):e0152570. doi: 10.1371/journal.pone.0152570.

Lipnizki, Frank. 2017. "4.10 Basic aspects and applications of membrane processes in agro-food and bulk biotech industries." In *Comprehensive Membrane Science and Engineering*, edited by Enrico Drioli, Lidietta Giorno and Enrica Fontananova, 2nd ed., 189–213. Elsevier.

Liu, Dasong, Mehdi Nikoo, Gökhan Boran, Peng Zhou, and Joe M Regenstein. 2015. "Collagen and gelatin." *Annu Rev Food Sci Technol* 6:527–557.

Liu, Kegang, Xiaohua Jiang, and Patrick Hunziker. 2016. "Carbohydrate-based amphiphilic nano delivery systems for cancer therapy." *Nanoscale* 8 (36):16091–16156.

Low, Liang Ee, Loh Teng-Hern Tan, Bey-Hing Goh, Beng Ti Tey, Boon Hoong Ong, and Siah Ying Tang. 2019. "Magnetic cellulose nanocrystal stabilized Pickering emulsions for enhanced bioactive release and human colon cancer therapy." *Int J Biol Macromol* 127:76–84.

Machałowski, Tomasz, Marcin Wysokowski, Mikhail V Tsurkan, Roberta Galli, Christian Schimpf, David Rafaja, Erica Brendler, Christine Viehweger, Sonia Żółtowska-Aksamitowska, and Iaroslav Petrenko. 2019. "Spider chitin: An ultrafast microwave-assisted method for chitin isolation from Caribena versicolor spider molt cuticle." *Molecules* 24 (20):3736.

Mansouri, M. Boroujeni, S. M. Barzi, M. Zafari, M. Chiani, M. Chehrazi, H. Nosrati, M. S. Shams Nosrati, S. Nayyeri, M. Khodaei, S. Bonakdar, and M. Shafiei. 2022. "Electrosprayed cefazolin-loaded niosomes onto electrospun chitosan nanofibrous membrane for wound healing applications." *J Biomed Mater Res B Appl Biomater* 110 (8):1814–1826. doi: 10.1002/jbm.b.35039.

Mathew, L., M. Burney, A. Gaikwad, P. Nyshadham, E. K. Nugent, A. Gonzalez, and J. A. Smith. 2017. "Preclinical evaluation of safety of fucoidan extracts from Undaria pinnatifida and fucus vesiculosus for use in cancer treatment." *Integr Cancer Ther* 16 (4):572–584. doi: 10.1177/1534735416680744.

Mendes, J. F., R. T. Paschoalin, V. B. Carmona, Alfredo R. Sena Neto, A. C. P. Marques, J. M. Marconcini, L. H. C. Mattoso, E. S. Medeiros, and J. E. Oliveira. 2016. "Biodegradable polymer blends based on corn starch and thermoplastic chitosan processed by extrusion." *Carbohydr Polym* 137:452–458.

Meng, Q., Y. Sun, H. Cong, H. Hu, and F. J. Xu. 2021. "An overview of chitosan and its application in infectious diseases." *Drug Deliv Transl Res* 11 (4):1340–1351. doi: 10.1007/s13346-021-00913-w.

Mišurcová, Ladislava, Jana Orsavová, and Jarmila Vávra Ambrožová. 2015. "Algal Polysaccharides and Health." In *Polysaccharides: Bioactivity and Biotechnology*, edited by Kishan Gopal Ramawat and Jean-Michel Mérillon, 109–144. Cham: Springer International Publishing.

Morin-Crini, Nadia, Eric Lichtfouse, Giangiacomo Torri, and Grégorio Crini. 2019. "Applications of chitosan in food, pharmaceuticals, medicine, cosmetics, agriculture, textiles, pulp and paper, biotechnology, and environmental chemistry." *Environmental Chemistry Letters* 17 (4):1667–1692. 10.1007/s10311-019-00904-x.

Mustafa, S., J. S. Pawar, and I. Ghosh. 2021. "Fucoidan induces ROS-dependent epigenetic modulation in cervical cancer HeLa cell." *Int J Biol Macromol* 181:180–192. doi: 10.1016/j.ijbiomac.2021.03.110.

Nam, Kyung-Soo, and Yun-Hee Shon. 2009. "Suppression of metastasis of human breast cancer cells by chitosan oligosaccharides." *J Microbiol Biotechnol* 19 (6):629–633.

Nandgude, Tanaji, and Roshani Pagar. 2021. "Plausible role of chitosan in drug and gene delivery against resistant breast cancer cells." *Carbohydr Res* 506:108357.

Nani, B. D., M. Franchin, J. G. Lazarini, I. A. Freires, M. G. da Cunha, B. Bueno-Silva, S. M. de Alencar, R. M. Murata, and P. L. Rosalen. 2018. "Isoflavonoids from Brazilian red propolis down-regulate the expression of cancer-related target proteins: A pharmacogenomic analysis." *Phytother Res* 32 (4):750–754. doi: 10.1002/ptr.6016.

Naz, S., M. Imran, A. Rauf, I. E. Orhan, M. A. Shariati, Haq Iahtisham Ul, IqraYasmin, M. Shahbaz, T. B. Qaisrani, Z. A. Shah, S. Plygun, and M. Heydari. 2019. "Chrysin: Pharmacological and therapeutic properties." *Life Sci* 235:116797. doi: 10.1016/j.lfs.2019.116797.

Ngwuluka, Ndidi C. 2018. "19 – Responsive polysaccharides and polysaccharides-based nanoparticles for drug delivery." In *Stimuli Responsive Polymeric Nanocarriers for Drug Delivery Applications, Volume 1*, edited by Abdel Salam Hamdy Makhlouf and Nedal Y. Abu-Thabit, 531–554. Woodhead Publishing.

Onnainty, Renée, Barbara Onida, Paulina Páez, Marcela Longhi, Antonello Barresi, and Gladys Granero. 2016. "Targeted chitosan-based bionanocomposites for controlled oral mucosal delivery of chlorhexidine." *Int J Pharm* 509 (1–2):408–418.

Orasugh, Jonathan Tersur, Nayan Ranjan Saha, Dipak Rana, Gunjan Sarkar, Md Masud Rahaman Mollick, Atiskumar Chattoapadhyay, Bhairab Chandra Mitra, Dibyendu Mondal, Swapan Kumar Ghosh, and Dipankar Chattopadhyay. 2018. "Jute cellulose nano-fibrils/hydroxypropylmethylcellulose nanocomposite: A novel material with potential for application in packaging and transdermal drug delivery system." *Industrial Crops and Products* 112:633–643.

Palko-Łabuz, A., J. Maksymowicz, B. Sobieszczańska, A. Wikiera, M. Skonieczna, O. Wesołowska, and K. Środa-Pomianek. 2021. "Newly obtained apple pectin as an adjunct to irinotecan therapy of colorectal cancer reducing E. coli Adherence and β-Glucuronidase activity." *Cancers (Basel)* 13 (12). doi: 10.3390/cancers13122952.

Panonnummal, Rajitha, R. Jayakumar, and M. Sabitha. 2017. "Comparative anti-psoriatic efficacy studies of clobetasol loaded chitin nanogel and marketed cream." *Eur J Pharm Sci* 96:193–206.

Park, H. Y., S. H. Park, J. W. Jeong, D. Yoon, M. H. Han, D. S. Lee, G. Choi, M. J. Yim, J. M. Lee, D. H. Kim, G. Y. Kim, I. W. Choi, S. Kim, H. S. Kim, H. J. Cha, and Y. H. Choi. 2017. "Induction of p53-Independent Apoptosis and G1 cell cycle arrest by fucoidan in HCT116 human colorectal carcinoma cells." *Mar Drugs* 15 (6). doi: 10.3390/md15060154.

Park, Hye Young, Il-Whan Choi, Gi-Young Kim, Byung Woo Kim, Wun-Jae Kim, and Yung Hyun Choi. 2015. "Fucoidan induces G1 arrest of the cell cycle in EJ human bladder cancer cells through down-regulation of pRB phosphorylation." *Revista Brasileira de Farmacognosia* 25 (3):246–251. https://doi.org/10.1016/j.bjp.2015.03.011.

Picot-Allain, Carene, and Vidushi Shradha Neergheen. 2023. "Pectin a multifaceted biopolymer in the management of cancer: A review." *Heliyon* 9 (11):e22236.

Piredda, M., G. Facchinetti, V. Biagioli, D. Giannarelli, G. Armento, G. Tonini, and M. G. De Marinis. 2017. "Propolis in the prevention of oral mucositis in breast cancer patients receiving adjuvant chemotherapy: A pilot randomised controlled trial." *Eur J Cancer Care (Engl)* 26 (6). doi: 10.1111/ecc.12757.

Queiroz, V. M., I. C. S. Kling, A. E. Eltom, B. S. Archanjo, M. Prado, and R. A. Simão. 2020. "Corn starch films as a long-term drug delivery system for chlorhexidine gluconate." *Mater Sci Eng C Mater Biol Appl* 112:110852. doi: 10.1016/j.msec.2020.110852.

Roshan, S., Y. Y. Liu, A. Banafa, H. J. Chen, K. X. Li, G. X. Yang, G. Y. He, and M. J. Chen. 2014. "Fucoidan induces apoptosis of HepG2 cells by down-regulating p-Stat3." *J Huazhong Univ Sci Technolog Med Sci* 34 (3):330–336. doi: 10.1007/s11596-014-1278-0.

Scaffaro, R., F. Lopresti, A. Marino, and A. Nostro. 2018. "Antimicrobial additives for poly(lactic acid) materials and their applications: Current state and perspectives." *Appl Microbiol Biotechnol* 102 (18):7739–7756. doi: 10.1007/s00253-018-9220-1.

Seferian, Peter G., and Mitzi L. Martinez. 2000. "Immune stimulating activity of two new chitosan containing adjuvant formulations." *Vaccine* 19 (6):661–668.

Senthilkumar, K., P. Manivasagan, J. Venkatesan, and S. K. Kim. 2013. "Brown seaweed fucoidan: biological activity and apoptosis, growth signaling mechanism in cancer." *Int J Biol Macromol* 60:366–374. doi: 10.1016/j.ijbiomac.2013.06.030.

Sioud, M., and A. Mobergslien. 2012. "Selective killing of cancer cells by peptide-targeted delivery of an anti-microbial peptide." *Biochem Pharmacol* 84 (9):1123–1132. doi: 10.1016/j.bcp.2012.08.002.

Sirivat, Anuvat, and Nophawan Paradee. 2019. "Facile synthesis of gelatin-coated Fe3O4 nanoparticle: Effect of pH in single-step co-precipitation for cancer drug loading." *Mater Des* 181:107942.

Sung, H., J. Ferlay, R. L. Siegel, M. Laversanne, I. Soerjomataram, A. Jemal, and F. Bray. 2021. "Global cancer statistics 2020: GLOBOCAN estimates of incidence and mortality worldwide for 36 cancers in 185 countries." *CA: A Cancer Journal for Clinicians, 71*(3), 209–249. https://doi.org/10.3322/caac.21660

Tan, Marselina Irasonia, and Irham Hayati. 2017. "Inhibition of mammary gland cancer development by propolis and mangostin in female mice Balb/C." *J Math Fundam Sci* 49:40–50.

Teruya, T., H. Tatemoto, T. Konishi, and M. Tako. 2009. "Structural characteristics and in vitro macrophage activation of acetyl fucoidan from Cladosiphon okamuranus." *Glycoconj J* 26 (8):1019–1028. doi: 10.1007/s10719-008-9221-x.

Tian, Bingren, Yumei Liu, and Jiayue Liu. 2021. "Chitosan-based nanoscale and non-nanoscale delivery systems for anticancer drugs: A review." *Eur Polym J* 154:110533.

Toshio, Yoshioka, Hashida Mitsuru, Muranishi Shozo, and Sezaki Hitoshi. 1981. "Specific delivery of mitomycin c to the liver, spleen and lung: Nano-and m1crospherical carriers of gelatin." *Int J Pharm* 8 (2):131–141.

Tsai, H. L., C. J. Tai, C. W. Huang, F. R. Chang, and J. Y. Wang. 2017. "Efficacy of low-molecular-weight fucoidan as a supplemental therapy in metastatic colorectal cancer patients: A double-blind randomized controlled trial." *Mar Drugs* 15 (4). doi: 10.3390/md15040122.

Usman, Adil, Sundas Khalid, Atif Usman, Zakir Hussain, and Yanmei Wang. 2017. "Chapter 5 - Algal Polysaccharides, Novel Application, and Outlook." In *Algae Based Polymers, Blends, and Composites*, edited by Khalid Mahmood Zia, Mohammad Zuber, and Muhammad Ali, 115–153. Elsevier.

Wang, H., K. Zhang, J. Liu, J. Yang, Y. Tian, C. Yang, Y. Li, M. Shao, W. Su, and N. Song. 2021. "Curcumin regulates cancer progression: Focus on ncRNAs and molecular signaling pathways." *Front Oncol* 11:660712. doi: 10.3389/fonc.2021.660712.

Wang, W., J. Wu, X. Zhang, C. Hao, X. Zhao, G. Jiao, X. Shan, W. Tai, and G. Yu. 2017. "Inhibition of influenza a virus infection by fucoidan targeting viral neuraminidase and cellular EGFR pathway." *Sci Rep* 7:40760. doi: 10.1038/srep40760.

Wei, S., P. Xu, Z. Yao, X. Cui, X. Lei, L. Li, Y. Dong, W. Zhu, R. Guo, and B. Cheng. 2021. "A composite hydrogel with co-delivery of antimicrobial peptides and platelet-rich plasma to enhance healing of infected wounds in diabetes." *Acta Biomater* 124:205–218. doi: 10.1016/j.actbio.2021.01.046.

Widhiantara, I. G., and I. M. Jawi. 2021. "Phytochemical composition and health properties of Sembung plant (Blumea balsamifera): A review." *Vet World* 14 (5):1185–1196. doi: 10.14202/vetworld.2021.1185–1196.

Wong, T. W., G. Colombo, and F. Sonvico. 2011. "Pectin matrix as oral drug delivery vehicle for colon cancer treatment." *AAPS PharmSciTech* 12 (1):201–14. doi: 10.1208/s12249-010-9564-z.

Xu, Yongbin, Yaxin Zi, Jinfeng Lei, Xingyue Mo, Zilun Shao, Yuyao Wu, Ye Tian, Defu Li, and Changdao Mu. 2020. "pH-Responsive nanoparticles based on cholesterol/imidazole modified oxidized-starch for targeted anticancer drug delivery." *Carbohydr Polym* 233:115858.

Yang, G., Q. Zhang, Y. Kong, B. Xie, M. Gao, Y. Tao, H. Xu, F. Zhan, B. Dai, J. Shi, and X. Wu. 2015. "Antitumor activity of fucoidan against diffuse large B cell lymphoma in vitro and in vivo." *Acta Biochim Biophys Sin (Shanghai)* 47 (11):925–931. doi: 10.1093/abbs/gmv094.

Yao, Y., Y. Zhou, L. Liu, Y. Xu, Q. Chen, Y. Wang, S. Wu, Y. Deng, J. Zhang, and A. Shao. 2020. "Nanoparticle-based drug delivery in cancer therapy and its role in overcoming drug resistance." *Front Mol Biosci* 7:193. doi: 10.3389/fmolb.2020.00193.

Yazdi Rouholamini, Seyede Elmira, Saeid Moghassemi, Zahra Maharat, Amirhossien Hakamivala, Susan Kashanian, and Kobra Omidfar. 2018. "Effect of silibinin-loaded nano-niosomal coated with trimethyl chitosan on miRNAs expression in 2D and 3D models of T47D breast cancer cell line." *Artif Cells Nanomed Biotechnol* 46 (3):524–535.

Yoshimoto, M., K. Higaki, E. Nanba, and M. Ikeguchi. 2015. "Anti-proliferation activity of fucoidan in MKN45 gastric cancer cells and downregulation of phosphorylated ASK1, a cell cycle-regulated kinase." *Yonago Acta Med* 58 (1):1–7.

Zabaiou, N., A. Fouache, A. Trousson, S. Baron, A. Zellagui, M. Lahouel, and J. A. Lobaccaro. 2017. "Biological properties of propolis extracts: Something new from an ancient product." *Chem Phys Lipids* 207 (Pt B):214–222. doi: 10.1016/j.chemphyslip.2017.04.005.

Zhang, W., J. Y. Du, Z. Jiang, T. Okimura, T. Oda, Q. Yu, and J. O. Jin. 2014. "Ascophyllan purified from Ascophyllum nodosum induces Th1 and Tc1 immune responses by promoting dendritic cell maturation." *Mar Drugs* 12 (7):4148–64. doi: 10.3390/md12074148.

Zhang, W., T. Oda, Q. Yu, and J. O. Jin. 2015. "Fucoidan from Macrocystis pyrifera has powerful immune-modulatory effects compared to three other fucoidans." *Mar Drugs* 13 (3):1084–104. doi: 10.3390/md13031084.

Zhang, Z., K. Teruya, H. Eto, and S. Shirahata. 2011. "Fucoidan extract induces apoptosis in MCF-7 cells via a mechanism involving the ROS-dependent JNK activation and mitochondria-mediated pathways." *PLoS One* 6 (11):e27441. doi: 10.1371/journal.pone.0027441.

Zhao, M. and M. Liu. 2018. "New avenues for nanoparticle-related therapies." *Nanoscale Res Lett* 13 (136). https://doi.org/10.1186/s11671-018-2548-8

Zhao, Y., M. Jayachandran, and B. Xu. 2020. "In vivo antioxidant and anti-inflammatory effects of soluble dietary fiber Konjac glucomannan in type-2 diabetic rats." *Int J Biol Macromol* 159:1186–1196. doi: 10.1016/j.ijbiomac.2020.05.105.

Zhao, Yuanyuan, Jinfeng Bi, Jianyong Yi, Jian Peng, and Qiancheng Ma. 2022. "Dose-dependent effects of apple pectin on alleviating high fat-induced obesity modulated by gut microbiota and SCFAs." *Food Sci Hum Well* 11 (1):143–154. https://doi.org/10.1016/j.fshw.2021.07.015.

Zhu, Jianzhong, Li Zhong, Wenxue Chen, Yizhe Song, Zhengming Qian, Xianying Cao, Qiang Huang, Bin Zhang, Haiming Chen, and Weijun Chen. 2019. "Preparation and characterization of pectin/chitosan beads containing porous starch embedded with doxorubicin hydrochloride: A novel and simple colon targeted drug delivery system." *Food Hydrocolloids* 95:562–570.

Zingue, S., S. Maxeiner, J. Rutz, D. T. Ndinteh, F. K. Chun, F. T. Fohouo, D. Njamen, and R. A. Blaheta. 2020. "Ethanol-extracted Cameroonian propolis: Antiproliferative effects and potential mechanism of action in prostate cancer." *Andrologia* 52 (9):e13698. doi: 10.1111/and.13698.

10 Immunomodulation by Dietary Interventions

Bettadahalli Sadashivaiah[1], *R. Chandrakanth*[1], *and L. Sunil*[2]

[1]Department of Molecular Biology, Yuvaraja's College, University of Mysore, Mysuru, India
[2]Department of Pharmacological & Pharmaceutical Sciences, College of Pharmacy, University of Houston, Houston, TX, U.S.A.

10.1 INTRODUCTION

Living organisms are constantly exposed to pathogenic microorganisms from the external environment, but as they are inherently equipped with defense mechanisms to deal with such factors, they can overcome the problems and maintain homeostatic conditions. Immunity is typically defined as the opposition offered by an organism against invading pathogenic microbes or altered self-cells. However, impairment in the natural defense mechanism in an organism may result in the onset of various diseases. The immune system has two types of immunity, namely, innate and adaptive. The innate immune system is composed of more generalized physical and chemical barriers comprising skin, mucosal tissue, and cells like neutrophils, macrophages, and natural killer (NK) cells. The innate response is non-specific and offers an instant immune response. The innate immune system is supplemented with a highly specific adaptive immunity that mounts a pathogen-specific response by employing T lymphocytes and antibody-producing B-lymphocytes. Various cytokines and chemokines secreted by different immune effector cells play an important role in interconnecting innate and adaptive immune systems and thus enable the host to impart a robust and highly effective immune response against a plethora of pathogens and altered cells in the body. The immune system of an individual depends on various factors including diet. "Let food be thy medicine and medicine be thy food" is a much thought-provoking statement made by the great scientist Hippocrates, which emphasizes the importance of consuming a balanced diet.

An adequate and nutritious diet is known to have a profound impact on the immune system of an organism. Malnutrition, with micronutrient deficiency, and hypernutrition, such as overweight and obesity, are found to alter the immune response with an increased presence of inflammatory markers in systemic circulation (Cunningham

 DOI: 10.1201/9781003514244-13

et al., 2005). Nutraceuticals and functional foods are gaining significant attention concerning their potential to modulate the immune response and thereby the health and well-being of an organism. These foods regulate the functioning of the immune response at different stages, ranging from nonspecific innate to more specific adaptive responses. Different functional foods are reported to exert varied outcomes under different pathophysiological conditions, with some showing stimulatory activity and others having suppressive activity towards the immune system. Thus, they can be classified as immunostimulators and immunosuppressors based on their effects on the immune response. Preventive medicine and prophylactic approaches are gaining significance over therapeutic interventions, suggesting that exploring the immunomodulatory potentials of different functional foods for better health is the right choice. This chapter discusses the immunomodulatory properties of some important functional foods and nutraceuticals.

10.2 FLAVONOIDS

The positive effect of a plant-based diet on human health has been established for several centuries. Several epidemiological studies have established a firm foundation for the therapeutic benefits of a diet rich in fruits and vegetables compared to animal-based diets (Bruckdorfer, 2008). Some studies have reported that a Mediterranean-type diet with reduced amounts of saturated fats results in fewer mortalities from different diseases (Sofi et al., 2008). A large body of evidence has reported that people who consume a diet rich in phytochemicals are less prone to many diseases and infections compared to those whose diet is deficient in or devoid of phytonutrients.

Flavonoids are hydroxylated polyphenols having two or more aromatic rings bridged by a pyran moiety with one or more aromatic hydroxyl groups attached (Ullah et al., 2020). They have been consumed from different dietary sources for many centuries. They are primarily produced by different plants as defensive secondary metabolites against a wide variety of biotic and abiotic stress. Fruits and vegetables including plums, cherries, apples, beans, olives, onions, spinach, cocoa, red wine, and black and green tea are rich sources of flavonoids (Kopustinskiene et al., 2020). They are found either in conjugation with a glycan moiety or devoid of glycan (Lila, 2004). Based on differences in structures and the oxidation state of the central pyran ring, flavonoids have been classified into eight sub-groups, namely, flavones, flavonols, isoflavones, flavan-3-ols, flavanonols, anthocyanidins, chalcones, and flavanones, comprising more than 4000 chemical moieties (Madunic et al., 2018). Additionally, a new subgroup called aurones has been reported, which is present in wide variety of sources (Chirumbolo, 2010). Flavonoids have been reported to exhibit several health benefits against a large number of diseases through their inherent pharmacological properties such as antioxidant, anti-inflammatory, antimicrobial, anticancer, and neuroprotective properties, and many more (Rodríguez-García et al., 2019; Kopustinskiene et al., 2020). In addition, they have been shown to act as potent immunomodulatory compounds in in vitro and in vivo studies (Yahfoufi et al., 2018).

10.2.1 Flavonoids as Immunomodulators in In Vitro and Preclinical Models

Monocytes, macrophages, neutrophils, mast cells, and NK cells are some of the major types of cellular machinery that participate in the innate immune response as the first line of defense (Kumagai and Akira, 2010). The functioning of immune cells is critically regulated in homeostatic conditions, and deregulation in their function results in altered immune responses, thus rendering the host susceptible to different pathological conditions. For instance, the degranulation of mast cells and the release of histamine is the primary reason for causing allergic reactions. Several flavonoids have been reported to prevent the degranulation of mast cells and basophils, and thus suppress allergic reactions that otherwise may lead to severe conditions and, at times, to fatality if not regulated properly (Kabashima et al., 2018). Apigenin, fisetin, kaempferol, luteolin, myricetin, and quercetin are some of the flavonoids that exhibit antiallergic functions (Matsuda et al., 2016). The flavones (apigenin, fisetin, and luteolin), flavonols (kaempferol and quercetin), and the chalcone (isoliquiritigenin) have been shown to abrogate the release of TNF-α and IL-4 from mast cells and basophils of allergic organisms (Tanaka, 2014; Hagenlocher and Lorentz, 2015). Fisetin and quercetin are reported to inhibit the kinase activity of ERK and JNK and block the transcription activity of NF-κB, thereby down-regulating the expression of genes involved in allergic reactions (Che et al., 2018).

Neutrophils get activated by recognizing the receptors associated with pathogenic microbes and cause the secretion of an array of defensive chemicals. Reactive oxygen species (ROS), reactive nitrogen species (RNS), cytokines, chemokines, proteases, and other mediators of inflammation like neutrophil elastase are some of the important effector molecules that eliminate the pathogen and establish homeostasis. However, the pathogen elimination process might cause tissue injury if it is not regulated optimally (Hellebrekers et al., 2018). Many groups have reported the regulatory role of phytochemicals and flavonoids on neutrophil activation and secretion of inflammatory mediators (Ribeiro et al., 2015). For instance, quercetin inhibited superoxide anion production in stimulated neutrophils. Quercetin, morin, fisetin, and kaempferol were reported to inhibit NADPH oxidase. Further, quercetin and 2-epicatechin were also shown to inhibit the expression of NADPH and thereby decreased the production of ROS and RNS. Additionally, bioflavonoids like procyanidin, fukugetin, amentoflavone, and podocarpus flavone decreased ROS production in phorbol myristate acetate (PMA)-stimulated neutrophils (Saroni et al., 2015). Myeloperoxidase (MPO), a key defensive enzyme of neutrophils and macrophages, was shown to be inhibited by epigallocatechin gallate (EGCG), quercetin, and myricetin (Tian et al., 2017). Flavonoids like apigenin, luteolin, and wogonin were also shown to promote apoptosis in neutrophils through the activation of caspases (Lucas et al., 2013). Thus, several reports have been published highlighting the beneficial role of flavonoids on neutrophil-mediated activities in immune response.

Macrophages are mononuclear phagocytes widely distributed across the body and are involved in the recognition, processing, and elimination of pathogens through their phagocytic activity. They are key players in frontline defense involving both innate and adaptive immune responses (Collin and Bigley, 2016). Tissue macrophages are derived from resident tissue-specific macrophage colony-forming unit cells (Gordon and Taylor, 2005). The circulatory monocytes differentiate into specific macrophages upon their entry into different tissues from the blood and take part in the inflammatory response. Some flavonoids have been reported to modulate the functions of monocytes and macrophages by suppressing the secretion of pro-inflammatory mediators such as MCP-1, TNF-α, and IL-6, down-regulating TLR4 expression, inhibiting the transcription of iNOS and cyclooxygenase-2 (COX-2) via NF-κB inhibition, preventing the activation of the NLRP3 inflammasome complex, etc. (Rahimifard et al., 2017; Jia et al., 2015; Kim et al., 2017).

The efficacy of the immune response is also mediated by another important group of cells called dendritic cells (DCs). These cells play a key role in antigen presentation to T lymphocytes. They stimulate T lymphocytes and MHC-II by expressing co-stimulatory receptors such as CD80 and CD86 (Dress et al., 2018). They also secrete IL-12, which further facilitates the proliferation and differentiation of T lymphocytes (Clark et al., 2019). Many phytochemicals, including flavonoids, have been proven to modulate the activities of DCs like differentiation, maturation, and antigen presentation and, in turn, the innate immune response in preclinical models (Del Cornò et al., 2016). The effect of daidzein, an isoflavone, on the maturation and activation of dendritic cells was examined in lipopolysaccharide (LPS)-stimulated DCs. Daidzein suppressed the expression of pro-inflammatory cytokines like IL-12, IL-6, and TNF-α (Yum et al., 2011). Further, fisetin was shown to decrease the expression of IL-12, IL-6, TNF-α, CD80, and CD86, along with lowering the endocytic activity of dendritic cells upon LPS-induction (Liu et al., 2010). Fisetin also abrogated LPS-driven NF-κB activation to suppress phagocytic activity in macrophages (Liu et al., 2010). Similarly, quercetin and apigenin have also been reported to reduce the expression of important receptors and some inflammatory mediators. Luteolin inhibited the nuclear translocation of NF-κB and thus down-regulated the genes involved in immune function in bone marrow-derived DCs (Kim et al., 2005). Also, in line with many previously reported findings, the flavone tangeretin was found to decrease the expression of inflammatory cytokines, compromise the expression of key receptors, and block the activation of NF-κB in LPS-stimulated bone marrow-derived DCs (Eun et al., 2017). Further, many flavonoids have been reported to exhibit anti-inflammatory activity by switching the macrophage phenotype from M1 to M2, thus helping to resolve the inflammatory effects (Feng et al., 2014). Various types of T lymphocytes (comprising cytotoxic, helper, and regulatory cells) and B-lymphocytes are the major components of adaptive immunity, where the former imparts a cell-mediated immune response and the latter offers an antibody-mediated immune response in eliminating the threat and thus restoring the normalcy of the system (Childs et al., 2019). Many phytochemicals, including flavonoids, are known to modulate the activities of these cells and thus the immune response.

Naive helper T lymphocytes get differentiated into different subsets such as Th1, Th2, and Th17, which are actively involved in immune responses (Kumar et al., 2018). Many flavonoids have been reported to modulate the differentiation of naive Th cells into specific subsets and thus regulate the immune response. Chrysin, quercetin, astilbin, and naringenin are known to modulate the Th cell activation and differentiation (Choi et al., 2017; Wei et al., 2017). Many flavonoids inhibit the proliferation and differentiation of Th cells through the suppression of the PI3K/Akt/mTOR axis (Granato et al., 2017). Regulatory T cells (Tregs) play a decisive role in autoimmune diseases. Alteration in their number or activity results in a deregulated immune response. Several flavonoids have been proven to positively modulate Tregs activity (Xu et al., 2019). Antibody types and titer values decide their efficacy in combating different types of pathogens. Flavonoids like rutin, catechin, and quercetin have been reported to enhance antibody titers (Ganeshpurkar and Saluja, 2017; Alam et al., 2017). However, some reports indicate the suppressive effects of flavonoids on antibody-dependent immune functions. Further, flavonoids have also been implicated in the therapeutic regimen against autoimmune diseases (Rengasamy et al., 2019). Apigenin, quercetin, and naringenin are reported to exhibit anti-inflammatory and immune modulatory properties by enhancing the activity of NK cells and T lymphocytes (Kilani-Jaziri et al., 2016). Modulation of the activity of NF-κB, activation protein-1, and mitogen-activated protein kinase (MAPK) are the common mechanisms exhibited by many flavonoids to impart their immunomodulatory activities (Rahman et al., 2006; Serafini et al., 2010; Kuo et al., 2014; Haghmorad et al., 2017). Flavonoids like quercetin were reported to prevent necroptosis of oligodendrocytes by inhibition of macrophage or microglia polarization from M2 to M1 phenotype through the inhibition of STAT1 and NF-κB pathways in in vitro and in vivo rat models (Fan et al., 2019). A large body of evidence has reported the positive immune modulation of flavonoids by suppressing the secretion of pro-inflammatory cytokines and alleviating anti-inflammatory mediators through inhibition of the NLRP3 inflammasome, NF-κB, MAPK, JAK/STAT pathways, and activating Nrf2, AMPK, and PI3k/Akt signals (Chen et al., 2022). Flavonoids like cryptotanshinone (CPT) and naringenin were shown to shift the immune response towards Th1 and alter the T regulatory cells via STAT3 pathways in breast cancer treatment (Noori et al., 2022). Naringenin, a citrus fruit-derived flavonoid, is a potent modulator of immune response. It was shown to down-regulate T cell proliferation and T cell-derived cytokines like IFN-γ, IL-17, IL-6, and TNF-α. These T cell suppressive effects of naringenin were found to be operative via IL2- and IL2R-mediated pathways and STAT5 phosphorylation in T cells. This function of naringenin is being explored to be used as a preventive or therapeutic agent in many autoimmune diseases (Niu et al., 2018).

10.2.2 Flavonoids as Immunomodulators in Humans

A large number of studies have reported the health benefits and immunomodulatory properties of flavonoids and their derivatives in in vitro and in vivo rodent models. Based on preclinical findings, many studies were performed to explore the immunomodulatory effect of flavonoids in human subjects. The effect of a flavonoid-rich diet on immunomodulation was measured using the expression of cytokines (such as IL-6, IL-1β, and TNF-α), ex vivo immune function markers (including lymphocyte proliferation, macrophage phagocytosis), and cytotoxicity of NK cells. Flavonoid-rich foods like green tea, red wine, mixed fruit juices, dark chocolate, capsules filled with flavonoid concentrates, soy supplements, and quercetin in pure form are some of the prominent sources used in many human dietary studies.

The collective outcome of these studies was neither consistent nor conclusive, with few groups reporting positive implications, while others recorded negative changes, and even some documenting negligible or no significant changes, with respect to the parameters studied (Bub et al., 2003; Fanti et al., 2006; Heinz et al., 2010). Many of the studies on healthy people did not find any notable changes in the examined parameters, while varied results were observed in people with obesity, diabetes, and cardiovascular diseases (Egert et al., 2009). Although many in vitro studies have demonstrated the positive effect of flavonoids in immune modulation, their efficacy in human studies was relatively low. Some studies have reported differences in the rate of absorption, bioavailability, and plasma levels of different flavonoids when consumed crudely and in purified form.

The correlation between flavonoid consumption and observed effects was also not consistent. There was a substantial increase in the levels of plasma quercetin upon its supplementation (500 and 1000 mg/day for 12 weeks) in adult female subjects. However, there was no change in granulocyte respiratory burst activity, phagocytosis, serum levels of IL-6 and TNF-α, and circulating counts of B/T lymphocytes or NK cells (Heinz et al., 2010). Poor bioavailability, faster metabolic clearance, alteration in chemical structure, and activity upon metabolism in physiological fluids are some major limiting factors associated with flavonoid administration in human subjects. Comprehensive data reporting the effects of different flavonoid interventions (including quercetin and isoquercetin, EGCG, flavonoid-rich fruits and vegetables, soy isoflavones, tea catechins, wine, and cocoa) on immune modulations in humans have been documented (Peluso et al., 2015). The immunomodulatory properties of flavonoids have been depicted in Figure 10.1.

To summarize, the primary limitation in harnessing the health benefits of flavonoids is their poor absorption and bioavailability, along with faster metabolic clearance. This is further stressed by the lower plasma levels of pure flavonoids, which are far lower compared to the concentration used in in vitro systems. Thus, if attempts are made to address the above-mentioned problems, numerous benefits could be attained from these plant bioactives.

10.3 FATTY ACIDS

Apart from being energy sources, fatty acids as components of cell membranes are known to influence the structure and functions of cells. Some classes of fatty acids have been reported to exert different immunomodulatory properties. Among the various fatty acids, polyunsaturated fatty acids (particularly ω-3 [also called n-3] fatty acids) have demonstrated a broad range of health benefits against a plethora of illnesses. It has been extensively demonstrated that the immune response of an organism is also dependent on lipids, due to the presence of lipids in the phospholipid bilayer of the cell membrane, which serve as precursors for the synthesis of different lipid mediators, eicosanoids, secondary messengers, etc., and which have a profound influence on different aspects of

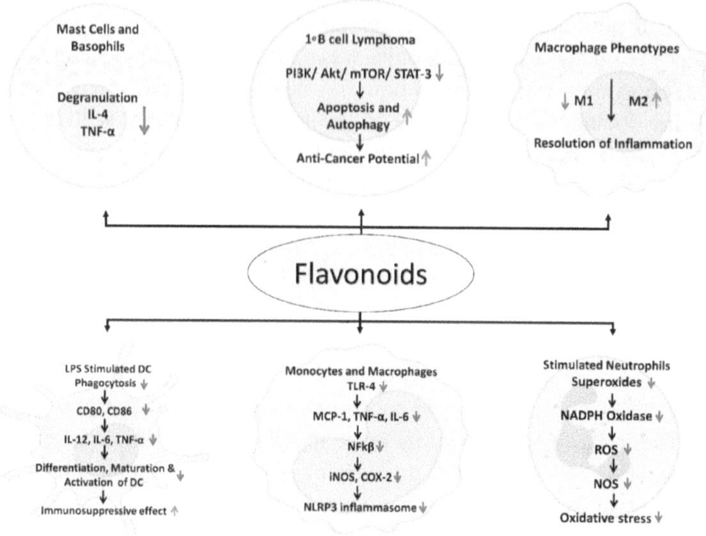

FIGURE 10.1 Schematic diagram representing regulatory effects of flavonoids on immune cells. LPS: lipopolysaccharides; CD80/86: cluster of differentiation 80/86; IL: interleukin; TLR: toll-like receptor; TNF-α: tumor necrosis factor-alpha; MCP-1: monocyte chemoattractant protein; iNOS: inducible nitric oxide synthase; COX-2: cyclooxygenase-2; NLRP-3: NOD-like receptor protein-3; ROS: reactive oxygen species; PI3K: phosphoinositide 3 kinase; mTOR: mammalian target of rapamycin; STAT-3: signal transducer and activator of transcription-3; NF-κB: nuclear factor kappa B; NADPH oxidase: nicotinamide adenine dinucleotide phosphate oxidase; M1: macrophage 1 phenotype; M2: macrophage 2 phenotype.

immune functions. The type and amount of fatty acids in the diet are known to alter the membrane phospholipids, specifically at the sn-2 position, and influence eicosanoid biosynthesis.

Immunomodulating properties of fatty acids are mediated by the change in membrane lipid composition and fluidity in the specialized regions of membranes called lipid rafts of immune cells and their associated downstream cell signaling pathways (Peck, 1995). Based on the effects of lipid mediators on the outcome of inflammation, fatty acids have been primarily classified as pro-inflammatory and anti-inflammatory. Saturated fatty acids and n-6 fatty acids belong to the former class, and n-3 fatty acids to the latter class.

10.3.1 Unsaturated Fatty Acids

Several reports have documented the effects of n-3 fatty acids on cytokine production in inflammatory diseases and demonstrated the role of n-3 fatty acids in immune modulation in health and disease (Calder, 2003). Most early studies demonstrated that fatty acids primarily mediate their effect on the immune system by altering membrane composition and fluidity, changing eicosanoid synthesis, and altering cytokine profiles. Further studies also revealed that fatty acids and their metabolites can alter gene expression patterns by binding to a class of nuclear receptors called peroxisomal proliferator-activated receptors (PPARs). Immune effector cells like myeloid-derived monocytes, macrophages, neutrophils, NK cells, and lymphocytes are reported to express PPARs, illustrating the modulatory potentials of fatty acids being exhibited through these receptors (Xu et al., 1999). N-6 fatty acid-derived eicosanoids like prostaglandins (PGE_2) and leukotrienes (LTA_4, LTB_4, and LTC_4) produced via the 5-lipoxygenase pathway function as powerful chemoattractants and vasodilators. They facilitate the infiltration of immune effector cells like neutrophils, monocytes, and macrophages to the site of infection or tissue injury, and thus act as potent pro-inflammatory mediators.

N-3 fatty acid-derived mediators like resolvins and neuroprotectins are reported to function as inhibitors of neutrophil and macrophage infiltration, and thus diminish immune response, thereby facilitating the resolution of inflammation (Serhan et al., 2000; Serhan, 2005). Cell adhesion molecules play an important role in the communication between antigen-presenting cells and T lymphocytes, which is necessary for the optimal priming of the adaptive immune system (Hogg and Landis, 1993). N-3 fatty acids, DHA in particular, have been shown to reduce the cytokine-stimulated expression of adhesion molecules (such as CD106) that are involved in the intercellular communication of immune effector cells (De Caterina et al., 1994). This feature is shown to be beneficial in many autoimmune disorders like type-I diabetes mellitus, rheumatoid arthritis, and systemic lupus erythematosus, where suppression of the adaptive immune response is known to ameliorate the complications of autoimmune disorders and thus establish homeostasis. Dietary DHA supplementation has been shown to down-regulate the expression of the nitric oxide synthase gene in macrophages stimulated with LPS and IFNγ (Khair-El-Din et al. 1996). Further, n-3 fatty acid supplementation in mice revealed that EPA and DHA were shown to suppress the proliferation of mitogen-primed T lymphocytes, which are accompanied by decreased secretion of IL-2 along with the reduction in vital lipid-based secondary messengers like diacylglycerol and ceramide, which is essential for cell signaling, further highlighting the anti-inflammatory properties of n-3 fatty acids in immune cells.

Unlike the Western diet, high in fat and sugar, and which is known to promote obesity accompanied by chronic pro-inflammatory situations, a diet enriched with omega-3 fatty acids is reported to tweak immune response towards an anti-inflammatory nature. This is accompanied by an increased prevalence of anti-inflammatory $CD4^+Foxp3^+$ and $CD4^+Foxp3^+CD25^+$ Tregs subtypes in lymphoid organs along with the decreased expression of ceramide synthase 2 (Camacho-Muñoz et al., 2022). In another study, omega-3 fatty acids were found to decrease the levels of effector memory $CD4^+$T cells in adipose tissue. They also changed the profiles of bioactive mediators, resulting in decreased chemotactic triggers and thus limiting the motility of $CD4^+$T cells in association with cytoskeletal elements, leading to decreased adaptive immune response and ameliorating the complications of cardiovascular metabolic disorders (Cucchi et al., 2020).

10.3.2 N-3 Fatty Acids in Human Clinical Trials

Although the benefits of n-3 fatty acids over n-6 and saturated fatty acids on various parameters of immune response and inflammation under different disease conditions are good in preclinical models, their effects in clinical studies are controversial. Results from the randomized clinical trials (RCT) on the effect of n-3 fatty acids on RA reported mixed responses. The majority of the studies presented negligible or no effect. Only a few of them reported beneficial effects, while one of the studies reported worsening of the RA symptoms upon n-3 supplementation (Kremer et al., 1990; Nielsen et al., 1992; Cleland et al., 1988). Another clinical study also indicated no beneficial effects of n-3 fatty acids in RA patients (Fortin et al., 1995). Similarly, the results of clinical trials on the effects of n-3 fatty acids on asthma were reported to contradict the results of preclinical models (Wong, 2005). These results were in agreement with the results of another review reporting no betterment in alleviating the symptoms of asthma with n-3 fatty acids supplementation (Schachter et al., 2004). The results from the Agency for Healthcare Research and Quality (AHRQ) differ from many other previously reported studies owing to some important factors like statistically underpowered numbers considered in the RCT, differences in experimental designs including patient profile, genetic setup, nature of intervention including control from other dietary consumption, and inadequate masking of patients about the treatment, which are the reasons for the controversial findings of the report. More accurate and critically performed trials need to be carried out before making conclusions about the effects of n-3 fatty acids on autoimmune disorders like RA and asthma in adults and children.

The reasons attributed to not realizing the potential reflected in in vitro and animal models in clinical trials are the use of comparable amounts of n-3 fatty acids and also the role played by the genotypes of the patients enrolled, offering much more heterogeneity than the laboratory animals, along with limited control over diet and nutrient patterns in human studies compared to strict diet control employed in animal studies. Further, studies in clinical trials explore therapeutic effects, while most in vitro and animal models measure prophylactic roles, thus offering much difference. In addition, results analyzed in in vitro and animal models are in terms of easily measurable parameters, while the clinical trials assess actual improvement in disease symptoms, which are not easily quantifiable. The absence of sensitive and precise assays quantifying immune functions in in vivo systems makes it further complicated. Thus, many factors, as mentioned above, contribute to not realizing the benefits of n-3 fatty acids in clinical trials as expected in in vitro models. Hence, results from nutrition immunology experiments should be carefully interpreted to harness maximum benefits.

10.4 MICRONUTRIENTS

Vitamins are an important class of organic compounds present in different foods and exhibit many important functions in the body. Although they are present in trace amounts, they are involved in numerous activities in different parts of the body.

Their absence or deficiency is known to compromise several functions and lead to the onset of many diseases. Based on their chemical structure and functions, micronutrients have been classified into different types. According to the solubility criteria, they are grouped into two classes: water-soluble and fat-soluble vitamins. Different vitamins, like vitamin C, E, and members of the B complex, have been reported to be involved in different aspects of immune response in a more generalized manner, such as acting as potent antioxidants. Vitamins A and D are known to be involved in many intricate functions of the immune system. They are reported to act more like hormones, exhibiting their biological functions through their metabolites rather than themselves by binding to their respective nuclear-hormone receptors and modulating a wide range of genes involved in immune modulation by these receptors.

10.4.1 Vitamin A

Vitamin A (VA) is an important micronutrient, well known for its primary role in maintaining the vision of an organism. Retinol, retinal, and retinoic acid (RA) are the three different forms of VA that exhibit an array of health benefits beyond their role in visual function (Pino-Lagos et al., 2010). VA is reported to modulate the immune response of an organism involving both innate and adaptive systems (Stephensen, 2001). 9-*cis*-retinoic acid and all-*trans*-retinoic acid (ATRA) are the two principal forms of retinoic acid exerting many biological activities. They are known to bring about their effects by binding to receptors called retinoic acid receptors (RAR) in the nucleus and modulating the expression of VA-responsive genes (Rochette-Egly and Germain, 2009).

VA was shown to regulate the development and differentiation of dendritic cells, which are the principal antigen-presenting cells essential for the optimal integration of innate and adaptive immune responses (Beijer et al., 2013). Retinoic acid was shown to be essential for the optimal migration of leukocytes into mucosal tissues, mediated by proper expression of different integrin molecules (Kang et al., 2011). Studies by Villamor demonstrated that VA supplementation to preschool children protected them from common diseases like measles, diarrhea, human immunodeficiency virus (HIV) infection, and malaria. This is attributed to the role played by VA in functions like maintaining an intact mucosal barrier, enhanced antibody production, and increased lymphocyte proliferation, indicating the protective function of VA against common ailments (Villamor and Fawzi, 2005).

VA has been directly linked to the activation, proliferation, and differentiation of many cells of the innate immune system, such as neutrophils, macrophages, and NK cells (Ramakrishnan et al., 2004). Innate immune functions, such as phagocytosis, pathogen processing, production of cytokines for recruitment of other immune effector cells, and antigen presentation for further amplification and orchestration of adaptive immune response by specialized lymphocytes, are directly linked to the VA status of the host (Chang and Hou, 2015). VA deficiency was shown to dramatically increase myeloid cell proliferation, which is seen in many myeloproliferative disorders, and this abnormality was found to be reversed on VA restoration in the diet, indicating the critical role of retinoids in the signaling pathway involved in myeloid cell homeostasis (Kuwata et al., 2000). In vivo, a study in rabbits revealed that beta-cryptoxanthin (b-Cr), a pro-vitamin A supplementation in the diet, has been shown to up-regulate the production of antibodies and thus has a role in B cell-mediated adaptive immune response (Ghodratizadeh et al., 2014).

RA signaling via RARα in B cells has been shown to be essential for mounting an optimal gut humoral response and for maintaining healthy gut microflora (Pantazi et al., 2015). VA has been shown to down-regulate the IL-4-induced synthesis of IgE in in vitro experiments and thus functions as an anti-sensitive agent which can be explored to ameliorate IgE-mediated allergic responses (Seo et al., 2017). In addition, VA is known to stimulate the differentiation of naive B cells into Tregs and also induce the synthesis of Tregs, which are known to perform immune regulatory functions in many diseases. Many studies have demonstrated the effect of RA in regulating T cell homeostasis by inhibiting the conversion of T cells into pro-inflammatory Th17 cells and promoting their conversion to regulatory Treg cells. Thus, vitamin A has a protective immune function in both innate and adaptive immune responses. ATRA was shown to induce a macrophage phenotype shift from M1 to M2 and thus down-regulate the macrophage-mediated immunity against Leishmania parasites. This was accompanied by decreased secretion of pro-inflammatory cytokines and lowered expression of iNOS in bone marrow-derived macrophages. Thus, care should be taken while using ATRA as a therapeutic tool targeting the innate immune response (Vellozo et al., 2017). VA was shown to play both developmental and regulatory roles in cells of the immune system at different steps and stages of the immune response, indicating its essentiality as a regulatory and modulatory micronutrient of the immune system.

10.4.2 Vitamin D

Vitamin D (VD) is synthesized in the skin upon exposure to sunlight containing UV-B light from its precursor 7-dehydrocholesterol. VD undergoes metabolic changes in the liver and kidney before converting into its biologically active form referred to as vitamin D3 (VD3) (Holick, 2007). Apart from the classical role of VD in calcium and phosphate metabolism and thus in bone homeostasis, VD is reported to be involved in the immunomodulatory functions of both the innate and adaptive immune systems in the body. The importance of VD in immune cells like dendritic cells, macrophages, T cells, and B cells is reflected by the presence of the vitamin D receptor (VDR) and the enzymes necessary for the conversion of VD to its biologically active form VD3 in addition to primary tissues like the liver and kidney.

The protective effect of VD3 against many infections is mainly attributed to its effect on the innate immune system. Studies have reported that upon the exposure of lipopolysaccharides present in many bacterial cell walls to macrophages, the TLR becomes activated and initiates a series of signal transduction events culminating in enhanced expression of genes for anti-bacterial proteins like cathelicidin and beta-defensin, which displays significant bactericidal activities and thus limits their proliferation and infection (Liu et al., 2006; Wang et al., 2004). VD is known to intervene in the growth, differentiation, and functions of dendritic cells, the principal antigen-presenting cells, rendering them more immune-tolerant by decreasing the expression of MHC II necessary for mounting an effective adaptive immune response (Adorini and Penna, 2009). VD was shown to prevent vascular dysfunction by preserving the integrity of the endothelium, up-regulating cadherin-based adhesion molecules, and preventing local inflammatory reactions with tissue injury (Vila Cuenca et al., 2018). Many cross-sectional studies have reported decreased incidence of respiratory tract infections and other infectious diseases in people with elevated levels of VD3 and vice versa (Ginde et al., 2009). Other studies have shown VD to intervene in the proliferation and differentiation of B cells and their ability to produce antibodies and autoantibodies against B cells when they undergo apoptosis. The role of VD in autoimmune disease is well documented. Many reports from animal experiments have highlighted the requirement of VD for the prevention, progression, and effective treatment of many autoimmune diseases (Hewison, 2012). In addition, one study reported that lower levels of VD3 consumption during pregnancy led to increased incidences of early onset islet autoimmunity in children (Fronczak et al., 2003). VD3 was shown to inhibit the adaptive immune system.

VD has been reported to inhibit the proliferation and differentiation of B cells along with decreased secretion of immunoglobulin (Chen et al., 2007). VD was shown to modulate activities like T cell proliferation, shifting the helper T cell phenotype from Th1 to Th2, reducing the maturation of inflammatory Th17 phenotypes, induction of regulatory T cell proliferation, and many more (Boonstra et al., 2001; Bhalla et al., 1984). These actions have been reported to create an anti-inflammatory milieu from pro-inflammatory conditions with corresponding changes in cytokine and chemokine profiles. Animal experiments have demonstrated decreased incidences of T1D upon VD supplementation, highlighting the role of VD in the management of T1D symptoms. The production of autoantibodies and the formation of auto-reactive Th1 and CTL are the main reasons attributed to the onset of T1D. This is reported to be due to enhanced stimulation of Tregs along with inhibition of Th1 cells, thus creating an immune tolerance microenvironment (Gregori et al., 2002). VD was also found to stimulate pancreatic β cells by binding to VDR and thus increasing the secretion of insulin.

VD levels are positively correlated in patients with RA, which is found to be due to the inhibitory effect of VD on Th1 and Th17 cells and the stimulation of Tregs cells. Some clinical trials have reported an improvement in the symptoms of RA with VD intervention. However, some reports indicate no definitive benefit from VD3 supplementation; hence, care should be taken before firmly recommending its beneficial effects. This discrepancy may be attributed to many indirect factors resulting in altered outcomes. In one of the studies, VD levels were found to be negatively associated in patients with unexplained recurrent spontaneous abortion (URSA). Peripheral blood lymphocyte subsets and serum cytokine from Th1 and Th2 cells were found to be inversely associated with VD levels before and after VD supplementation, implicating the beneficial role of VD in alleviating the complications associated with URSA patients (Yang and Lu, 2024).

A cross-sectional analysis study revealed the importance of VD and physical exercise concerning NK cell activity, an important component of immune response, highlighting the essentiality of VD for optimal immune function (Oh et al., 2021). In another study, VD supplementation revealed a positive association between VD levels and gut microbiome composition and stability, which led to a significant increase in the count of beneficial bacteria along with a drastic decrease in opportunistic pathogens, suggesting the role of VD in the positive modulation of the gut microbiome and its associated functions in healthy individuals (Wyatt et al., 2024). Various immunomodulatory functions of VD have been depicted in Figure 10.2.

10.4.3 Vitamin E

Vitamin E (VE) is an important lipid-based antioxidant that exerts an array of health benefits. It is represented by a group of organic molecules referred to as tocopherols and tocotrienols. Of all the different subtypes, alpha-tocopherol is the most abundant and effective form of VE. It is an important constituent of cell membranes of all cells but is present in significantly higher concentrations in immune cells, thus highlighting its importance in these cells (Coquette et al., 1986). Its deficiency has been linked with the impairment of both innate and adaptive immune responses in various animal and human systems. Dietary intervention studies with VE supplementation were found to restore and enhance immune functions, suggesting the involvement of VE

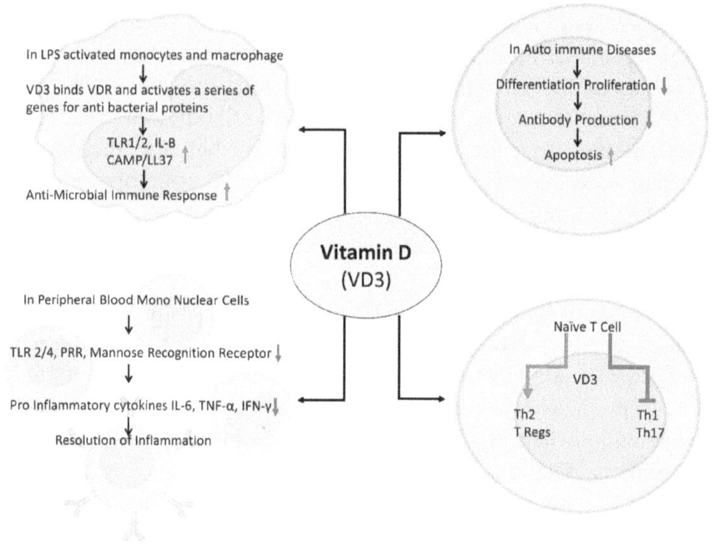

FIGURE 10.2 Schematic diagram depicting immunomodulatory effects of vitamin D (VD3) on different effector cells of the innate and adaptive immune systems. TLR: toll-like receptors; PBMN: peripheral blood mononuclear cells; PRR: pathogen recognition receptor; MRR: mannose recognition receptor; Th1: T helper cell type 1; Th2: T helper cell type 2; Tregs: regulatory T cells; IL-6: interleukin 6; IFNγ: interferon-gamma; TNF-α: tumor necrosis factor-alpha; CAMP: cathelicidin antimicrobial peptide.

in immune response (Han and Meydani, 2006). Optimal maintenance of membrane structure and composition is crucial for several membrane-dependent functions. VE is reported to maintain the integrity of cell membranes mainly in regions like lipid rafts, which are rich in membrane receptors essential for many membrane-associated functions like ligand binding, signal transduction, and production of proteins and mediators essential for different cell functions. VE in immune cells complements this function specifically by scavenging free radicals produced by their high metabolic activity and increased oxidation of PUFA, a predominant type of fatty acid in immune cells, which would otherwise hamper their function (Wang and Quinn, 1999).

VE was reported to down-regulate the proliferation of monocytes, macrophages, and neutrophils through the inhibition of PKC activity mediated by the action of protein phosphatase 2B, thus decreasing the production of superoxides in macrophages (Zingg, 2015). High dietary supplementation of VE was found to increase the numbers and activities of splenic lymphocytes, alveolar macrophages, and NK cells, and activities like phagocytosis, opsonization, and response to concanavalin A (ConA) stimulation were found to be enhanced in rats (Moriguchi et al., 1990). VE supplementation was shown to suppress PGE2 production by inhibiting COX2 enzyme activity in LPS-stimulated macrophages from old rats, thus reversing the age-associated production of PGE2 in murine macrophages (Kalinski, 2012). Lower antibody response to red blood cells (RBCs), decreased lymphocyte response to mitogen stimulus, and altered chemotaxis are some common features reported under VE deficiency in many animal studies.

Defects in CD4^{+}T cell homeostasis, reduced secretion of IL-2 and its receptors, phosphorylation of key proteins involved in signal transduction, and synapse formation, etc., are cardinal signs of the age-associated T cell specific defects observed. However, most of these defects were found to be reversed with the supplementation of VE, indicating the essential role of VE in augmenting age-specific immune response and thus protecting the elderly from frequent infections (Molano and Meydani, 2012). Further, VE was found to enhance the proliferation of T cells and secretion of IL-2 in T cells from old mice compared to young, indicating the role of VE to be more sensitive against T cells that have become more responsive to oxidative stress in older mice than younger ones (Adolfsson et al., 2001). In vivo, supplementation of VE was found to enhance the production of IL-2 and IFN-γ via the activation of Th1 cells in mice infected with the influenza virus and mice with colorectal cancer (Han et al., 2000). VE was found to stimulate humoral immune response too, with a corresponding increase in antibody titers in both human and animal models. However, whether VE is exerting this effect directly through B cells or through its indirect effect on T cells needs further confirmation (De Waart et al., 1997).

10.4.4 Vitamin C

Vitamin C (VC) is an important micronutrient essential for the optimal health of an organism. VC has been implicated in the regulation of various aspects of immune cell functions. The absence of the key enzyme necessary for the biosynthesis of VC makes it an essential nutrient to be obtained through dietary sources (Burns, 1957). Multiple factors influence VC levels in the body; being water soluble in nature, VC should be consumed consistently in high amounts to harness its benefits. Much of its immune-modulating function is attributed to its inherent non-enzymatic antioxidant properties. VC has been directly linked with the neutralization of excess ROS generated by activated immune cells during the phagocytosis of pathogens. VC is known to prevent ROS-induced lipid peroxidation, protein modifications, and disruption of membrane fluidity and integrity. Although ROS is necessary for optimal defense, if produced in excess, it can create havoc in the cells and tissues and thus disrupt normal homeostasis, leading to many degenerative diseases by chronic oxidative stress (Calder et al., 2002).

Studies have reported reduced incidences of upper respiratory infections with increased supplementation of dietary VC. Leukocytes are known to possess VC in adequate quantities and thus become primed to be activated during infections. Neutrophils and monocytes in particular are forerunners during the innate immune response. VC deficiency is known to cause a severe disease called scurvy, characterized by poor wound healing, weakened collagenous tissue, and a weakened immune system, rendering the host more susceptible to fatal respiratory infections like pneumonia (Hemila, 2017). It is reported to be present in keratinocytes and fibroblasts, which are important cells involved in collagen biosynthesis, and it is essential for proper skin development and enabling protection from external insults, including pathogens (Pullar et al., 2017).

Rapid collagen synthesis, enhanced keratinocyte differentiation, increased fibroblast proliferation, faster wound healing, and protection against free radical-induced damage are some important therapeutic properties associated with VC treatment (Savini et al., 2001; Kishimoto et al., 2013). As a cofactor, VC is involved in reactions catalyzed by enzymes like hydroxylases, monooxygenases, and dioxygenases, which are are involved in different biosynthetic and metabolic reactions (Mandl et al., 2009). VC was shown to modulate neutrophil functions like phagocytosis, chemotaxis, cytokine production, pathogen killing, and macrophage clearance. Rapid accumulation of VC by leukocytes indicates the role of VC in their immune modulation functions (Bergsten et al., 1990).

VC was shown to inhibit NF-κB activation in neutrophils and decrease oxidative and inflammatory stress. VC was shown to regulate neutrophil apoptosis by macrophages, thereby avoiding the necrosis of neutrophils, which otherwise enhance persistent chronic inflammation. The effect was found to be reversed in VC intervention studies. Similarly, through the enhanced phagocytic activity of macrophages and neutrophils, VC facilitates the innate immune response to be more effective in its actions, such as chemotaxis mediated by specific chemokines and cytokines secreted by activated macrophages, which, in turn, can induce stimuli for initiation of adaptive immune response (Lämmermann, 2016). The same kinds of results have also been observed in human studies where subjects were supplemented with adequate amounts of VC (Bozonet et al., 2015). Similar to the results observed in the innate immune response, VC supplementation has also been reported to be beneficial in the adaptive immune response.

VC is represented in both T and B lymphocytes exhibiting elevated functions such as activation, proliferation, and differentiation, resulting in a protective immune response against many ailments. VC-incubated lymphocytes are known to secrete anti-inflammatory cytokines with a concomitant decrease in pro-inflammatory cytokine production (Molina et al., 2014). Many human clinical studies have demonstrated an inverse correlation between serum and tissue VC levels concerning the onset and progression of numerous oxidative and inflammatory stress-induced diseases like type 2 diabetes, obesity, hypertension, and other complicated conditions like metabolic syndrome (Harding et al., 2008). Furthermore, VC has also been linked inversely with the development of many infectious diseases like respiratory tuberculosis, asthma, and pneumonia. Supplementation of VC was found to reverse many of these diseases to a large extent. Most of these results are primarily attributed to the ability of VC to positively modulate both innate and adaptive immune responses (Bakaev and Duntau, 2004; Hunt et al., 1994). The diverse benefits of VC are mainly attributed to its inherent antioxidant properties, protecting against an array of diseases resulting from oxidative stress.

A pharmacological dose of VC was shown to enhance oxidative stress in tumor cells, resulting in an enhanced expression of cell death markers like calreticulin and HMGB1, thus marking them to be targeted by cytotoxic cells like NK cells and activated $CD8^+$ T lymphocytes, indicating the immune-mediated antitumor properties of VC (Cavaliere et al., 2023). Many clinical studies have revealed the immunomodulating properties of VC in alleviating complications in sepsis patients; the molecular mechanism of this effect was found to be exhibited by inhibition of STAT1 phosphorylation and down-regulating the expression of PD-L1 in an experimental septic mouse model administered with a high dose of 250 mg/kg body weight intravenously (Zhao et al., 2024).

Complications associated with sepsis were found to be inversely linked with systemic VC levels in these patients. In connection with this, an ex vivo study of LPS-induced cytokine response in peripheral blood mononuclear cells from healthy individuals revealed decreased levels of inflammatory cytokines like IL-10, IL-23, and MCP-1 in VC supplementation, and vice versa, indicating the beneficial role of VC in the management of sepsis (Lauer et al., 2021). In another tumor immunotherapy experiment, targeted VC delivery through lipid nanoparticles was found to elevate oxidative toxicity and induce macrophage polarization from M2 to M1, resulting in the infiltration of active cytotoxic T lymphocytes into the tumor microenvironment, thus indicating VC's antitumor properties (Ma et al., 2022). Different immunomodulatory functions of VC are represented schematically in Figure 10.3.

To conclude, VC, through its inherent antioxidant properties, is well known to exert many health benefits in immune effector cells like neutrophils, macrophages, and T and B lymphocytes. Apart from this, VC as a cofactor is involved in key enzymes of biosynthetic and gene regulatory pathways in immune cells.

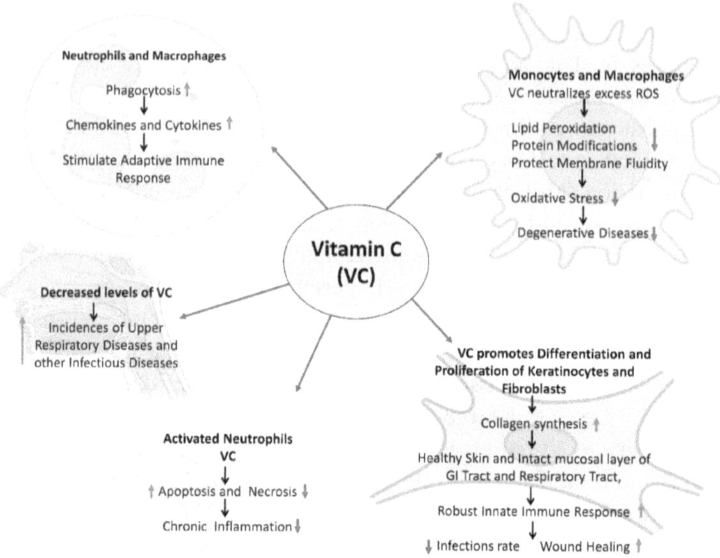

FIGURE 10.3 Schematic diagram depicting immunomodulatory effects of vitamin C on different cells of the immune system. VC: vitamin C; ROS: reactive oxygen species; GI tract: gastrointestinal tract.

10.4.5 Zinc

Although micronutrients are present in minute concentrations, their absence or deficiency results in the manifestation of many diseases with compromised body functions. Zinc is shown to be essential for maintaining the optimal integrity of mucous membranes in the respiratory, gastrointestinal, and urinary tracts. It is shown to be essential for tight junction proteins in the epithelial layer, which helps to maintain an intact physical barrier against many pathogens. It is also shown to be essential for the functions of neutrophils, macrophages, and NK cells of the innate immune response. Further, it is reported to be crucial for the functioning of both B and T cells of the adaptive immune response (Shankar and Prasad, 1998). It participates as a cofactor in hundreds of key enzymes involved in immune response (Vallee and Falchuk, 1993).

Thymic atrophy, lymphopenia, and compromised cell-mediated and antibody-mediated immune responses are some cardinal signs exhibited during zinc deficiency (Fraker and King, 2004). Studies have reported that plasma zinc levels are directly linked to oxidative stress mediators and inflammatory cytokines. Their levels were found to be significantly high in zinc deficiency but returned to normal levels upon zinc supplementation in elderly people (Prasad et al., 2007). Zinc facilitates the growth and development of lymphocytes, which are necessary for mounting an effective adaptive immune response. Compared to T cells, B cells are more dependent on zinc for their optimal proliferation and differentiation. Mice deficient in zinc were shown to exhibit decreased phagocytosis of pathogens. However, the restoration of zinc in their diets resulted in improved phagocytosis and neutralization of parasites (Wirth et al., 1989). Zinc deficiency was shown to affect the mitogenesis of T cells with no effect on B cell mitogenesis, indicating the essentiality of zinc for optimal T cell function and the differential sensitivity of zinc for B cell and T cell mitogenesis (Zanzonico et al., 1981). In addition to B cell lymphopoiesis, zinc deficiency has resulted in a drastic decrease in antibody production, highlighting its critical role in humoral-mediated immunity. The function of thymulin, an important hormone secreted by thymic cells essential for optimal T cell differentiation, is reported to be decreased during zinc deficiency. Zinc supplementation restores its function and induces the differentiation and development of T cells, resulting in an efficient cell-mediated immune response (Prasad et al., 1988). Zinc deficiency was also reported to alter the balance between Th1 and Th2 levels, resulting in decreased Th1-dependent cytokines like IFN-γ, IL-2, and tumor necrosis factor (TNF)-α, with no significant changes in Th2-dependent cytokines, thus indicating a differential regulatory control over T cell subtypes in the in vivo system (Beck et al., 1997).

Trace element supplementation studies with zinc and selenium have documented increased antibody titers against influenza vaccination, indicating enhanced functioning of the humoral immune response and decreased morbidity from infectious diseases. This emphasizes the role of trace elements like zinc in improving public health concerns, particularly in the elderly population (Girodon et al., 1999). Results from many studies indicate that zinc acts as an indispensable ion necessary for the optimal functioning of the immune system and thus offers protection against many pathogenic organisms. In addition, with zinc status in the aged population being suboptimal, increased dietary supplementation of zinc was shown to provide elevated protection for these individuals.

10.4.6 Selenium

Selenium is an important micronutrient which is incorporated into proteins cotranslationally (Hatfield et al., 2011). It is an important constituent of an antioxidant enzyme, glutathione peroxidase, involved in the neutralization of free radicals. It thereby regulates the redox status in many cells and tissues including immune effector cells. Selenium exists principally as a component of many important selenoproteins. There are around 25 different families of selenoproteins, of which glutathione peroxidases and thioredoxin reductases are two well-known ones (Gladyshev et al., 2016). Many animal and human studies have indicated diminished immune functions against many viral infections under selenium deficiency, which were restored to normal levels upon selenium supplementation, indicating the essentiality of selenium for an effective defense response (Huang et al., 2012). Lower selenium status in pregnant women was found to be associated with higher incidences of bacterial and parasitic infections in their neonates up to six months, indicating decreased immune potential of neonates born to mothers with decreased selenium levels during pregnancy (Varsi et al., 2017). Selenium deprivation was shown to lower the proliferation potential of lymphocytes upon mitogen stimulation. Further, neutrophil chemotaxis was also compromised due to lowered secretion of leukotriene B4 in macrophages. In addition, important antibody titers comprising IgG, IgM, and IgA were lowered during infection under selenium deficiency. Both of the above-mentioned events were found to be reversed upon the supplementation of selenium, suggesting the role of selenium in both innate and adaptive immune responses (Jackson et al., 2003). Under selenium deficiency, the cytolytic potential of neutrophils in the phagosomes was reported to be compromised, which was attributed to decreased activity of cytosolic glutathione peroxidase necessary for the generation of an oxidative burst for the elimination of pathogens in the phagolysosome complexes (Boyne and Arthur, 1986).

A novel selenium-containing protein called selenoprotein K (Sel K), present in the endoplasmic reticulum membrane, is found to be expressed in higher amounts in many immune cells. In Sel K-depleted mice, receptor-mediated Ca^{2+} flux was reported to be decreased in T cells, neutrophils, and macrophages, resulting in lowered functions of Ca^{2+}-dependent activities like the proliferation and migration of T cells and macrophages. In addition, Fcγ receptor-mediated oxidative burst in macrophages was decreased. The potential for viral clearance and survival against viral infection was also lowered in Sel K-depleted mice compared to the control, implicating the necessity of selenium for optimum immune function (Verma et al., 2011). Selenium was shown to protect against ultraviolet (UV)-induced chromosome damage and gene mutation through functional BRCA1 activity in human MCF-7 breast carcinoma cells and mouse fibroblasts. Further, overexpression of GPx-1, a selenoprotein with antioxidant activity, was reported to decrease UV-induced micronuclei formation (Baliga et al., 2007).

Similar to many other key micronutrients, selenium is also implicated in enhanced immune competence in the aged population, who otherwise would be prone to the development of many infections and diseases due to a weakened immune system. This has been demonstrated in many animal and human studies involving various parameters of immunity against the dietary supplementation of selenium (Ventura et al., 1994; Wood et al., 2000; Berg et al., 2005). Thus results from in vitro, in vivo, and human interventional studies indicate the essentiality of selenium, although in minimal concentrations, for optimal functioning of the immune system in addition to many other associated functions for leading a healthy life.

10.5 PROBIOTICS

According to world bodies like WHO and FAO, probiotics are defined as a set of live microorganisms that, when consumed or administered in adequate quantities to an organism, have the potential to bring about several health benefits in the host. With progressive advances in probiotic research, the definition is being continuously modified to meet the latest findings (Hill et al., 2014). Microbial strains having the potential to modulate the immune response under altered conditions are being explored for use as probiotics in many hosts.

The most commonly used organisms as probiotics belong to the genera Lactobacillus, Enterococcus, and Bifidobacterium, and multiple species within them. However, *Escherichia coli*, a Gram-negative bacteria, and Saccharomyces fungi are less frequently used as probiotics (Nath et al., 2018). These probiotics and their metabolites are known to exhibit several health benefits and can significantly modulate the health status of their host. Probiotics are known to colonize the mucosal layers of the gastrointestinal tract, respiratory tract, and urogenital tract, and compete with harmful bacteria for space, nutrients and other essentials, thus establish themselves as more efficient microorganisms. Probiotics and their metabolites are known to influence a variety of activities ranging from metabolism to pathophysiological functions.

Advancements in proteomic and genomic studies have resulted in the identification of different proteins and metabolites that have a profound impact on immune modulation (Yan et al., 2010). Studies related to the biological significance of probiotics and their metabolites have identified that they can modulate different components and arms of the immune system. This ranges from physical barriers like the skin and mucosal layer, generalized innate immune responses with neutrophils, NK cells, and macrophages, to highly specific T cell- and B cell-mediated adaptive immune responses (López-Varela et al., 2002). Epithelial cells and dendritic cells play an important role in the regulation of the innate immune system.

Dendritic cells, the professional antigen-presenting cells, are known to be distributed in peripheral tissues. Intestinal dendritic cells, the main immune sensor cells present in gut-associated lymphoid tissue (GALT), recognize microbe-specific ligands through Toll-like receptors (TLR) and c-type lectin receptors, and activate signaling pathways causing altered expression of genes, resulting in the production of a different set of cytokines and chemokines specific to the kind of microbe recognized. Studies with probiotic supplementation in many inflammatory diseases have revealed that dendritic cells, upon recognition of pathogen-specific ligands, activate and mediate the production of anti-inflammatory cytokines and chemokines, accompanied by suppression of pro-inflammatory cytokines via TLR-mediated signaling (Dasgupta et al., 2014).

Probiotics are known to induce immune tolerance with substantial stimulation of Treg cells along with reduced secretion of macrophage inflammatory proteins (Scully et al., 2013). Further, some probiotics are reported to augment the production of anti-bacterial peptides having broad-spectrum antibiotic potential by epithelial cells, Paneth cells, macrophages, and neutrophils, thereby reinforcing mucosal barrier immunity (Habil et al., 2014).

A few probiotics belonging to the genus Bifidobacterium are known for inducing autophagy of epithelial cells infected with pathogenic microbes so that they will not divide and proliferate further (Lin et al., 2014). A few other probiotics are reported to induce elevated secretion of mucin, which impedes the anchorage of pathogens, and many other probiotics are reported to down-regulate the production of pro-inflammatory mediators like cytokines and chemokines and thus alleviate the symptoms of chronic inflammatory insults (Elian et al., 2015). The activity and expression of a specific receptor on NK cells are seen to be up-regulated upon the supplementation of probiotics, thus rendering the host more immunocompetent (Aziz and Bonavida, 2016). Concerning adaptive immunity, probiotic consumption is reported to increase the proportion of Treg cells through the up-regulation of FoxP3, which facilitates the synthesis of anti-inflammatory mediators and lowers pro-inflammatory mediators, thus alleviating individuals from the effects of many inflammatory diseases (Zhang et al., 2017).

Short-chain fatty acids like butyrate and propionate, the key metabolites of many probiotics, are known to stimulate the production of Treg cells. Further, through the activation of GPR 43, they are known to alleviate the complications of inflammatory diseases by the suppression of Th2- and Th17-based activities (Kim et al., 2014). They are also known to augment the production of IgA, the principal secretory antibody playing an important role in binding pathogens and neutralizing their toxins at the mucosal surface. They are reported to act via TLR2 signaling in the Peyer's patches and lamina propria of mouse small intestine (Sakai et al., 2014).

Another study concerning the *Saccharomyces cerevisiae* strain showed decreased secretion of pro-inflammatory cytokines and chemokines through down-regulation of ERK1/2 and p38 MAPK phosphorylation in enterotoxigenic *E. coli* (ETEC)-induced studies (Zanello et al., 2011). Transcriptomic analysis of gene expression profiles against different probiotics in humans revealed that probiotic modulation of mucosal immunity is more specific concerning host genetics and strain, each having varied roles under different conditions (van Baarlen et al., 2011).

Evidence from different experiments involving the use of probiotics shows a promising future in the exploration of probiotics and their metabolites as important modulators of gut microbiota and immune response. Further insights into the underlying mechanisms, types and interactions of probiotic strains used, host dietary practices, genetic set-up, etc. will unravel additional information that will assist in the proper translation of the therapeutic potential of probiotics to prevent and treat many ailments, including immune system-dependent diseases, effectively.

10.6 CONCLUSION AND FUTURE PERSPECTIVES

Dietary molecules are known to modulate several functions, including the immune response of an organism. Modulation of the immune response has been implicated in the pathogenesis and therapeutic regimen of many diseases and disorders. Dietary components like polyphenols, flavonoids, omega-3 fatty acids, certain vitamins and minerals, and probiotics are key dietary molecules shown to exhibit a profound influence on immune functions in animal experiments and clinical studies. Different dietary molecules are reported to influence different aspects of immune response involving both innate and adaptive immune responses. Some are shown to modulate innate immune functions like phagocytosis and NK cell activity, while others are reported to alter B cell- and T cell-mediated adaptive immune response and thus bring about immune modulation. Many of these functions are brought out by regulating the levels of oxidative and inflammatory stress mediators like reactive oxygen and nitrogen species, cytokines, chemokines, and other key signaling molecules. Although most results from preclinical studies have reported the positive effects of dietary molecules with respect to immune modulation, some have reported contradictory findings, and others have shown insignificant results. Thus, additional clinical studies taking all the variable factors into consideration should be carried out to generate ample evidence-based outcomes in terms of their efficacy, specificity, and robustness in immune-modulating properties. Hence, the consumption of a diet abundant in bioactive molecules might be beneficial for their prophylactic effects, but their use as therapeutic molecules for immune modulation in many diseases cannot be firmly advocated.

REFERENCES

Adolfsson, O., Huber, B. T., & Meydani, S. N. (2001). Vitamin E-enhanced IL-2 production in old mice: Naive but not memory T cells show increased cell division cycling and IL-2-producing capacity. *Journal of Immunology (Baltimore, Md.: 1950)*, *167*(7), 3809–3817.

Adorini, L. (2005). Intervention in autoimmunity: The potential of vitamin D receptor agonists. *Cellular Immunology*, *233*(2), 115–124.

Adorini, L., & Penna, G. (2009). Induction of tolerogenic dendritic cells by vitamin D receptor agonists. *Dendritic Cells*, *188*, 251–273.

Alam, R. T., Zeid, E. H., & Imam, T. S. (2017). Protective role of quercetin against hematotoxic and immunotoxic effects of furan in rats. *Environmental Science and Pollution Research International*, *24*(4), 3780–3789.

Aziz, N., & Bonavida, B. (2016). Activation of natural killer cells by probiotics. *Forum on Immunopathological Diseases and Therapeutics*, *7*(1–2), 41–55.

Bakaev, V. V., & Duntau, A. P. (2004). Ascorbic acid in blood serum of patients with pulmonary tuberculosis and pneumonia. *The International Journal of Tuberculosis and Lung Disease*, *8*(2), 263–266.

Baliga, M. S., Wang, H., Zhuo, P., Schwartz, J. L., & Diamond, A. M. (2007). Selenium and GPx-1 overexpression protect mammalian cells against UV-induced DNA damage. *Biological Trace Element Research*, *115*, 227–241.

Beck, F. W., Prasad, A. S., Kaplan, J., Fitzgerald, J. T., & Brewer, G. J. (1997). Changes in cytokine production and T cell subpopulations in experimentally induced zinc-deficient humans. *American Journal of Physiology-Endocrinology and Metabolism*, *272*(6), E1002–E1007.

Beijer, M. R., Molenaar, R., Goverse, G., Mebius, R. E., Kraal, G., & den Haan, J. M. (2013). A crucial role for retinoic acid in the development of Notch-dependent murine splenic CD8- CD4- and CD4+ dendritic cells. *European Journal of Immunology*, *43*(6), 1608–1616.

Berg, B. M., Godbout, J. P., Chen, J., Kelley, K. W., & Johnson, R. W. (2005). α-Tocopherol and selenium facilitate recovery from lipopolysaccharide-induced sickness in aged mice. *The Journal of Nutrition*, *135*(5), 1157–1163.

Bergsten, P., Amitai, G., Kehrl, J., Dhariwal, K. R., Klein, H. G., & Levine, M. (1990). Millimolar concentrations of ascorbic acid in purified human mononuclear leukocytes. Depletion and reaccumulation. *Journal of Biological Chemistry*, *265*(5), 2584–2587.

Bhalla, A. K., Amento, E. P., Serog, B., & Glimcher, L. H. (1984). 1, 25-Dihydroxyvitamin D3 inhibits antigen-induced T cell activation. *Journal of Immunology* (Baltimore, Md.: 1950), *133*(4), 1748–1754.

Boonstra, A., Barrat, F. J., Crain, C., Heath, V. L., Savelkoul, H. F., & O'Garra, A. (2001). 1alpha, 25-Dihydroxyvitamin d3 has a direct effect on naive CD4(+) T cells to enhance the development of Th2 cells. *Journal of Immunology* (Baltimore, Md.: 1950), *167*(9), 4974–4980.

Boyne, R., & Arthur, J. R. (1986). The response of selenium-deficient mice to Candida albicans infection. *The Journal of Nutrition*, *116*(5), 816–822. https://doi.org/10.1093/jn/116.5.816

Bozonet, S. M., Carr, A. C., Pullar, J. M., & Vissers, M. C. (2015). Enhanced human neutrophil vitamin C status, chemotaxis and oxidant generation following dietary supplementation with vitamin C-rich SunGold kiwifruit. *Nutrients*, *7*(4), 2574–2588.

Bruckdorfer, K. R. (2008). Antioxidants and CVD. *The Proceedings of the Nutrition Society*, *67*(2), 214–222.

Bub, A., Watzl, B., Blockhaus, M., Briviba, K., Liegibel, U., Müller, H., Pool-Zobel, B. L., & Rechkemmer, G. (2003). Fruit juice consumption modulates antioxidative status, immune status and DNA damage. *The Journal of Nutritional Biochemistry*, *14*(2), 90–98.

Burns, J. J. (1957). Missing step in man, monkey and guinea pig required for the biosynthesis of L-ascorbic acid. *Nature*, *180*(4585), 553–553.

Calder P. C. (2003). N-3 polyunsaturated fatty acids and inflammation: From molecular biology to the clinic. *Lipids*, *38*(4), 343–352. https://doi.org/10.1007/s11745-003-1068-y). N-3 polyunsaturated fatty acids and inflammation: from molecular biology to the clinic. *Lipids*, *38*(4), 343–352.

Calder, P. C., Field, C. J., & Gill, H. S. (2002). *Nutrition and immune function*. CABI.

Camacho-Muñoz, D., Niven, J., Kucuk, S., Cucchi, D., Certo, M., Jones, S. W., & Nicolaou, A. (2022). Omega-3 polyunsaturated fatty acids reverse the impact of western diets on regulatory T cell responses through averting ceramide-mediated pathways. *Biochemical Pharmacology*, *204*, 115211.

Cavaliere, A., Macagno, M., Maione, F., Amodio, V., Chilà, R., Lamba, S., & Di Nicolantonio, F. (2023). 18P Exposure of calreticulin is required for vitamin C immunomediated cancer surveillance. *Annals of Oncology*, *34*, S192.

Chang, H. K., & Hou, W. S. (2015). Retinoic acid modulates interferon-γ production by hepatic natural killer T cells via phosphatase 2A and the extracellular signal-regulated kinase pathway. *Journal of Interferon & Cytokine Research*, *35*(3), 200–212.

Che, D. N., Cho, B. O., Shin, J. Y., Kang, H. J., Kim, Y. S., & Jang, S. I. (2018). Fisetin inhibits IL-31 production in stimulated human mast cells: Possibilities of fisetin being exploited to treat histamine-independent pruritus. *Life Sciences*, *201*, 121–129.

Chen, S., Sims, G. P., Chen, X. X., Gu, Y. Y., Chen, S., & Lipsky, P. E. (2007). Modulatory effects of 1, 25-dihydroxyvitamin D3 on human B cell differentiation. *Journal of Immunology* (Baltimore, Md.: 1950), *179*(3), 1634–1647.

Chen, Y., Peng, F., Xing, Z., Chen, J., Peng, C., & Li, D. (2022). Beneficial effects of natural flavonoids on neuroinflammation. *Frontiers in Immunology*, *13*, 1006434.

Childs, C. E., Calder, P. C., & Miles, E. A. (2019). Diet and immune function. *Nutrients*, *11*(8), 1933.

Chirumbolo S. (2010). The role of quercetin, flavonols and flavones in modulating inflammatory cell function. *Inflammation & Allergy Drug Targets*, *9*(4), 263–285.

Choi, J. K., Jang, Y. H., Lee, S., Lee, S. R., Choi, Y. A., Jin, M., Choi, J. H., Park, J. H., Park, P. H., Choi, H., Kwon, T. K., Khang, D., & Kim, S. H. (2017). Chrysin attenuates atopic dermatitis by suppressing inflammation of keratinocytes. *Food and Chemical Toxicology: An International Journal Published for the British Industrial Biological Research Association*, *110*, 142–150.

Clark, G. J., Silveira, P. A., Hogarth, P. M., & Hart, D. N. J. (2019). The cell surface phenotype of human dendritic cells. *Seminars in Cell & Developmental Biology*, *86*, 3–14.

Cleland, L. G., French, J. K., Betts, W. H., Murphy, G. A., & Elliott, M. J. (1988). Clinical and biochemical effects of dietary fish oil supplements in rheumatoid arthritis. *The Journal of Rheumatology*, *15*(10), 1471–1475.

Collin, M., & Bigley, V. (2016). Monocyte, macrophage, and dendritic cell development: The human perspective. *Microbiology Spectrum*, *4*(5), 10.1128/microbiolspec.MCHD-0015-2015.

Coquette, A., Vray, B., & Vanderpas, J. (1986). Role of vitamin E in the protection of the resident macrophage membrane against oxidative damage. *Archives internationales de physiologie et de biochimie*, *94*(5), S29–S34.

Cucchi, D., Camacho-Munoz, D., Certo, M., Niven, J., Smith, J., Nicolaou, A., & Mauro, C. (2020). Omega-3 polyunsaturated fatty acids impinge on CD4+ T cell motility and adipose tissue distribution via direct and lipid mediator-dependent effects. *Cardiovascular Research*, *116*(5), 1006–1020.

Cunningham-Rundles, S., McNeeley, D. F., & Moon, A. (2005). Mechanisms of nutrient modulation of the immune response. *The Journal of Allergy and Clinical Immunology*, *115*(6), 1119–1129.

Dasgupta, S., Erturk-Hasdemir, D., Ochoa-Reparaz, J., Reinecker, H. C., & Kasper, D. L. (2014). Plasmacytoid dendritic cells mediate anti-inflammatory responses to a gut commensal molecule via both innate and adaptive mechanisms. *Cell Host & Microbe*, *15*(4), 413–423. https://doi.org/10.1016/j.chom.2014.03.006

De Caterina, R., Cybulsky, M. I., Clinton, S. K., Gimbrone, M. A., Jr, & Libby, P. (1994). The omega-3 fatty acid docosahexaenoate reduces cytokine-induced expression of proatherogenic and proinflammatory proteins in human endothelial cells. *Arteriosclerosis and Thrombosis: A Journal of Vascular Biology*, *14*(11), 1829–1836.

De Waart, F. G., Portengen, L., Doekes, G., Verwaal, C. J., & Kok, F. J. (1997). Effect of 3 months vitamin E supplementation on indices of the cellular and humoral immune response in elderly subjects. *The British Journal of Nutrition*, *78*(5), 761–774.

Del Cornò M, Scazzocchio B, Masella R, Gessani S. Regulation of dendritic cell function by dietary del Cornò, M., Scazzocchio, B., Masella, R., & Gessani, S. (2016). Regulation of dendritic cell function by dietary polyphenols. *Critical Reviews in Food Science and Nutrition*, *56*(5), 737–747.

Dress, R. J., Wong, A. Y., & Ginhoux, F. (2018). Homeostatic control of dendritic cell numbers and differentiation. *Immunology and Cell Biology*, *96*(5), 463–476.

Egert, S., Bosy-Westphal, A., Seiberl, J., Kürbitz, C., Settler, U., Plachta-Danielzik, S., Wagner, A. E., Frank, J., Schrezenmeir, J., Rimbach, G., Wolffram, S., & Müller, M. J. (2009). Quercetin reduces systolic blood pressure and plasma oxidised low-density lipoprotein concentrations in overweight subjects with a high-cardiovascular disease risk phenotype: a double-blinded, placebo-controlled cross-over study. *The British Journal of Nutrition*, *102*(7), 1065–1074.

Elian, S. D., Souza, E. L., Vieira, A. T., Teixeira, M. M., Arantes, R. M., Nicoli, J. R., & Martins, F. S. (2015). Bifidobacterium longum subsp. infantis BB-02 attenuates acute murine experimental model of inflammatory bowel disease. *Beneficial Microbes*, *6*(3), 277–286.

Eun, S. H., Woo, J. T., Kim, D. H. (2017). Tangeretin inhibits IL-12 expression and NF-κB activation in dendritic cells and attenuates colitis in mice. *Planta Medica*, *83*(6), 527–533.

Fan, H., Tang, H. B., Shan, L. Q., Liu, S. C., Huang, D. G., Chen, X., Chen, Z., Yang, M., Yin, X. H., Yang, H., & Hao, D. J. (2019). Quercetin prevents necroptosis of oligodendrocytes by inhibiting macrophages/microglia polarization to M1 phenotype after spinal cord injury in rats. *Journal of Neuroinflammation*, *16*(1), 206.

Fanti, P., Asmis, R., Stephenson, T. J., Sawaya, B. P., & Franke, A. A. (2006). Positive effect of dietary soy in ESRD patients with systemic inflammation–correlation between blood levels of the soy isoflavones and the acute-phase reactants. *Nephrology, Dialysis, Transplantation: Official Publication of the European Dialysis and Transplant Association - European Renal Association*, *21*(8), 2239–2246.

Feng, X., Qin, H., Shi, Q., Zhang, Y., Zhou, F., Wu, H., Ding, S., Niu, Z., Lu, Y., & Shen, P. (2014). Chrysin attenuates inflammation by regulating M1/M2 status via activating PPARγ. *Biochemical Pharmacology*, *89*(4), 503–514.

Fortin, P. R., Lew, R. A., Liang, M. H., Wright, E. A., Beckett, L. A., Chalmers, T. C., & Sperling, R. I. (1995). Validation of a meta-analysis: the effects of fish oil in rheumatoid arthritis. *Journal of Clinical Epidemiology*, *48*(11), 1379–1390.

Fraker, P. J., & King, L. E. (2004). Reprogramming of the immune system during zinc deficiency. *Annual Review of Nutrition*, *24*, 277–298.

Fronczak, C. M., Barón, A. E., Chase, H. P., Ross, C., Brady, H. L., Hoffman, M., ... Norris, J. M. (2003). In utero dietary exposures and risk of islet autoimmunity in children. *Diabetes Care*, *26*(12), 3237–3242. VD3 has been reported to be used as a protective supplement against many ailments from many years

Ganeshpurkar, A., & Saluja, A. K. (2017). Protective effect of rutin on humoral and cell mediated immunity in rat model. *Chemico-biological Interactions*, *273*, 154–159.

Ghodratizadeh, S., Kanbak, G., Beyramzadeh, M., Dikmen, Z. G., Memarzadeh, S., & Habibian, R. (2014). Effect of carotenoid β-cryptoxanthin on cellular and humoral immune response in rabbit. *Veterinary Research Communications*, *38*, 59–62.

Ginde, A. A., Mansbach, J. M., & Camargo, C. A. (2009). Association between serum 25-hydroxyvitamin D level and upper respiratory tract infection in the third national health and nutrition examination survey. *Archives of Internal Medicine*, *169*(4), 384–390.

Girodon, F., Galan, P., Monget, A. L., Boutron-Ruault, M. C., Brunet-Lecomte, P., Preziosi, P., ... Hercberg, S. (1999). Impact of trace elements and vitamin supplementation on immunity and infections in institutionalized elderly patients: A randomized controlled trial. *Archives of Internal Medicine*, *159*(7), 748–754.

Gladyshev, V. N., Arnér, E. S., Berry, M. J., Brigelius-Flohé, R., Bruford, E. A., Burk, R. F., ... Zhang, Y. (2016). Selenoprotein gene nomenclature. *Journal of Biological Chemistry*, *291*(46), 24036–24040.

Gordon, S., & Taylor, P. R. (2005). Monocyte and macrophage heterogeneity. *Nature Reviews: Immunology*, *5*(12), 953–964.

Granato, M., Rizzello, C., Gilardini Montani, M. S., Cuomo, L., Vitillo, M., Santarelli, R., Gonnella, R., D'Orazi, G., Faggioni, A., & Cirone, M. (2017). Quercetin induces apoptosis and autophagy in primary effusion lymphoma cells by inhibiting PI3K/AKT/mTOR and STAT3 signaling pathways. *The Journal of Nutritional Biochemistry*, *41*, 124–136.

Gregori, S., Giarratana, N., Smiroldo, S., Uskokovic, M., & Adorini, L. (2002). A 1α, 25-dihydroxyvitamin D3 analog enhances regulatory T-cells and arrests autoimmune diabetes in NOD mice. *Diabetes*, *51*(5), 1367–1374.

Habil, N., Abate, W., Beal, J., & Foey, A. D. (2014). Heat-killed probiotic bacteria differentially regulate colonic epithelial cell production of human β-defensin-2: Dependence on inflammatory cytokines. *Beneficial Microbes*, *5*(4), 483–495. https://doi.org/10.3920/BM2013.0061

Hagenlocher, Y., & Lorentz, A. (2015). Immunomodulation of mast cells by nutrients. *Molecular Immunology*, *63*(1), 25–31.

Haghmorad, D., Mahmoudi, M. B., Salehipour, Z., Jalayer, Z., Momtazi Brojeni, A. A., Rastin, M., Kokhaei, P., & Mahmoudi, M. (2017). Hesperidin ameliorates immunological outcome and reduces neuroinflammation in the mouse model of multiple sclerosis. *Journal of Neuroimmunology*, *302*, 23–33. https://doi.org/10.1016/j.jneuroim.2016.11.009

Han, S. N., & Meydani, S. N. (2006). Impact of vitamin E on immune function and its clinical implications. *Expert Review of Clinical Immunology*, *2*(4), 561–567.

Han, S. N., Wu, D., Ha, W. K., Beharka, A., Smith, D. E., Bender, B. S., & Meydani, S. N. (2000). Vitamin E supplementation increases T helper 1 cytokine production in old mice infected with influenza virus. *Immunology*, *100*(4), 487–493.

Harding, A. H., Wareham, N. J., Bingham, S. A., Khaw, K., Luben, R., Welch, A., & Forouhi, N. G. (2008). Plasma vitamin C level, fruit and vegetable consumption, and the risk of new-onset type 2 diabetes mellitus: The European prospective investigation of cancer–Norfolk prospective study. *Archives of Internal Medicine, 168*(14), 1493–1499.

Hatfield, D. L., Berry, M. J., & Gladyshev, V. N. (Eds.). (2011). *Selenium: Its molecular biology and role in human health.* Springer Science & Business Media.

Heinz, S. A., Henson, D. A., Nieman, D. C., Austin, M. D., & Jin, F. (2010). A 12-week supplementation with quercetin does not affect natural killer cell activity, granulocyte oxidative burst activity or granulocyte phagocytosis in female human subjects. *The British Journal of Nutrition, 104*(6), 849–857.

Hellebrekers, P., Vrisekoop, N., & Koenderman, L. (2018). Neutrophil phenotypes in health and disease. *European Journal of Clinical Investigation, 48* (Suppl. 2), e12943.

Hemila, H. (2017). Vitamin C and infections. *Nutrients, 9*(4), 339. https://doi.org/10.3390/nu9040339

Hewison, M. (2012). An update on vitamin D and human immunity. *Clinical Endocrinology, 76*(3), 315–325.

Hill, C., Guarner, F., Reid, G., Gibson, G. R., Merenstein, D. J., Pot, B., & Sanders, M. E. (2014). Expert consensus document: The International Scientific Association for Probiotics and Prebiotics consensus statement on the scope and appropriate use of the term probiotic. *Nature Reviews Gastroenterology & Hepatology, 11*(8), 506–514. https://doi.org/10.1038/nrgastro.2014.66

Hogg, N., & Landis, R. C. (1993). Adhesion molecules in cell interactions. *Current Opinion in Immunology, 5*(3), 383–390.

Holick, M. F. (2007). Vitamin D deficiency. *New England Journal of Medicine, 357*(3), 266–281.

Huang, Z., Rose, A. H., & Hoffmann, P. R. (2012). The role of selenium in inflammation and immunity: From molecular mechanisms to therapeutic opportunities. *Antioxidants & Redox Signaling, 16*(7), 705–743.

Hunt, C., Chakravorty, N. K., Annan, G., Habibzadeh, N., & Schorah, C. J. (1994). The clinical effects of vitamin C supplementation in elderly hospitalised patients with acute respiratory infections. *International Journal for Vitamin and Nutrition Research, 64*(3), 212–219.

Jackson, M. J., Broome, C. S., & McArdle, F. (2003). Marginal dietary selenium intakes in the UK: Are there functional consequences? *The Journal of Nutrition, 133*(5), 1557S–1559S.

Jia, Z., Nallasamy, P., Liu, D., Shah, H., Li, J. Z., Chitrakar, R., Si, H., McCormick, J., Zhu, H., Zhen, W., & Li, Y. (2015). Luteolin protects against vascular inflammation in mice and TNF-alpha-induced monocyte adhesion to endothelial cells via suppressing IKBα/NF-κB signaling pathway. *The Journal of Nutritional Biochemistry, 26*(3), 293–302.

Kabashima, K., Nakashima, C., Nonomura, Y., Otsuka, A., Cardamone, C., Parente, R., De Feo, G., & Triggiani, M. (2018). Biomarkers for evaluation of mast cell and basophil activation. *Immunological Reviews, 282*(1), 114–120.

Kalinski, P. (2012). Regulation of immune responses by prostaglandin E2. *The Journal of Immunology, 188*(1), 21–28.

Kang, S. G., Park, J., Cho, J. Y., Ulrich, B., & Kim, C. H. (2011). Complementary roles of retinoic acid and TGF-β1 in coordinated expression of mucosal integrins by T cells. *Mucosal immunology, 4*(1), 66–82.

Khair-El-Din, T., Sicher, S. C., Vazquez, M. A., Chung, G. W., Stallworth, K. A., Kitamura, K., Miller, R. T., & Lu, C. Y. (1996). Transcription of the murine iNOS gene is inhibited by docosahexaenoic acid, a major constituent of fetal and neonatal sera as well as fish oils. *The Journal of Experimental Medicine, 183*(3), 1241–1246.

Kilani-Jaziri, S., Mustapha, N., Mokdad-Bzeouich, I., El Gueder, D., Ghedira, K., & Ghedira-Chekir, L. (2016). Flavones induce immunomodulatory and anti-inflammatory effects by activating cellular anti-oxidant activity: A structure-activity relationship study. *Tumour Biology: The Journal of the International Society for Oncodevelopmental Biology and Medicine*, *37*(5), 6571–6579.

Kim, H. J., Kim, Y. J., Lee, S. H., Yu, J., Jeong, S. K., & Hong, S. J. (2014). Effects of lactobacillus rhamnosus on allergic march model by suppressing Th2, Th17, and TSLP responses via CD4+ CD25+ Foxp3+ tregs. *Clinical Immunology*, *153*(1), 178–186.

Kim, J. K., Jin, H. S., Suh, H. W., & Jo, E. K. (2017). Negative regulators and their mechanisms in NLRP3 inflammasome activation and signaling. *Immunology and Cell Biology*, *95*(7), 584–592.

Kim, J. S., & Jobin, C. (2005). The flavonoid luteolin prevents lipopolysaccharide-induced NF-kappaB signalling and gene expression by blocking IkappaB kinase activity in intestinal epithelial cells and bone-marrow derived dendritic cells. *Immunology*, *115*(3), 375–387.

Kishimoto, Y., Saito, N., Kurita, K., Shimokado, K., Maruyama, N., & Ishigami, A. (2013). Ascorbic acid enhances the expression of type 1 and type 4 collagen and SVCT2 in cultured human skin fibroblasts. *Biochemical and Biophysical Research Communications*, *430*(2), 579–584.

Kopustinskiene, D. M., Jakstas, V., Savickas, A., & Bernatoniene, J. (2020). Flavonoids as anticancer agents. *Nutrients*, *12*(2), 457.

Kremer, J. M., Lawrence, D. A., Jubiz, W., DiGiacomo, R., Rynes, R., Bartholomew, L. E., & Sherman, M. (1990). Dietary fish oil and olive oil supplementation in patients with rheumatoid arthritis. Clinical and immunologic effects. *Arthritis and Rheumatism*, *33*(6), 810–820.

Kumagai, Y., & Akira, S. (2010). Identification and functions of pattern-recognition receptors. *The Journal of Allergy and Clinical Immunology*, *125*(5), 985–992.

Kumar, B. V., Connors, T. J., & Farber, D. L. (2018). Human T cell development, localization, and function throughout life. *Immunity*, *48*(2), 202–213.

Kuo, C. L., Chen, T. S., Liou, S. Y., & Hsieh, C. C. (2014). Immunomodulatory effects of EGCG fraction of green tea extract in innate and adaptive immunity via T regulatory cells in murine model. *Immunopharmacology and Immunotoxicology*, *36*(5), 364–370.

Kuwata, T., Wang, I. M., Tamura, T., Ponnamperuma, R. M., Levine, R., Holmes, K. L., Morse, H. C., De Luca, L. M., & Ozato, K. (2000). Vitamin A deficiency in mice causes a systemic expansion of myeloid cells. *Blood*, *95*(11), 3349–3356.

Lämmermann, T. (2016). In the eye of the neutrophil swarm—navigation signals that bring neutrophils together in inflamed and infected tissues. *Journal of Leucocyte Biology*, *100*(1), 55–63.

Lauer, A., Burkard, M., Niessner, H., Leischner, C., Renner, O., Vollbracht, C., & Venturelli, S. (2021). Ex vivo evaluation of the sepsis triple therapy high-dose vitamin C in combination with vitamin B1 and hydrocortisone in a human peripheral blood mononuclear cells (PBMCs) model. *Nutrients*, *13*(7), 2366.

Lila, M. A. (2004). Anthocyanins and human health: An in vitro investigative approach. *Journal of Biomedicine & Biotechnology*, *2004*(5), 306–313.

Lin, R., Jiang, Y., Zhao, X. Y., Guan, Y., Qian, W., Fu, X. C., Ren, H. Y., & Hou, X. H. (2014). Four types of Bifidobacteria trigger autophagy response in intestinal epithelial cells. *Journal of Digestive Diseases*, *15*(11), 597–605. https://doi.org/10.1111/1751-2980.12179

Liu, P. T., Stenger, S., Li, H., Wenzel, L., Tan, B. H., Krutzik, S. R., & Modlin, R. L. (2006). Toll-like receptor triggering of a vitamin D-mediated human antimicrobial response. *Science*, *311*(5768), 1770–1773.

Liu, S. H., Lin, C. H., Hung, S. K., Chou, J. H., Chi, C. W., & Fu, S. L. (2010). Fisetin inhibits lipopolysaccharide-induced macrophage activation and dendritic cell maturation. *Journal of Agricultural and Food Chemistry*, *58*(20), 10831–10839. https://doi.org/10.1021/jf1017093

López-Varela, S., González-Gross, M., & Marcos, A. (2002). Functional foods and the immune system: A review. *European Journal of Clinical Nutrition*, *56*(Suppl. 3), S29–S33. https://doi.org/10.1038/sj.ejcn.1601481

Lucas, C. D., Allen, K. C., Dorward, D. A., Hoodless, L. J., Melrose, L. A., Marwick, J. A., Tucker, C. S., Haslett, C., Duffin, R., & Rossi, A. G. (2013). Flavones induce neutrophil apoptosis by down-regulation of Mcl-1 via a proteasomal-dependent pathway. *FASEB Journal: Official Publication of the Federation of American Societies for Experimental Biology*, *27*(3), 1084–1094.

Ma, Z., Yang, M., Foda, M. F., Zhang, K., Li, S., Liang, H., & Han, H. (2022). Polarization of tumor-associated macrophages promoted by vitamin C-loaded liposomes for cancer immunotherapy. *ACS Nano*, *16*(10), 17389–17401.

Madunic, J., Madunic, I. V., Gajski, G., Popic, J., & Garaj-Vrhovac, V. (2018). Apigenin: A dietary flavonoid with diverse anticancer properties. *Cancer Letters*, *413*, 11–22.

Mandl, J., Szarka, A., & Banhegyi, G. (2009). Vitamin C: Update on physiology and pharmacology. *British Journal of Pharmacology*, *157*(7), 1097–1110.

Matsuda, H., Nakamura, S., & Yoshikawa, M. (2016). Degranulation Inhibitors from medicinal plants in antigen-stimulated Rat Basophilic Leukemia (RBL-2H3) cells. *Chemical & Pharmaceutical Bulletin*, *64*(2), 96–103.

Molano, A., & Meydani, S. N. (2012). Vitamin E, signalosomes and gene expression in T cells. *Molecular Aspects of Medicine*, *33*(1), 55–62.

Molina, N., Morandi, A. C., Bolin, A. P., & Otton, R. (2014). Comparative effect of fucoxanthin and vitamin C on oxidative and functional parameters of human lymphocytes. *International Immunopharmacology*, *22*(1), 41–50.

Moriguchi, S., Kobayashi, N., & Kishino, Y. (1990). High dietary intakes of vitamin E and cellular immune functions in rats. *The Journal of Nutrition*, *120*(9), 1096–1102.

Nath, A., Haktanirlar, G., Varga, Á. Molnár, M. A., Albert, K., Galambos, I., Koris, A., & Vatai, G. (2018). Biological activities of lactose-derived prebiotics and symbiotic with probiotics on gastrointestinal system. *Medicina (Kaunas, Lithuania)*, *54*(2), 18.

Nielsen, G. L., Faarvang, K. L., Thomsen, B. S., Teglbjaerg, K. L., Jensen, L. T., Hansen, T. M., Lervang, H. H., Schmidt, E. B., Dyerberg, J., & Ernst, E. (1992). The effects of dietary supplementation with n-3 polyunsaturated fatty acids in patients with rheumatoid arthritis: A randomized, double blind trial. *European Journal of Clinical Investigation*, *22*(10), 687–691.

Nieman, D. C., Henson, D. A., Davis, J. M., Angela Murphy, E., Jenkins, D. P., Gross, S. J., Carmichael, M. D., Quindry, J. C., Dumke, C. L., Utter, A. C., McAnulty, S. R., McAnulty, L. S., Triplett, N. T., & Mayer, E. P. (2007). Quercetin's influence on exercise-induced changes in plasma cytokines and muscle and leukocyte cytokine mRNA. *Journal of Applied Physiology (Bethesda, Md.: 1985)*, *103*(5), 1728–1735.

Niu, X., Wu, C., Li, M., Zhao, Q., Meydani, S. N., Wang, J., & Wu, D. (2018). Naringenin is an inhibitor of T cell effector functions. *The Journal of Nutritional Biochemistry*, *58*, 71–79.

Noori, S., Nourbakhsh, M., Imani, H., Deravi, N., Salehi, N., & Abdolvahabi, Z. (2022). Naringenin and cryptotanshinone shift the immune response towards Th1 and modulate T regulatory cells via JAK2/STAT3 pathway in breast cancer. *BMC Complementary Medicine and Therapies*, *22*(1), 145.

Oh, S., Chun, S., Hwang, S., Kwack, K., & Choi, S. W. (2021). Vitamin D and exercise are major determinants of natural killer cell activity, which is age-and gender-specific. *Frontiers in Immunology, 12*, 594356.

Pantazi, E., Marks, E., Stolarczyk, E., Lycke, N., Noelle, R. J., & Elgueta, R. (2015). Cutting edge: Retinoic acid signaling in B cells is essential for oral immunization and microflora composition. *Journal of immunology (Baltimore, Md.: 1950), 195*(4), 1368–1371.

Peck MD. 1995. Interaction of lipids with immune function I: Biochemical effects of dietary lipids on plasma membranes. *Journal of Nutritional Biochemistry, 5*, 466–478

Peluso, I., Miglio, C., Morabito, G., Ioannone, F., & Serafini, M. (2015). Flavonoids and immune function in human: A systematic review. *Critical Reviews in Food Science and Nutrition, 55*(3), 383–395.

Pino-Lagos, K., Guo, Y., & Noelle, R. J. (2010). Retinoic acid: A key player in immunity. *Biofactors, 36*(6), 430–436.

Prasad, A. S., Beck, F. W., Bao, B., Fitzgerald, J. T., Snell, D. C., Steinberg, J. D., & Cardozo, L. J. (2007). Zinc supplementation decreases incidence of infections in the elderly: Effect of zinc on generation of cytokines and oxidative stress. *The American Journal of Clinical Nutrition, 85*(3), 837–844.

Prasad, A. S., Meftah, S., Abdallah, J., Kaplan, J., Brewer, G. J., Bach, J. F., & Dardenne, M. (1988). Serum thymulin in human zinc deficiency. *The Journal of Clinical Investigation, 82*(4), 1202–1210.

Pullar, J. M., Carr, A. C., & Vissers, M. (2017). The roles of vitamin C in skin health. *Nutrients, 9*(8), 866.

Radzikowska, U., Rinaldi, A. O., Çelebi Sözener, Z., Karaguzel, D., Wojcik, M., Cypryk, K., Akdis, M., Akdis, C. A., & Sokolowska, M. (2019). The influence of dietary fatty acids on immune responses. *Nutrients, 11*(12), 2990.

Rahimifard, M., Maqbool, F., Moeini-Nodeh, S., Niaz, K., Abdollahi, M., Braidy, N., Nabavi, S. M., & Nabavi, S. F. (2017). Targeting the TLR4 signaling pathway by polyphenols: A novel therapeutic strategy for neuroinflammation. *Ageing Research Reviews, 36*, 11–19.

Rahman, I., Biswas, S. K., & Kirkham, P. A. (2006). Regulation of inflammation and redox signaling by dietary polyphenols. *Biochemical Pharmacology, 72*(11), 1439–1452.

Ramakrishnan, U., Web, A. L., Ologoudou, K., Gershwin, N. E., Nestel, P., & Keen, C. L. (2004). *Handbook of nutrition and immunity.* Springer Science & Business Media.

Rengasamy, K. R. R., Khan, H., Gowrishankar, S., Lagoa, R. J. L., Mahomoodally, F. M., Khan, Z., Suroowan, S., Tewari, D., Zengin, G., Hassan, S. T. S., & Pandian, S. K. (2019). The role of flavonoids in autoimmune diseases: Therapeutic updates. *Pharmacology & Therapeutics*, 194, 107–131.

Ribeiro, D., Freitas, M., Tomé, S. M., Silva, A. M., Laufer, S., Lima, J. L., & Fernandes, E. (2015). Flavonoids inhibit COX-1 and COX-2 enzymes and cytokine/chemokine production in human whole blood. *Inflammation, 38*(2), 858–870.

Rochette-Egly, C., & Germain, P. (2009). Dynamic and combinatorial control of gene expression by nuclear retinoic acid receptors (RARs). *Nuclear Receptor Signaling, 7*(1), nrs-07005.

Rodríguez-García, C., Sánchez-Quesada, C., & J Gaforio, J. (2019). Dietary flavonoids as cancer chemopreventive agents: An updated review of human studies. *Antioxidants (Basel, Switzerland), 8*(5), 137.

Sakai, F., Hosoya, T., Ono-Ohmachi, A., Ukibe, K., Ogawa, A., Moriya, T., Kadooka, Y., Shiozaki, T., Nakagawa, H., Nakayama, Y., & Miyazaki, T. (2014). Lactobacillus gasseri SBT2055 induces TGF-β expression in dendritic cells and activates TLR2 signal to produce IgA in the small intestine. *PLoS One, 9*(8), e105370. https://doi.org/10.1371/journal.pone.0105370

Saroni Arwa, P., Zeraik, M. L., Ximenes, V. F., da Fonseca, L. M., Bolzani, V. S., & Siqueira Silva, D. H. (2015). Redox-active biflavonoids from Garcinia brasiliensis as inhibitors of neutrophil oxidative burst and human erythrocyte membrane damage. *Journal of Ethnopharmacology*, *174*, 410–418.

Savini, I., Rossi, A., Duranti, G., Avigliano, L., Catani, M. V., & Melino, G. (2001). Characterization of keratinocyte differentiation induced by ascorbic acid: protein kinase C involvement and vitamin C homeostasis. *Journal of Investigative Dermatology*, *118*(2), 372–379.

Schachter, H. M., Reisman, J., Tran, K., Dales, B., Kourad, K., Barnes, D., Sampson, M., Morrison, A., Gaboury, I., & Blackman, J. (2004). Health effects of omega-3 fatty acids on asthma. *Evidence Report/Technology Assessment (Summary)*, *91*, 1–7.

Scully, P., Macsharry, J., O'Mahony, D., Lyons, A., O'Brien, F., Murphy, S., Shanahan, F., & O'Mahony, L. (2013). Bifidobacterium infantis suppression of Peyer's patch MIP-1α and MIP-1β secretion during Salmonella infection correlates with increased local CD4+CD25+ T cell numbers. *Cellular Immunology*, *281*(2), 134–140.

Seo, G. Y., Lee, J. M., Jang, Y. S., Kang, S. G., Yoon, S. I., Ko, H. J., ... Kim, P. H. (2017). Mechanism underlying the suppressor activity of retinoic acid on IL4-induced IgE synthesis and its physiological implication. *Cellular Immunology*, *322*, 49–55.

Serafini, M., Peluso, I., & Raguzzini, A. (2010). Flavonoids as anti-inflammatory agents. *The Proceedings of the Nutrition Society*, *69*(3), 273–278.

Serhan, C. N. (2005). Novel omega – 3-derived local mediators in anti-inflammation and resolution. *Pharmacology & Therapeutics*, *105*(1), 7–21.

Serhan, C. N., Clish, C. B., Brannon, J., Colgan, S. P., Chiang, N., & Gronert, K. (2000). Novel functional sets of lipid-derived mediators with antiinflammatory actions generated from omega-3 fatty acids via cyclooxygenase 2-nonsteroidal antiinflammatory drugs and transcellular processing. *The Journal of Experimental Medicine*, *192*(8), 1197–1204.

Shankar, A. H., & Prasad, A. S. (1998). Zinc and immune function: The biological basis of altered resistance to infection. *The American Journal of Clinical Nutrition*, *68*(2), 447S–463S.

Sofi, F., Cesari, F., Abbate, R., Gensini, G. F., & Casini, A. (2008). Adherence to Mediterranean diet and health status: Meta-analysis. *BMJ* (Clinical research ed.), *337*, a1344.

Stephensen, C. B. (2001). Vitamin A, infection, and immune function. *Annual Review of Nutrition*, 21(1), 167–192.

Tanaka T. (2014). Flavonoids for allergic diseases: Present evidence and future perspective. *Current Pharmaceutical Design*, *20*(6), 879–885.

Tian, R., Ding, Y., Peng, Y. Y., & Lu, N. (2017). Inhibition of myeloperoxidase- and neutrophil-mediated hypochlorous acid formation in vitro and endothelial cell injury by (-)-epigallocatechin gallate. *Journal of Agricultural and Food Chemistry*, *65*(15), 3198–3203.

Ullah, A., Munir, S., Badshah, S. L., Khan, N., Ghani, L., Poulson, B. G., Emwas, A. H., & Jaremko, M. (2020). Important flavonoids and their role as a therapeutic agent. *Molecules (Basel, Switzerland)*, *25*(22), 5243.

Vallee, B. L., & Falchuk, K. H. (1993). The biochemical basis of zinc physiology. *Physiological Reviews*, *73*(1), 79–118.

van Baarlen, P., Troost, F., van der Meer, C., Hooiveld, G., Boekschoten, M., Brummer, R. J., & Kleerebezem, M. (2011). Human mucosal in vivo transcriptome responses to three lactobacilli indicate how probiotics may modulate human cellular pathways. *Proceedings of the National Academy of Sciences*, *108*(Suppl. 1), 4562–4569.

Varsi, K., Bolann, B., Torsvik, I., Rosvold Eik, T. C., Høl, P. J., & Bjørke-Monsen, A. L. (2017). Impact of maternal selenium status on infant outcome during the first 6 months of life. *Nutrients*, *9*(5), 486.

Vellozo, N. S., Pereira-Marques, S. T., Cabral-Piccin, M. P., Filardy, A. A., Ribeiro-Gomes, F. L., Rigoni, T. S., DosReis, G. A., & Lopes, M. F. (2017). All-*Trans* retinoic acid promotes an M1- to M2-Phenotype shift and inhibits macrophage-mediated immunity to *Leishmania major*. *Frontiers in Immunology*, *8*, 1560.

Ventura, M. T., Serlenga, E., Tortorella, C., & Antonaci, S. (1994). In vitro vitamin E and selenium supplementation improves neutrophil-mediated functions and monocyte chemoattractant protein-1 production in the elderly. *Cytobios*, *77*(311), 225–232.

Verma, S., Hoffmann, F. W., Kumar, M., Huang, Z., Roe, K., Nguyen-Wu, E., & Hoffmann, P. R. (2011). Selenoprotein K knockout mice exhibit deficient calcium flux in immune cells and impaired immune responses. *The Journal of Immunology*, *186*(4), 2127–2137

Vila Cuenca, M., Ferrantelli, E., Meinster, E., Pouw, S. M., Kovačević, I., de Menezes, R. X., & Vervloet, M. G. (2018). Vitamin D attenuates endothelial dysfunction in uremic rats and maintains human endothelial stability. *Journal of the American Heart Association*, *7*(17), e008776.

Villamor, E., & Fawzi, W. W. (2005). Effects of vitamin A supplementation on immune responses and correlation with clinical outcomes. *Clinical Microbiology Reviews*, *18*(3), 446–464.

Wang, T. T., Nestel, F. P., Bourdeau, V., Nagai, Y., Wang, Q., Liao, J., & White, J. H. (2004). Cutting edge: 1, 25-dihydroxyvitamin D3 is a direct inducer of antimicrobial peptide gene expression. *The Journal of Immunology*, *173*(5), 2909–2912.

Wang, X., & Quinn, P. J. (1999). Vitamin E and its function in membranes. *Progress in Lipid Research*, *38*(4), 309–336.

Wei, C. B., Tao, K., Jiang, R., Zhou, L. D., Zhang, Q. H., & Yuan, C. S. (2017). Quercetin protects mouse liver against triptolide-induced hepatic injury by restoring Th17/Treg balance through Tim-3 and TLR4-MyD88-NF-κB pathway. *International Immunopharmacology*, *53*, 73–82.

Wirth, J. J., Fraker, P. J., & Kierszenbaum, F. (1989). Zinc requirement for macrophage function: effect of zinc deficiency on uptake and killing of a protozoan parasite. *Immunology*, *68*(1), 114.

Wong K. W. (2005). Clinical efficacy of n-3 fatty acid supplementation in patients with asthma. *Journal of the American Dietetic Association*, *105*(1), 98–105.

Wood, S. M., Beckham, C., Yosioka, A., Darban, H., & Watson, R. R. (2000). β-Carotene and selenium supplementation enhances immune response in aged humans. *Integrative Medicine*, *2*(2–3), 85–92.

Wyatt, M., Choudhury, A., Dohlen, G. V., Heileson, J. L., Forsse, J. S., & Greathouse, L. (2024). Vitamin D3 supplementation alters gut microbiome composition and increases stability in healthy individuals. *International Journal of Exercise Science: Conference Proceedings*, *2*(16), 9.

Xu, H. E., Lambert, M. H., Montana, V. G., Parks, D. J., Blanchard, S. G., Brown, P. J., Sternbach, D. D., Lehmann, J. M., Wisely, G. B., Willson, T. M., Kliewer, S. A., & Milburn, M. V. (1999). Molecular recognition of fatty acids by peroxisome proliferator-activated receptors. *Molecular Cell*, 3(3), 397–403.

Xu, S., Kong, Y. G., Jiao, W. E., Yang, R., Qiao, Y. L., Xu, Y., Tao, Z. Z., & Chen, S. M. (2019). Tangeretin promotes regulatory T cell differentiation by inhibiting Notch1/Jagged1 signaling in allergic rhinitis. *International Immunopharmacology*, *72*, 402–412.

Yahfoufi, N., Alsadi, N., Jambi, M., & Matar, C. (2018). The immunomodulatory and anti-inflammatory role of polyphenols. *Nutrients*, *10*(11), 1618. https://doi.org/10.3390/nu10111618

Yan, F., & Polk, D. B. (2010). Probiotics: Progress toward novel therapies for intestinal diseases. *Current Opinion in Gastroenterology*, *26*(2), 95.

Yang, P., & Lu, F. (2024). Study on the immunomodulatory mechanism of vitamin D in patients with unexplained recurrent spontaneous abortion. *Heliyon*, *10*(6), e27280. https://doi.org/10.1016/j.heliyon.2024.e27280

Yum, M. K., Jung, M. Y., Cho, D., & Kim, T. S. (2011). Suppression of dendritic cells' maturation and functions by daidzein, a phytoestrogen. *Toxicology and Applied Pharmacology*, *257*(2), 174–181.

Zanello, G., Berri, M., Dupont, J., Sizaret, P. Y., d'Inca, R., Salmon, H., & Meurens, F. (2011). Saccharomyces cerevisiae modulates immune gene expressions and inhibits ETEC-mediated ERK1/2 and p38 signaling pathways in intestinal epithelial cells. *PLoS One*, *6*(4), e18573.

Zanzonico, P., Fernandes, G., & Good, R. A. (1981). The differential sensitivity of T-cell and B-cell mitogenesis to in vitro zinc deficiency. *Cellular Immunology*, *60*(1), 203–211.

Zhang, M., Zhang, S., Hua, Z., & Zou, X. (2017). Long-term use of bifidobacterium longum alleviates colorectal colitis in rats by regulating inflammatory cytokines and Treg cells. *International Journal of Clinical and Experimental Medicine*, *10*, 7543–7552.

Zhao, X., Liu, M., Li, C., Liu, X., Zhao, J., Ma, H., & Qu, J. (2024). High dose Vitamin C inhibits PD-L1 by ROS-pSTAT3 signal pathway and enhances T cell function in TNBC. *International Immunopharmacology*, *126*, 111321.

Zingg, J. M. (2015). Vitamin E: A role in signal transduction. *Annual Review of Nutrition*, *35*, 135–173.

Index

FSC
www.fsc.org
MIX
Papier | Fördert gute Waldnutzung
FSC® C083411

Kolibri 360 GmbH
Ferdinand-Jühlke-Straße 7
99095 Erfurt, Deutschland
produktsicherheit@kolibri360.de